The Scientific Exploration of Mars

What do we know about Mars? What remains to be understood? Is there evidence of life there? Will humans ever travel there?

The dream of exploring Mars has been around since the early days of human civilisation and still forms part of our vision of the future for the human race. Today, we send unmanned spacecraft to explore this neighbouring world to examine its climate, search for evidence of past or present life, and learn how conditions there relate to those on Earth. Plans for a manned mission to Mars recur regularly, set against an uncertain background of political, practical, technical and financial considerations.

This unique book provides a complete description of the past, present and possible future of Mars exploration. Written by a scientist intimately involved with missions to Mars, it provides a personal first-hand account. It will appeal to anyone interested in this fascinating planet.

FREDRIC W. TAYLOR is Halley Professor of Physics at Oxford University, and has spent many years in the Space Science Division of the NASA Jet Propulsion Laboratory, California Institute of Technology. In addition to Mars projects, he has worked on missions to Venus, Jupiter, Saturn and Titan and on Earth observation space experiments. He has received numerous awards and honours, including the Medal for Exceptional Scientific Achievement from NASA and the Bates Medal of the European Geophysical Society for Excellence in the Planetary Sciences.

The Scientific Exploration of Mars

Fredric W. Taylor

Halley Professor of Physics,
University of Oxford

CAMBRIDGE
UNIVERSITY PRESS

CAMBRIDGE UNIVERSITY PRESS
Cambridge, New York, Melbourne, Madrid, Cape Town, Singapore, São Paulo, Delhi

Cambridge University Press
The Edinburgh Building, Cambridge CB2 8RU, UK

Published in the United States of America by Cambridge University Press, New York

www.cambridge.org
Information on this title: www.cambridge.org/9780521829564

First published 2010

Printed in the United Kingdom at the University Press, Cambridge

A catalogue record for this publication is available from the British Library

Library of Congress Cataloguing in Publication data
Taylor, F. W. The scientific exploration of Mars / F. W. Taylor.
p. cm.
ISBN 978-0-521-82956-4 (hardback)
1. Mars (Planet) – Exploration.   2. Mars probes.   3. Space flight to Mars.   I. Title.
QB641.T285   2009
523.43–dc22
2009039347

ISBN 978-0-521-82956-4 Hardback

2|11

Contents

The plates are situated between pages 50 and 51.

Prologue

Mars is another world in more senses than one. A protective and victorious god, representing both husbandry and war, and the celestial father of Rome; an alternative domain, a 'kind of mythic arena onto which we have projected our Earthly hopes and fears'[1]; a planetary sibling of the Earth that has suffered catastrophic climate change through processes that uncomfortably resemble those now causing concern nearer to home. This book is about the long-term, wide-ranging international scientific research programme that continues to seek to understand Mars, written for non-specialists, such as amateur astronomers or general science readers with an interest in the endeavours, successes and goals of an undertaking that most of us find endlessly fascinating. The aim is to give interested scientists and educated laypersons an intimate view of the science of Mars exploration and its implementation that is as realistic and comprehensive as possible in a readable book of modest size.

The main focus is the current and recent exploration of Mars using spacecraft, which has been pursued by four different national agencies, NASA in the USA, ESA in Europe, RKA in Russia and JAXA in Japan. Others, most notably India and China, say they are preparing to join in. The chapters that follow are a personal perspective on what motivates the scientific community and what it is trying to do, tempered by practical considerations that are often far from obvious to anyone working outside the programmes, not excluding those who report on them in the newspapers and other media. In the later chapters, I seek to take a realistic view of where Mars exploration is heading in the years ahead, in the light particularly of NASA and ESA's plans. These are less vague than those of the other nations who nevertheless will probably participate, or even lead, when the time comes.

There are many books about Mars already, of course, and some are very good. This book aims to be rather different from those, focussing on the *process* of exploring Mars, seeking to pin down the key short- and long-term objectives, and

---

[1] Carl Sagan, *Cosmos* (1980). Random House, New York.

how progress is made in practical terms, hopefully without being too speculative and unrealistic, nor too technical and specialised. The motivation to explore Mars and expend large sums of money doing so are aired, not to provide justification, but because views on this topic are many and varied, even among supporters and participants. The way in which these are reconciled by governments and agencies determines what exploration shall take place, and when. Few outsiders realise how much work goes into designing missions to Mars and elsewhere that never happen. The politics of space is every bit as hard as the technology when it comes to putting everything together successfully.

An account of such a complex subject is bound to be at least partly subjective. A different author, tackling the same topic, would write a different book. This one is based on a lifetime of studying Mars exploration, from boys' books of science fact and fiction to varying degrees of participation in innumerable NASA and ESA committees and design teams, and in some of the missions that actually flew to Mars, including one that is still active at the time of writing. The events, decisions, missions and scientific progress represented by these, plus the plans and discoveries discussed at conferences and workshops, and the guesses about the future, are based on the work of thousands of people, but are seen through the author's eyes. This is a necessary disclaimer – like all complex activities, from sport to wars, no two people involved will have seen it the same way, only the outcome is beyond dispute.

The perspective in this book comes from nearly four decades of working with and within the NASA programme, starting with ten years at the Jet Propulsion Laboratory, where the Mars work is based, in the 1970s when it all came of age with the hugely successful *Mariner 9* and *Viking* missions. Later, I was there when Europe's first venture to Mars, *Mars Express*, was planned, and for the selection of the payload, including the brilliant but unfortunate *Beagle 2* lander. More importantly for this story, hardware built in our lab in Oxford from 1980 on has been to Mars, no less than three times. The first, on *Mars Observer*, is orbiting the Sun somewhere in the vicinity of Mars, but lifelessly and silently, since that fateful day in 1982 when the spacecraft exploded in an abortive attempt to go into orbit around the Red Planet. The second is on the surface of Mars – most people think that *Beagle 2* was the first British hardware to land there; in fact, the Oxford-built part of the *Pressure Modulator Infrared Radiometer*, an instrument for measuring atmospheric temperatures on NASA's *Mars Climate Orbiter*, was first by nearly five years. Like *Beagle 2*, the *Climate Orbiter* crashed and did not survive. Today there is a new version, *Mars Climate Sounder*, happily circling Mars and taking excellent data at a great rate.

A crucial aspect, little discussed outside specialised meetings and study documents, is the fight to develop the pathways or chains of reasoning that must be followed by the scientific community and the space agencies in setting up a programme of exploration that has the maximum chance of success in achieving stated objectives. The objectives themselves have to be carefully thought through and clearly enunciated, and conflicting opinions and interests resolved. Cost and resource limitations have to be fully assessed and unexpected difficulties and failures resolved. The logical processes involved are as complex and essential a part of a mission to Mars as the engineering and project management that sends it on its way; I have tried to give the reader a feeling for these topics without getting too much into the (almost literally, sometimes) gory details. Simplified answers to questions about why and how Mars is being explored may not always do justice to a very complex subject, and there is no account that is accurate and complete that is not long, involved and sometimes uncertain. The current situation is evolving rapidly, and all plans for the future are subject to change (indeed, will almost certainly change – there has been considerable evolution, some progressive and some regressive, in the time it has taken to write this book).

The narrative is in three parts. Part I reviews the history of Mars exploration, up to the present day. Then Part II takes stock, asking what is known about Mars as a result of more than a century of serious scientific exploration, and what it is that makes the paymasters – all of us – want to continue. The outstanding scientific questions are conveniently considered under three headings: (1) the origin and evolution of the planetary system, of which Mars and Earth are family members; (2) the processes that regulate the climate and underlie climate change on Earth-like planets; and (3) the search for life outside the Earth. Origins, survival and, for want of a better word, loneliness are the themes. Given these ongoing motivations, Part III looks at the plans for the next steps to be taken that are in the pipeline at space agencies around the world. These are sure to include surface and orbital reconnaissance, subsurface and polar studies, the return of geological samples, at least thinking about manned expeditions, and the omnipresent theme of searching for evidence of life.

Acknowledgements

The story of the exploration of Mars so far, and of ideas and plans for the future, is based on the work of a large number of planetary scientists, aerospace engineers, mission planners and all-round Mars experts toiling in countless study and planning exercises, particularly those of NASA (the American National Aeronautics and Space Administration) and ESA (the European Space Agency). It is impossible to acknowledge everyone involved individually, and if I tried the result would be unreadable, but I do want to express deep thanks to Dr Dan McCleese and the NASA Mars Program team at the Jet Propulsion Laboratory, who hosted much of the work that went into the writing of this book. I would also like to thank colleagues from the various experiment teams and study groups I have worked with for their insights and inspiration: the main ones are listed in an appendix at the end of the book. Several colleagues read all or part of the manuscript in various versions and offered advice; I should like particularly to thank Dr Crofton B. Farmer, Professor Richard Moxon and Vince Higgs for their detailed comment and criticism. Any bias or emphasis placed on facts, probabilities and controversies is of course my own responsibility, as are any errors or exclusions.

Acknowledgements are also due to the originators of figures and artwork reproduced here, and especially to Dr D. J. Taylor for so skilfully drawing the original material used in many of the figures. The remaining illustrations are largely drawn from the projects and planning of the stunningly successful American and European space agencies, as identified from the context and accompanying captions and text. Figures and plates representing futuristic art by commercial artists contain individual acknowledgements and copyright information. Nearly every one of the illustrations in the book is taken or derived from a collection I have made over the last forty years of close observation or involvement in Mars programmes, and is included to invoke what happened then and to add interest and information to the narrative. In a few places where my text may appear to change or embellish the information content as expressed in the original usage, this reflects my own perspective and the responsibility is again mine.

The book is dedicated with respect and admiration to the Mars programmes at NASA and JPL, ESA and ESTEC, to all past and present colleagues in the planetary scientific community, and everyone involved in the history of Mars exploration so far. I feel very fortunate to have been around when they made so much of it happen.

Part I

Views of Mars, from the beginning to the present day

To put present and future Mars exploration in context we must first review the history of mankind's aspirations, investigations and knowledge regarding our planetary neighbour. The most basic facts about Mars, which are summarised in Appendix A, have been obtained as a result of observations spanning hundreds of years, during most of which researchers were limited to observations from the Earth, although latterly through telescopes of considerable size and sophistication. For the last half century, however, it has been possible to study Mars close up using instruments on space probes, and these too have been gradually increasing in size and complexity. Around 35 spacecraft have been dispatched (see the summary in Appendix B) and, while less than half of the total have been successful, no fewer than five are currently operational on and around Mars.

These current missions include the *Mars Exploration Rovers*, true robot geologists that, at the time of writing, have been crawling over the surface of Mars for more than two and a half years, covering more than ten miles between them. The discovery of sedimentary rocks and salt deposits has confirmed what was long suspected: liquid water played a major role on early Mars. Climate change has dried and cooled the planet, and depleted the Martian atmosphere to a thin remnant of its former, more Earth-like state. What may have been a good environment for life is no longer so benign. Any life that once existed on the surface may have been driven below ground, where explorers from Earth have yet to look.

Chapter 1
The dawn of Mars exploration

1.1 Early observations

Mars was known to the ancients as a planet, or wanderer, in the night sky. What they thought about it is the stuff of myths and legends: usually, the baleful blood-red planet personified the God of War. Mars became recognised as a world, a companion to the Earth in the cosmos, about the time of the ancient Greeks, and as a member of the Sun's family of orbiting planets when Kepler and Copernicus worked out its orbit in the Middle Ages.

Nicolaus Copernicus proposed in *De revolutionibus orbium coelestium* in 1543 that Mars, like Earth and the other planets, followed an orbit around the Sun, presumed on philosophical grounds to be a perfect circle. Stimulated by Copernicus' thinking, and some ideas of his own in which the orbits of the planets were related to the five regular 'Platonic' polyhedrons,[1] in 1600 Johannes Kepler joined Tycho Brahe in Prague. Here, and before that in his native Denmark, Brahe had been observing the movement of Mars in the sky with unprecedented precision. After Brahe died suddenly in 1601 (poisoned by Kepler, according to some accounts), Kepler spent nearly a decade with Tycho's data attempting to find a mathematical explanation for the apparent meanderings of the red planet. A true wanderer, Mars actually reverses its apparent direction of travel twice every two years, tracing a loop against the backdrop of 'fixed' stars as seen by an observer on the Earth. The Earth-centred plot of Mars' motion in Figure 1.1 shows how this occurs.

Kepler concluded that the orbit of Mars is not only not related to any of the five Platonic solids, it is not even circular, but must follow instead an elliptical

---

[1] Probably discovered by Pythagoras, but forming a cornerstone of Plato's philosophy, there are only five of these 'perfect' solids, ranging from four to twenty surfaces forming completely regular shapes. In his early thinking, Kepler proposed that they determined the geometry that separated the spheres containing the six planets. The shapes are, in the order Kepler placed them in distance from the Sun, the octahedron, the icosahedron, dodecahedron, tetrahedron and finally, framing the orbit of Saturn, the cube.

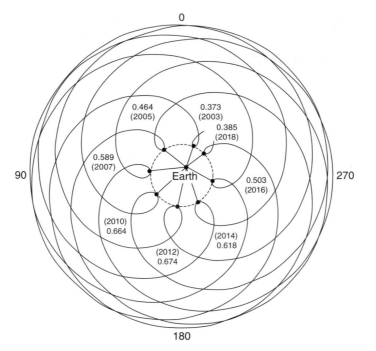

Figure 1.1 The elliptical orbit of Mars, plotted in a coordinate system that has the Earth at the centre. The dates, 2003, 2005 etc., are the years in which the closest approaches of Mars to Earth occur. The minimum distances are also shown; the closest is in 2003, when Mars was only 0.373 astronomical units (54.42 million kilometres) from Earth. This representation (based on plots by Tom Ruen using *Full Sky* software) makes it easier to understand why Mars periodically changes direction across the sky as viewed from Earth's surface. This apparently bizarre behaviour stimulated Kepler to develop his laws of planetary motion, which he published in 1609.

orbit with the Sun at one of its foci. This, and his second law of planetary motion, which states that a line from Mars to the Sun sweeps out equal areas in equal times as the planet moves, were published in *Astronomia Nova* in 1609. Three-quarters of a century later, in *Philosophiae Naturalis Principia Mathematica*, Isaac Newton showed that Kepler's laws were a specific application of more general laws of motion and of gravitation that applied to all objects at all times throughout the observable universe. This was one of the foundation stones of modern science, remaining essentially unchallenged until Einstein, with the Theory of Relativity, found deeper truths behind Newton's laws in the early 1900s. Today, the recent discoveries of apparently non-Newtonian behaviour in galaxies has led to concepts such as dark matter and dark energy that are understood by no one.

Most professional astronomers these days are cosmologists, wrestling with concepts of the nature and origin of the Universe at large, even discussing 'multiverses' ruled more by mathematics than observations. Planetary scientists are now a separate breed, often closer to Earth scientists than mainstream twenty-first-century astronomers because of a common interest in aspects of Solar System formation, geophysics and atmospheric science. Their tools are now more likely to be robotic spacecraft than telescopes on mountains, and data still rule over theory, just as in the time of Kepler and Tycho.

1.2 Another planet with an atmosphere

The scientific exploration of Mars as a planet like the Earth with a surface and an atmosphere started with the foundation of the great observatories, following the invention of the telescope *circa* 1610. In 1659 Christiaan Huygens made the sketch of Mars reproduced in Figure 1.2, which shows a glimpse of some features on the surface, while Giovanni Cassini was the first to record the bright polar cap at the Martian north pole in 1666. In 1672 Huygens saw its equivalent in the south. That same year, working with a colleague observing Mars from South America at the same time as he made his own observations in France, Cassini obtained a value for the parallax[2] of Mars and hence for its distance from the Earth, and that of the Earth from the Sun. This was the first reasonably accurate determination of these distances, less than ten per cent too small compared to modern values.

In 1781 William Herschel, observing both polar caps, wrote that he believed they were ice sheets like those on the Earth. In an address to the Royal Society in 1784, he went further:

It appears that this planet is not without considerable atmosphere; for besides the permanent spots on the surface, I have often noticed occasional changes of partial bright belts; and also once a darkish one. These alterations we can hardly ascribe to any other cause

Figure 1.2 Huygens' 1659 sketch of Mars; on the left, is the earliest known record of markings on the surface. Thirteen years later, on 13 August 1672, he drew what seem to be the same features, but added the polar cap. The third drawing is dated 17 May 1683.

---

[2] Parallax is the apparent displacement of an object when viewed from different directions. In this case it refers to the apparent change in the position of Mars relative to the background stars when viewed from two locations on the Earth, or in principle when the Earth is at the two extremes of its orbit, either side of the Sun. This small angle is not simple to determine (particularly since Mars moves along its own orbit during the time of the measurement) but once it is available, the distance to Mars can be determined from simple triangulation.

than the variable disposition of clouds and vapours floating in the atmosphere of the planet. Mars has a considerable but modest atmosphere, so that its inhabitants probably enjoy a situation in many respects similar to ours.

Along with the idea that Mars had a substantial atmosphere came the assumption that it was likely to be more or less like that on the Earth, a not unreasonable starting point given the absence of quantitative measurements. Even less was known then than now about atmospheric evolution on the terrestrial planets, and today it is generally believed that the atmospheres of Earth, Mars, and indeed Venus, used to resemble each other more than they do now. It was not until 1908 that the first scientific attempt to quantify the surface pressure on Mars was made by Percival Lowell in the USA. As discussed below, he came up with less than one-tenth of the terrestrial value, but it was still ten times too large.

1.3   The Mountains of Mitchel

Mars studies surged in Victorian times, with the work of many keen amateur and professional observers using telescopes that had improved rapidly in availability and performance with the flourishing of technology that characterised this period. Along with the new observations came a rash of interpretations and speculations. In 1846, Ormsby Mitchel was the director of the recently completed observatory (shown in Figure 1.3), which still stands at what is now the University of Cincinnati in Ohio. He made drawings of Mars like those in Figure 1.4, in which he noted that a region of the south polar cap remained snow covered in the spring after the ice had receded all around it. He reasonably assumed that this meant a high elevation, and the region is still known as the Mountains of Mitchel, although observations from the recent *Mars Global Surveyor* spacecraft in orbit around Mars show that the terrain Mitchel saw, although rough, is only moderately hilly. Its long-lived brightness is probably due to the alignment of the outcrops of rock keeping much of the surface in shadow in the early spring, when the Sun is low in the sky, so that the winter frost is slower to clear than in surrounding regions at the same latitude. However, some other regions that look topographically quite similar do not stand out in this way, and the *Global Surveyor* camera team noted as recently as 1999 that the exceptional brightness of the region remains something of a mystery.

The infrared spectroscopy team on the same mission found that the carbon dioxide frost covering the 'Mountains' appeared to be unusually small-grained, making it very reflective and hence slower to sublime away in the spring (and, incidentally, easier for Mitchel himself to observe with his relatively primitive equipment). It seems that something about the local meteorological conditions provides the region with particularly heavy deposits of fresh, highly reflective snow each winter season. Whatever the reason, the phenomenon is a remarkable example of a very early study of localised behaviour on Mars that has stood the test of time without being shown to be illusory, even if the interpretation was wrong.

Figure 1.3 Ormsby M. Mitchel, in uniform as a general during the US Civil War, and the observatory in Cincinnati that housed his twelve-inch refracting telescope. (University of Cincinnati)

Mitchel made regular observations of Mars and wrote a book, *The Planetary and Stellar Worlds*, in which he said:

The reddish tint which marks the light of Mars has been attributed by Sir John Herschel to the prevailing colour of its soil, which he considers the greenish hue of certain tracts to

Figure 1.4 Early observers had to draw by hand what they saw through the telescope. The sketches of Mars on the left are by Ormsby Mitchel, who was one of the more reliable nineteenth-century observers. His discovery in 1846 of an isolated bright region as the polar cap receded in the Martian southern spring was confirmed by spacecraft observations more than a century and a half later. The 'Mountains of Mitchel' can be seen just to the left of the polar cap in both Mitchel's drawing and the modern photograph (taken through the twenty-four-inch telescope at Lowell Observatory by C. F. Capen) on the right. South is at the top of both pictures, since telescopic observers traditionally draw or photograph planets upside-down, which is the way they appear through their eyepieces. (University of Cincinnati, courtesy Mrs V. Grohe)

distinguish them as covered with water. This is all pure conjecture, based upon analogy and derived from our knowledge of what exists in our own planet. If we did not know of the existence of seas on the Earth, we could never conjecture or surmise their existence in any neighbouring world. Under what modification of circumstances sentient beings may be placed, who inhabit the neighbouring worlds, it is vain for us to imagine. It would be most incredible to assert, as some have done, that our planet, so small and insignificant in its proportions when compared with other planets with which it is allied, is the only world in the whole universe filled with sentient, rational, and intelligent beings capable of comprehending the grand mysteries of the physical universe.

1.4 The wave of darkening

Several nineteenth century astronomers observed a dark front that travelled from the pole towards the equator in the springtime, which became known as the 'wave of darkening'. The popular view, which persisted until relatively recently, attributed this the seasonal advance of vegetation – 'herbage and plants' according to Camille Flammarion in 1873. In 1909 another French astronomer, Georges Fournier, wrote: 'As the spring advances, the dark shading progressively encroaches from the pole towards the equator across the seas. This advance, across wide expanses and along channels, takes place at a variable speed, but nearly always very rapidly; a few weeks only suffice to change the landscape completely.' E. M. Antoniadi in his book *La Planete Mars 1659–1929* wrote: 'It was almost exactly the colour of leaves which fall from trees in summer and autumn in our latitudes.'

In fact, leaves were ruled out fairly early by infrared observations of Mars. The dark areas were found to lack the high reflectivity at infrared wavelengths that is characteristic of the chlorophyll in green vegetation. Those who wanted to save the life-based explanation proposed some sort of dark cactus or lichens instead; others considered that the non-biological possibilities needed to be ruled out first. The latter included the Swedish chemist Svante Arrhenius, who pointed out in 1918 that certain substances that could be present in the Martian soil might change colour as they absorbed moisture seasonally released from the polar caps. In 1954, Dean McLaughlin suggested drifting volcanic ash from active volcanoes, and in 1957 Gerard Kuiper proposed windblown dust. Carl Sagan and James Pollack, writing in the journal *Nature* in 1969, developed the dust hypothesis in convincing detail.

The phenomenon is real and still regularly observed by amateurs, but the greenish colours are illusory. Kuiper, Sagan and Pollack were right: the space missions of the 1970s observed vast dust storms moving across the planet, most frequently in the spring. The early optical observers, with small telescopes that delivered very limited spatial resolution, were detecting Martian meteorology, not biology.

1.5 Lowell and the canals

Today, when Mars explorers of the early telescopic era are discussed, the names that everyone remembers are those of Giovanni Schiaparelli and Percival Lowell. Lowell in particular is memorable as a serious observer who interpreted what he saw in dramatic and eloquent terms that turned out to be wrong. Schiaparelli was an astronomer at the Observatory of Milan and he was the first, in 1877, to observe channels or 'canali' on the surface of Mars. He did not claim that these were artificial, but many others thought they were evidence of large-scale engineering on Mars and hence must be the work of intelligent life. Lowell, who held a distinction in mathematics from Harvard University, had an early career as a diplomat and author working mainly in the Far East. In 1894, inspired in part by Schiaparelli's discovery, he decided to dedicate his time and his considerable personal wealth to building an observatory in Flagstaff, Arizona, specifically for the study of these and other interesting features on Mars. Figure 1.5 shows Lowell at work, and an example of his observations, recorded, as was still the practice at the time even for professional astronomers, as drawings.

Lowell's dedicated campaign, using instrumentation carefully selected and designed to be the best available at the time, revealed to him a network of artificial canals on Mars, the purpose of which was to irrigate the planet by supplying water from the melting polar caps. Lowell knew that he would not be

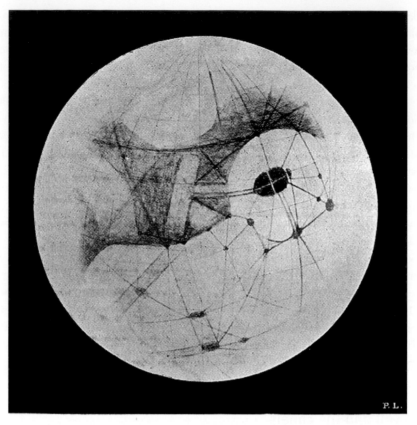

MARS
LONGITUDE 60° ON THE MERIDIAN

Figure 1.5 Percival Lowell in his observatory, and one of his drawings of Mars. (From *Mars and its Canals*, 1906)

able to see canals of any sensible width from the Earth, even if they existed, but he further surmised that the water would be bordered by fertile zones several miles wide, similar to those seen on Earth along the River Nile as it winds through otherwise arid desert.

Although 'unburdened with advanced degrees', Lowell was already an experienced author. He published, among others, books about Japanese culture, having lived in Japan as a young man. He was keen to communicate his insights on Mars to a receptive public, and wrote extensively about his findings in the scientific and the popular literature. Two of his books are still famous today: *Mars and its Canals* (1906) and *Mars as the Abode of Life* (1909), and they still convey the sense of excitement that Lowell felt over his discoveries:

In steady air the canals are perfectly distinct lines, not unlike the Fraunhofer ones of the Spectrum, pencil lines or gossamer filaments according to size. All the observers at Flagstaff concur in this. The photographs of them taken there also confirm it up to the limit of their ability. Careful experiments by the same observers on artificial lines show that if the canals had breaks amounting to 16 miles across, such breaks would be visible. None are; while the lines themselves are thousands of miles long and perfectly straight. Between expert observers representing the planet at the same epoch the accordance is striking; differences in drawings are differences of time and are due to seasonal and secular changes in the planet itself. These seasonal changes have been carefully followed at Flagstaff, and the law governing them detected. They are found to depend upon the melting of the polar caps. After the melting is under way the canals next to the cap proceed to darken, and the darkening thence progresses regularly down the latitudes. Twice this happens every Martian year, first from one cap and then six Martian months later from the other. The action reminds one of the quickening of the Nile valley after the melting of the snows in Abyssinia; only with planet-wide rhythm. Some of the canals are paired. The phenomenon is peculiar to certain canals, for only about one-tenth of the whole number, 56 out of 585, ever show double and these do so regularly. Each double has its special width; this width between the pair being 400 mi. in some cases, only 75 in others. Careful plotting has disclosed the fact that the doubles cluster round the planet's equator, rarely pass 40° Lat., and never occur at the poles, though the planet's axial tilt reveals all its latitudes to us in turn. They are thus features of those latitudes where the surface is greatest compared with the area of the polar cap, which is suggestive. Space precludes mention of many other equally striking peculiarities of the canals' positioning and development. At the junctions of the canals are small, dark round spots, which also wax and wane with the seasons. These facts and a host of others of like significance have led Lowell to the conclusion that the whole canal system is of artificial origin, first because of each appearance and secondly because of the laws governing its development. Every opposition has added to the assurance that the canals are artificial; both by disclosing their peculiarities better and better and by removing generic doubts as to the planet's habitability. The warmer temperature disclosed from Lowell's investigation on the subject, and the spectrographic detection[3] by Slipher[4] of water-vapour in the Martian air, are among the latest of these confirmations.

---

[3] 'Spectrography' is an early name for what is usually now called spectroscopy, see Figure 1.6.

[4] Vesto Slipher worked with Lowell at the Flagstaff Observatory. His work on Mars is discussed further below.

While Lowell convinced himself and enthused the public by this kind of rhetoric, other astronomers were generally much less certain that there were massive civil engineering works on Mars, not least because most of them could not see the canals through their own telescopes. In 1894, Edward Emerson Barnard[5] wrote to a friend:

To save my soul I can't believe in the canals as Schiaparelli draws them. I see details where some of his canals are, but they are not straight lines at all. When best seen these details are very irregular and broken up – that is, some of the regions of his canals; I verily believe – for all the verifications – that the canals as depicted by Schiaparelli are a fallacy and that they will be so proved before many oppositions are past.

Alfred Russel Wallace, the renowned English naturalist and proponent with Darwin of natural selection, had argued in his book *Man's Place in the Universe*, published in 1902, that Mars was not habitable. He further advanced this view in 1907 with another book, *Is Mars Habitable? A Critical Examination of Professor Percival Lowell's Book, Mars and its Canals*. Although he was not a physicist or astronomer, he had friends who were, and he argued convincingly and, as we now know, correctly, that the temperatures and pressures on Mars were too low for water-filled canals, pointing out that even the relatively warm Earth has 'arctic climates and perpetual snow at heights where the air is still far denser than it is on the surface of Mars'.[6]

The notion of observable global-scale hydraulic engineering on Mars lost further support with new observations by Eugene Antoniadi at the thirty-three-inch telescope at Meudon, near Paris, then the largest in Europe and superior, in aperture at least, to the twenty-four-inch used by Lowell. During the close approach of Mars in 1909, when the skies over Europe were exceptionally clear and still, he concluded, 'the geometrical canal network is an optical illusion', describing the true appearance of features on Mars as 'overwhelmingly natural'. Other observers, including leading figures in the USA such as George Ellery Hale, reached similar conclusions from their own studies.

The optical illusion that produced Schiaparelli and Lowell's canals is one that comes from working at the limits of resolution and atmospheric 'seeing', possibly with a contribution from the network of veins in the retina of the observer's eye. (Lowell also recorded and published drawings of linear features on cloud-covered Venus, which ought to have produced more of a reappraisal of his Martian revelations than it apparently did at the time.) The seasonally varying dark areas are now known to be the effect of natural surface features and migrating wind-borne dust and not any form of plant life. Nowadays Lowell's

---

[5] Barnard was an outstanding observational astronomer, whose achievements included the last optical discovery (in 1892) of a new moon of Jupiter, Amalthea.

[6] Wallace initially took Lowell's estimate of 'not more than about four inches of barometric pressure as we reckon it' for the mean surface pressure on Mars. In a footnote, Wallace adds: 'In a paper written since the book appeared the density of air at the surface of Mars is said to be 1/12 of the Earth's.' This is still more than 10 times too high, relative to the modern value, although this fact just makes Wallace's argument stronger.

passionate over-interpretation of his data is often considered comical. He was a very committed and generally competent observer, however, and succeeded in encouraging others, whether as supporters or protagonists. The net effect was to give rise to a wave of public and professional enthusiasm for the further study of Mars that is still palpable today, and has driven the application of gradually improving techniques, including space probes, to the study of the true nature of conditions on the Martian surface.

1.6 The detection of water

In Lowell's time, as now, a key question was how much water was to be found on Mars. Early attempts to find the spectral signature of water vapour in the Martian atmosphere using spectroscopic techniques (see Figure 1.6 for a basic description of what spectroscopy is) were unsuccessful. This was a frustrating situation for advocates of an open irrigation system on the surface since evaporation would be expected to lead to a readily observable level of atmospheric humidity. After a campaign using 'improved spectroscopic apparatus', which still showed no trace of water vapour, William Campbell of Lick Observatory wrote that Mars must be too dry for 'life as we know it', and that the atmosphere might be much thinner than Lowell had supposed. Then in 1909 Lowell's colleague Vesto Slipher, who would later succeed him as director of the observatory in Flagstaff,[7] claimed a tentative detection that quickly became one of Lowell's 'confirmations' of the existence of canals. In fact, the breakthrough was another delusion: Mars is colder than Slipher knew, and even though it is, in fact, close to saturation most of the time, the atmosphere there cannot hold enough water to be detected with any of the equipment available to him a century ago.

1.6.1 Spectroscopy

Decades later, although techniques had improved, a big problem for Earth-based observers remained the intervening, much larger, amount of water vapour in the Earth's atmosphere. This produces spectral lines that overlap and tend to obliterate the weak lines that originate on Mars. Skilled observers such as Adams and Dunham, who had in 1932 reported the detection of carbon dioxide in the atmosphere of Venus, searched for water on Mars when the Doppler shift produced by the relative velocities of the two planets in their orbits was large enough to separate the lines.[8] Addressing a meeting of the Astronomical Society of the Pacific in June 1937, they said of their observations made in February of that year with the hundred-inch telescope on Mount Wilson, California:

---

[7] Slipher was Director from 1916, when Lowell died (and was entombed in a mausoleum that stands near the telescope dome where he made his observations), until he retired in 1952.

[8] This technique, now commonplace, was devised and applied originally by Percival Lowell.

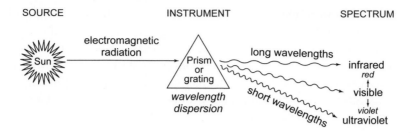

Figure 1.6 The principle of spectroscopy involves dispersing the light from an object, in our case a planet viewed through a telescope, into its component wavelengths by a device such as a prism or a diffraction grating. The separation of sunlight into the colours of the rainbow by Newton in 1665 was the earliest recorded spectroscopic experiment.

Different materials absorb or reflect more strongly at some wavelengths than others, and an analysis of the brightness, as a function of wavelength, of the dispersed light (called a spectrum from the Latin for a ghost or phantom) can show what materials are present. The technique works best with gases; for example water vapour in the air is easier to detect and quantify than water or ice on the surface, because the features produced in the spectrum by a gas are sharper and easier to identify conclusively than the broader, more diffuse features produced by solid or liquid materials.

Had Newton had better equipment, he would have seen narrow dark lines in his solar spectrum long before they were discovered by Fraunhofer in 1817, and he could have inferred some details of the composition of the Sun, and of the Earth's atmosphere through which the sunlight had passed. The most famous example of this is the discovery of helium in the Sun in 1868, long before it was first identified in a laboratory on the Earth in 1895.

The lines in the spectrum occur because atoms and molecules undergo discrete transitions in their internal electronic, rotational and vibrational energy by absorbing and emitting photons at wavelengths that are unique to that substance. The detailed shape and intensity of the spectral lines reveals further details of the physical conditions, such as temperature and pressure, of the gas or solid emitting the line. Thus, for example, the temperature and pressure of the atmosphere near the surface of Mars can be inferred from sufficiently sensitive high-resolution measurements of the spectrum of carbon dioxide.

Many of the important discoveries about Mars have been made using spectroscopy, either through Earth-based telescopes or from instruments on orbiting spacecraft. The modern technique that uses spectroscopy combined with imaging or mapping to cover all or part of a planet in detail with measurements made from a distance is often described as *remote sensing*.

... the radial velocity of Mars relative to the observer was large, about nineteen kilo-metres/sec of approach on the dates of the observations. Such a velocity, corresponding to a displacement of 0.45 Å, should completely separate lines of water vapor due to the planet's atmosphere from those of terrestrial origin. No evidence was found, however, of the presence of a planetary component either on the original negatives or on micropho-tometer tracings of the spectrum. ... any Martian lines cannot exceed five percent of the terrestrial lines and must probably be considerably less.

Despite the fact that they used a state-of-the art instrument on a large tele-scope at a high-altitude site, they still had insufficient sensitivity and spectral resolution to detect the Martian water. Limitations of 'seeing' meant that they could only observe the centre of the Martian disc, possibly over very dry desert

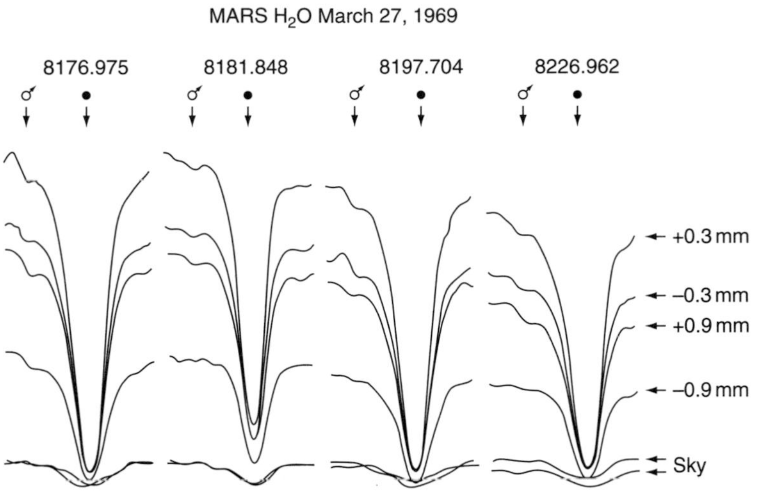

MARS H₂O March 27, 1969

Figure 1.7 These are the absorption lines of water vapour in the spectrum of Mars, as observed in 1969 by Robert Tull using the one hundred and seven-inch telescope at McDonald Observatory in Texas (reported in the journal of Solar System research, *Icarus*, in 1970). The Martian lines are the small bumps, marked ♂, shifted by the Doppler effect into the wings of the much stronger terrestrial lines. Four different lines are shown, with the wavelength (in angstroms, 1 Å = 10⁻¹⁰ metre) at the top. The strength of each line is shown as measured at four different places on the planet, with the (almost flat) spectrum of the empty sky near Mars shown at the bottom as a calibration.

terrain. They speculated that a positive result was more likely to be achieved if they could resolve the icy polar regions.

Many other unsuccessful attempts were made, including those by Kuiper in 1949, and Kiess and Korliss in 1957. In 1963, Spinrad and Richardson reported yet another negative result from observations made on 24 December 1962, with the forty eight inch reflector at Victoria in Canada. However, in a note added in proof, they rather bathetically reported: 'Several Mars water lines were detected by Spinrad, Munch and Kaplan from Mount Wilson in April 1963. A preliminary abundance estimate suggests about 5–10 μm precipitable H₂O for Mars.' Figure 1.7 shows the sort of thing that they saw – very weak Martian lines in the wings of their much stronger telluric[9] equivalents.

At about the same time, Audouin Dollfus of Paris Observatory detected Martian water from an observatory on the Jungfraujoch in the Swiss Alps. A glance at the data shows why the detection was nearly a century in coming: even at maximum elongation, when the Doppler shift is at a maximum, the lines are still strongly blended and the Martian feature is almost buried in the much stronger terrestrial feature.[10] The two lines have the same intrinsic strength, since they are both due to

---

[9] 'Telluric' refers to spectral lines or other features or artefacts in astronomic observations that originate in the Earth's atmosphere.

[10] The author attended a seminar by Dr Bill Sinton of the Institute for Astronomy in Hawaii in 1971, where Sinton showed spectra of Mars that he had used to determine the water abundance on the red planet. The audience squinted furiously at the spectra but claimed, to the speaker's very

the same quantum-mechanical transition, which has to be the same for every water molecule, whether on Earth or Mars. The apparent difference in strength is therefore due to the different amounts of water vapour in the two atmospheres. When they completed their analysis, the amount of water on Mars was inferred by Hyron Spinrad and colleagues to be only about fourteen precipitable micrometres. This means that, if the vapour were to condense on the surface, it would form a liquid layer only fourteen-millionths of a metre thick. The amount in Earth's atmosphere is at least a hundred times more than this. The problems of making this kind of measurement from the surface of the Earth, even on a high mountain, are obvious, as are the advantages of moving the spectrometer onto an Earth-orbiting telescope like *Hubble Space Telescope*, or, better still, onto a spacecraft in orbit around Mars.

1.7 Carbon dioxide

Water was not the first Martian gas to be detected by spectroscopy. Not surprisingly, since there is actually more of it on Mars than on the Earth, that distinction went to carbon dioxide, CO_2. In 1947, Gerard Kuiper, working at the McDonald Observatory in Texas, failed to find water or oxygen, but did find the signature of Martian CO_2 in his spectra. He estimated that the ratio of the column amounts of the gas on the two planets was about two to one. The contribution of CO_2 to the surface pressure on Earth is less than a tenth of one per cent, around one millibar,[11] so that on Mars might be only about two millibars. It seemed likely, however, that Mars would have a significant proportion of atmospheric nitrogen, like the Earth, in addition to the CO_2. It also was hard to accept that a surface pressure as low as a few millibars could be compatible with the presence of clouds and polar caps, which implies a substantial atmosphere. On the other hand, as early as the time of his advocacy of evidence for intelligent life on Mars, Lowell maintained that Mars had less of an atmosphere than Earth because of the lower gravity, which made it less capable of holding down a dense atmosphere. He also noted correctly that the low albedo (reflectivity) of Mars was a sign of limited backscattering by atmospheric gases and cloud particles. By what one historian called 'juggling some unreliable numbers', i.e. by what we would now call modelling, he came up with a value of eighty-seven millibars as early as 1908. This was refined by Donald Menzel of the Harvard College Observatory in 1925 to sixty-six millibars, and by the French astronomer Bernard Lyot to twenty-four millibars in 1929.

These were the best estimate available for the next thirty years, the problem being that nitrogen has no infrared spectral lines that could be observed to

---

obvious annoyance, that they could see no definite sign of the Martian lines in the wings of the much stronger features in the Earth's atmosphere. Sinton, an excellent observer, had previously published a spectroscopic detection of plant life on Mars, which turned out to be the lines of deuterated or heavy water, HDO, in Earth's atmosphere. He was right about the Mars water, however.

[11] 1 millibar (usually written mb) is 1/1000 of 1 bar, which is approximately the mean surface pressure on the Earth.

measure its contribution. However, the shapes of the spectral lines of other gases, including carbon dioxide and water, are affected by the total pressure, and with the very high resolution measurements that they used to find Martian water in 1963, Spinrad and his colleagues were able to show that the total pressure was less than twenty-five millibars, of which about four was carbon dioxide and the rest still presumed to be mostly nitrogen.

In 1971 a more exact value was obtained by Louise Gray Young, a theoretical planetary scientist working at the Jet Propulsion Laboratory in California. She used observations by Pierre Connes and colleagues and an improved theoretical calculation to come up with the correct value of between five and six millibars.[12] However, by this time the difficult task of deriving these numbers from telescopic observations of all or most of the disc of Mars, painstakingly acquired from mountain observatories on Earth, no longer represented the state of the art. Instead, miniaturised versions of the spectrometers and their telescopes were built and flown all the way to Mars on spacecraft launched by giant rockets, similar to those that had flown men to the Moon in 1969. Observing Mars close up meant far fewer problems with sensitivity and spatial resolution, while those with telluric atmospheric absorption and observing conditions ('seeing') were eliminated altogether. Along with more sensitive spectroscopic measurements came high-resolution pictures from television cameras mounted on the probes, revealing details of the features on the surface of Mars for the first time (and no canals). Science was at last beginning to catch up with human imagination, but it had quite a way to go.

1.8 Prelude to the space age: the von Braun Mars expeditions

Before the real space age began, there were many dreams of manned missions to Mars, including one published in 1956 in a book (arrestingly titled *There IS Life on Mars*) by the sixth Earl Nelson, descendant of the great admiral and victor of Trafalgar.[13] By far the best of these however was that of the rocketry pioneer Wernher von Braun, which he started in 1948, fresh from his move from defeated Nazi Germany to the USA, and the successful transfer of the German V-2 rocket project to New Mexico. In his spare time, von Braun developed plans for a human expedition to the red planet, originally in fictional form intended as the basis for a futuristic novel entitled *Project Mars, A Technical Tale*, set in the year 1980. This was shelved when he came to take the project seriously, and did not see the light of day until finally published thirty years after his death, when it revealed a stirring tale of the first expedition from Earth to Mars in which the human pioneers encounter an advanced humanoid civilisation on Mars. Its most important feature is an appendix sixty-two pages long, which presents in great technical detail the design and assembly in Earth orbit of the spaceships in which

---

[12] Her value was 5.16 ± 0.64 mb; from the most recent data, the true seasonal average is about 5.6 mb.

[13] The modern Earl Nelson concluded: 'when they succeed in getting there, men will find that Mars is a world inhabited only by low forms of vegetation, and probably bacterial life of a sort'.

the expedition travels to Mars and returns safely. In his personal notes, von Braun wrote that putting the project 'in simple, narrative form permits me to outline the scientific, financial and organizational efforts which will be necessary before space travel can actually be brought into being'.

By 1952, the fictional part had been set aside and 'Das Marsprojekt' had grown into a detailed technical proposal. To us in today's safety-obsessed culture, it seems astonishing that this experienced and talented engineer was able to sell the idea that manned flight to Mars was possible with the knowledge, resources and technology available at the time. However, not only did he convince himself, he inspired others to the point where the project could have (and, after *Apollo* succeeded in putting men on the Moon, nearly did) become a reality. He began to reveal his plans to the public, starting in 1952, in a series of magazine articles, books, and even a television programme sponsored by Walt Disney, another man with a powerful imagination, whom von Braun had embraced as a fellow enthusiast who would help to get his plans publicised. The mission to Mars that they unveiled to the world was not the tiny, instrumented probe called *Mariner* that, in the real world, finally got there in 1964, but a vast armada of ten ships each weighing nearly four thousand tons, with a total crew of seventy, shown in Plate 2. Every detail was worked out, and every known contingency allowed for, including screening to protect the crew from deadly solar and cosmic radiation, and provision for some limited exchange of personnel between the craft en route to Mars to alleviate the 'unbearable monotony' of the flight. Part of the reason for such a large expedition, although not discussed explicitly in the book, was that some of the ships and crew were expendable in order to guarantee overall success.

In 1956 von Braun published a revised version of the study in a book principally authored by Willy Ley, an experienced science author, and dramatically illustrated by the famous space artist Chesley Bonestell. Ley wrote about the history of Mars exploration, including an objective discussion of the canal and life questions, issues that were still being debated at the time. Recent developments in rocketry, he asserted, were sure to lead quite naturally to Moon landings and Mars expeditions sooner rather than later. Bonestell's dramatic pictures of these departing against the familiar backdrop of the Earth imprinted themselves on the psyche of an enthralled younger generation. Von Braun then continues the narrative to show how it will be done, again with plenty of practical detail including pages of tables giving the principal engineering parameters. Having got the message from reaction to his earlier publications that the concept needed to be scaled down to make it affordable in practical terms, and presumably realising even back then that treating humans as redundant was not a practical approach where the sentimental American public was concerned, the crew was reduced to just twelve people in two ships. The beautiful and seductive frontispiece, reproduced in Figure 1.8, shows the needle-nosed mother craft and its accompanying freighter homing in on the red planet, backwards of course so the main rocket motor can be used to decelerate. Mars fills the sky just eight thousand miles away.

Figure 1.8 *Approaching Mars* (c. 1955) by Chesley Bonestell. From *The Exploration of Mars*, by Willy Ley and Wernher von Braun. Reproduced courtesy of private collection.

The 'Landing Boat' that would take nine of the crew down to the surface of Mars was still a glider, one which landed on skids at a hundred and twenty-two miles per hour and which had a wingspan of a hundred and fifty metres (compare that to sixty metres for a Boeing 747 jumbo jet). The earlier plan had targeted the polar ice cap as the safest region to touch down; in the new plan, a low-latitude site was selected, which would require less fuel since it could be reached from an equatorial, rather than a polar, orbit around Mars.[14] In addition, the climate would be far more equitable for the landing party, with the daytime temperature

---

[14] Spacecraft flying from Earth to Mars usually travel close to the ecliptic plane, which contains the equator of the Sun and, approximately, the orbits of all of the planets. On arrival, the spacecraft passes over the Martian equator, and will enter a low-latitude (equatorial) orbit if slowed down by its rocket motor. To get from there to a high-latitude (polar) orbit requires a further burn of the

at least expected to rise above freezing. On the basis of the observations available at the time, von Braun chose the region known as Margaritifer Sinus, seen in a modern view not available to von Braun in Figure 1.9, as the best place to land. Twenty years later, the *Viking* orbiters photographed this region, and it does indeed look fairly smooth, and not only that, it is also one of the areas on Mars that shows clear signs of flowing water in times gone by. Whether any part of it is quite as smooth as Bonestell envisaged in his fine painting (seen in Plate 3) of the massive craft skidding to a halt under the cloudy Martian skies is quite another matter. Nevertheless, the concept was a technical tour de force and soon captured the public imagination. The time scale envisioned was thirty to fifty years in the future; it is noteworthy that the longer of these estimates has now passed and today's experts are still projecting similar periods before a manned landing will occur, so in schedule terms no progress at all has been made during the space age so far as manned flight to Mars is concerned.

Without unmanned precursor missions, which did not form part of von Braun's plans, the project would certainly have failed. For one thing, he assumed an atmospheric pressure that was more than ten times higher than its true value of less than one per cent of Earth's, and the landing craft would not have been able to glide in as planned and land at no more than a hundred and twenty miles per hour, even with its huge wings. True, the crew would have known about this problem before they commenced their descent, because they were to measure height profiles of the pressure, temperature and composition of the Martian atmosphere by means of instrumented probes fired downwards from their parking orbit, six hundred and twenty miles above the surface. The surface would also be telescopically surveyed (the instrument to be carried for this purpose was huge, comparable to an Earth-based astronomical telescope, and weighed over three thousand pounds) in order to ensure that the landing site was flat and free from dangerous obstructions. They might have stopped at that point and returned to Earth, and come again another day.

Had they pressed on, a total of one hundred and sixty-five tons would have landed on Mars, including the nine crew, the inflatable rubberised fabric tent in which they were to live, and two tractors for getting around during the four hundred days they were to spend on the surface. Six tons of fuel would be enough to drive a total of one thousand, six hundred miles across Mars, studying 'the meteorology and climate, rock formation and soil bacteria, plant life and seasonal changes on another planet'. Seismographs would be deployed and explosive charges detonated to probe the interior of the planet, and searches conducted for 'possible remnants of higher forms of life that might have populated Mars in past

---

motor to tilt the orbit through approximately ninety degrees. As an example, in 2003 *Mars Express* fired its motor for thirty-four minutes, consuming two hundred and sixty-five kilograms of propellant, to get into equatorial orbit, and then made a further four-minute burn to reach its final, polar, orbit.

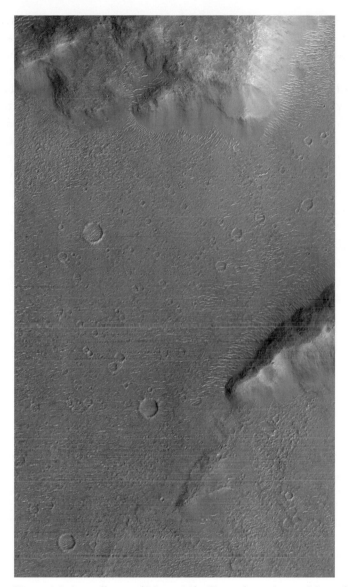

Figure 1.9 A small part of Margaritifer Sinus, the region on Mars where von Braun planned to land his manned Mars expedition (on skids, see Plate 3) and which today is one of the candidates for the landing of NASA's *Mars Science Laboratory*, an advanced robotic explorer. The site was photographed in January 2007 by the high-resolution camera on *Mars Reconnaissance Orbiter*, with which objects as small as fifty centimetres across can be identified on the surface. (NASA/JPL/University of Arizona)

geological ages, and for indications of whether Mars had ever been inhabited by intelligent beings'. The planned landing site was near the 'prominent canal' Lowell had named Thoth-Nepentes, with the intention that its true nature could be examined first-hand at last.

Von Braun revised his vision for Mars a final time in 1969, when the successful *Apollo* landings on the Moon had established a manned space programme in the USA and provided the technology, including the huge launch vehicles he needed for an *Apollo*-style landing on Mars. His dream at last within reach, von Braun moved to an administrative job in NASA Headquarters where he could muster support from within the agency. The 1969 scheme paid much more attention to testing, redundancy and back-up; von Braun had learned a lot of NASA politics since the early days with Walt Disney, and become even more convinced that detailed attention to risk reduction was essential. The Mars landing, now scheduled for 1982, would use *Mars Excursion Modules* that landed vertically under power, like those used so successfully by *Apollo*. The crew of three would now stay on the surface for only sixty days. The plan went on to envision a Mars base manned by fifty people, which was to be in place by 1989. By then, the programme would have run up a total cost of a hundred and forty billion dollars.

The plan was initially well received. Ironically, the largest factor in its demise was the success of the relatively modest *Mariner 7* mission, which photographed Mars close up just days after von Braun made his big pitch to NASA. The *Mariner* pictures showed a barren, Moon-like world, and the popular and political backing for an expensive and still hazardous manned project receded. Backing for the continuation of *Apollo* flights to the Moon was soon to evaporate, too. Disappointed, von Braun retired from NASA two years later, in 1972.

Further reading

The following historic books, with the possible exception of the one by Earl Nelson which I obtained for a few pounds from a used book shop, are available as modern facsimile editions.

The Planetary and Stellar Worlds: A Popular Exposition of the Great Discoveries and Theories of Modern Astronomy, by Ormsby M. Mitchel, W. L., Allison, 1848.

Mars and its Canals, by Percival Lowell, Macmillan, 1906.

Mars as the Abode of Life, by Percival Lowell, Macmillan, 1909.

Is Mars Habitable? A Critical Examination of Professor Percival Lowell's Book 'Mars and its Canals', with an Alternative Explanation, by Alfred Russel Wallace, Macmillan, 1907.

Lowell and Mars, by William G. Hoyt, University of Arizona Press, 1982.

There IS life on Mars, by The Earl Nelson, Werner Laurie (London), 1956.

The Exploration of Mars, by Willy Ley and Wernher von Braun, George, Allen and Unwin, 1960.

Chapter 2
The first space missions to Mars

2.1   Dreams are not enough

Just as rocketry had its origins in war and Hitler's V-2 led to the *Saturn V* rocket that put men on the Moon, the first unmanned missions to Mars had their origins in mankind's basest behaviour, rather than the noble aspirations expressed by von Braun and many others over time. When *Sputnik 1* and *Explorer 1* became the first artificial satellites of the Earth in 1957 and 1958, respectively, their scientific potential was much discussed, but it was their military implications that first drove the Soviet and US agencies to look to the Moon and beyond.

In the USA, the National Aeronautics and Space Administration was formed in 1958 after the discordant attempts by the Army, Navy and Air Force failed to beat the Soviet *Sputnik* satellite into Earth orbit. Work on an American national programme began by setting up various large laboratories around the country. The Jet Propulsion Laboratory in Pasadena, California, which had been founded during the Second World War to work on rocket development, including guided missiles, lobbied under its director William H. Pickering[1] to have responsibility for unmanned lunar and planetary flight projects and the development of the supporting science and technology. Within NASA, JPL (like NASA, often called by its initials) has retained this leading role, although not exclusively, to the present day. Its site in the foothills of the San Gabriel Mountains, originally chosen for its relative seclusion to protect the public from what was seen as (and sometimes really was) some quite dangerous work with rockets, is shown in Plate 4.

The new agency produced a programme for the development of launch vehicles and spacecraft that indicated it could send a probe to Mars or Venus by 1967, with the first planetary missions by humans possible by 1980. NASA's planning for manned spaceflight soon experienced further pressure from the

---

[1]   Dr Pickering was still the Director when the author joined the Jet Propulsion Laboratory as a post-doctoral researcher in 1970, retiring in 1976 after twenty-two years in the post. He was an inspirational gentleman to the end, and very kind even to the most junior of his colleagues.

Soviet Union, with the flight into Earth orbit of Yuri Gagarin on 12 April 1961. Having lost both the manned and unmanned races into Earth orbit, President John F. Kennedy made his famous speech to the US Congress on 25 May 1961 in which he committed the nation to putting Americans on the Moon 'before this decade is out'.

From 1964 to 1968 JPL developed the *Ranger* and *Surveyor* robotic missions to the Moon to carry out photo-reconnaissance in preparation for the manned *Apollo* landings. In parallel with these, and often in competition for vital resources, JPL drew up plans for an interplanetary spacecraft to be known as *Mariner*. The intention was to share launch vehicles and other infrastructure between the lunar and planetary programmes, although NASA's official policy meant that any conflict between the two had to be resolved in favour of the priority set by the President, which was to get men to the Moon and back. (The presidential 'vision' of a programme of Mars exploration via a permanent lunar base, which dominates NASA's planetary activities in the current decade, has had a similar effect.)

Nevertheless, Dr Pickering and his JPL colleagues were inspired by the idea of a planetary programme, and continued to make all the progress they could on the path to Venus and Mars. In 1960, teams from Pasadena started along what is now a well-worn groove in the road to Washington to present their plans to the NASA Administrator. At that time the incumbent was a Yale graduate called Keith Glennan who had, before the Second World War, been studio manager for Paramount Pictures. Pickering showed Glennan a ten-year plan for planetary exploration that included the development of nuclear and ion propulsion systems that could carry spacecraft to Jupiter and beyond.

A two-stage rocket, *Atlas-Centaur*, was being developed for the *Surveyor* program to land instruments on the Moon. Figure 2.1 shows how this vehicle combined the Air Force's *Atlas* Intercontinental Ballistic Missile with a new high-energy upper stage called *Centaur* from the Convair division of General Dynamics. The latter, powered by Pratt & Whitney engines fuelled by liquid oxygen and liquid hydrogen, was dauntingly complex and dangerous, but potentially capable of lifting large payloads. It also had the capability, not available to solid boosters, of starting and stopping several times to give multiple burns, another feature that would be very useful in the long journey to the planets. JPL calculated that an *Atlas-Centaur* could take an eight hundred and eighty-five kilogram payload to Venus, and five hundred and eighty-five kilograms to Mars. To gain experience in deploying spacecraft beyond Earth orbit, and to better understand the hazards there (such as unprotected exposure to the intense radiation environment produced by the Sun and by cosmic rays), they first developed plans to send two prototype *Mariners* to explore the interplanetary medium. Assuming these succeeded in 1963, they would be followed by another two aimed at the nearest and most accessible planet, Venus, in 1964. Shortly thereafter, an improved version called *Mariner B*, under parallel development, would be sent to Mars.

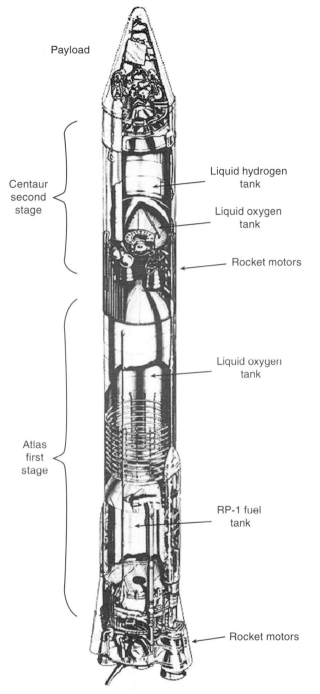

Payload

Centaur
second
stage

Liquid hydrogen
tank

Liquid oxygen
tank

Rocket motors

Liquid oxygen
tank

Atlas
first
stage

RP-1 fuel
tank

Rocket motors

Figure 2.1 The *Atlas-Centaur* launch vehicle, first launched in May 1962, and used to put NASA's unmanned *Surveyor 1* spacecraft on the Moon in May 1966. The RP-1 fuel for the first (Atlas) stage is refined petroleum. The second (*Centaur*) stage weighed about thirteen and a half tons and could put just over a ton in Earth orbit. An upgraded *Atlas-Centaur* took *Mariner 9* to become the first Mars orbiter in 1971; *Centaur* was later mated to the more powerful *Titan* first stage for the *Viking* launches in 1975.

Centaur remained an ambitious development, however, and there were long delays due primarily to problems with its rocket engines, which several times exploded on the test bed. At the same time, JPL experienced multiple failures with the early *Ranger* shots to the Moon,[2] which used the simpler *Atlas-Agena* launcher. Rather than wait for *Centaur* to be ready, JPL decided to make a lightweight planetary spacecraft, based on *Ranger* and known as *Mariner R*, that could be launched to Venus with the *Atlas-Agena* combination. The Lockheed *Agena* upper stage, fuelled with hydrazine and nitrogen tetroxide, was less powerful than *Centaur*, but had overcome its early teething troubles and performed successfully, delivering the *Rangers* successfully to the Moon in 1964–5. In an amazing eleven months from getting the go-ahead for *Mariner R*, JPL was ready to inaugurate the age of planetary exploration by spacecraft.

Thus, 22 July 1962 saw the launch of the first *Mariner*, bound for Venus. The flight lasted only five minutes before the spacecraft strayed off course and was destroyed by a command sent by the Range Safety Officer. However, its identical twin, *Mariner 2*, did not fail; launched a month later, it flew past Venus on 14 December 1962 and, amongst other achievements, confirmed that the planet has an improbably high surface temperature, higher than the melting point of soft metals such as lead and tin. Obviously, this profoundly changed the prospects for a manned mission to Venus, which previously had been the favourite choice of some planners, because it is the nearest planet to Earth, and had been expected to have a lush, water-rich environment, unlike dry, barren Mars.

2.2   The Mars *Mariners*

The *Mariner* Mars project was authorised by NASA Headquarters in November 1962, with the intention to launch twin spacecraft just two years later. Challenges that would have to be overcome in that time included making the mechanical and electronics systems more reliable to cope with the significantly longer flight time to Mars compared to Venus, which often meant adding redundancy.[3] In addition, more power and mass would be needed for communications and thermal control at Mars' much greater distance from the Sun and from the Earth. On top of these basic requirements, scientists were keen to add larger and more sophisticated instruments to make measurements that could start to address exciting goals like trying to detect the presence of life.

---

[2] The *Ranger* programme that preceded *Surveyor* was designed to take close-up pictures of selected regions on the Moon before it crashed into them. *Surveyor*, in contrast, was to soft-land instruments on the lunar surface, a much more ambitious goal that required a more powerful launcher.

[3] 'Redundancy' refers to the practice of doubling-up on systems, such as tape recorders for storing data onboard, in order to have a backup if one should fail. In the early planetary exploration programme, it was common to have not just redundant systems within a spacecraft, but to build two copies of the entire spacecraft, to be launched separately. Nowadays, redundancy as a policy has gone out of fashion, since, for example, reliable solid-state memories have replaced tape recorders, and launch vehicles fail less often.

Putting the spacecraft into orbit was obviously more desirable than flying rapidly past the planet, since much more time could be spent taking observations (typically months or years, versus a few hours) and extensive global coverage of the planet becomes possible. Getting into orbit means reducing the velocity of the spacecraft on arrival, so it can be captured by Mars' gravitational field. This would require carrying a heavy retro-rocket and its fuel all the way to Mars, which would not be possible until a reliable *Centaur* upper stage became available, and this was still several years away. Detaching a small, instrumented package from a spacecraft flying past Mars at twenty thousand miles per hour to parachute safely to the surface of the planet was another possibility that, though daunting, was seriously considered.

As they still do, these exciting plans attracted a very high calibre of engineer keen to take up the challenge. They found ingenious ways to make the flight hardware lighter and at the same time more capable. Most importantly, they worked fanatically to make sure the spacecraft and their science instruments were ready for launch on time. Any project that missed its narrow launch 'window', when Earth and Mars were favourably aligned, would not be able to reach its target without waiting for the next window more than two years later (Figure 2.2 explains about launch windows and why they occur with this separation). Such delays were not only frustrating for the team; they meant increased costs and possible cancellation.

The first *Mariner* flight to Mars failed, however. The housing, known as a shroud, at the top of the rocket, should have split and separated once it was out of the Earth's atmosphere, releasing the interplanetary spacecraft. When it failed to do so, the solar cell arrays that were meant to provide power to the spacecraft could not be unfolded. This left *Mariner 3* with only the current from its batteries to maintain communication with the control station on the ground, and these were soon depleted. NASA rushed to test a spare shroud in the laboratory to find the problem, and learned a salutary lesson. The shroud was a honeycomb structure made of fibreglass to reduce the weight, always a goal where space flight is concerned. When pumped down in a vacuum chamber to simulate the fall in pressure as the rocket climbed, the shroud actually exploded under the pressure from the air trapped inside the material. Undaunted, exercising the brilliant crisis management and engineering skill that has come (along with the inevitable mistakes and miscalculations) to characterise the space programme, the team designed, built and tested a metal shroud and had *Mariner 4* ready to go before the launch window closed, only three weeks after the disaster to its sibling. And the replacement shroud was actually lighter than the original.

Mariner 4, pictured in Figure 2.3, took off successfully on 28 November 1964, but soon encountered some fresh problems. Like its early namesakes plying the oceans, *Mariner* was to navigate using a bright star, in this case the yellow giant Canopus, as a reference to maintain its orientation and direction. For the first ten

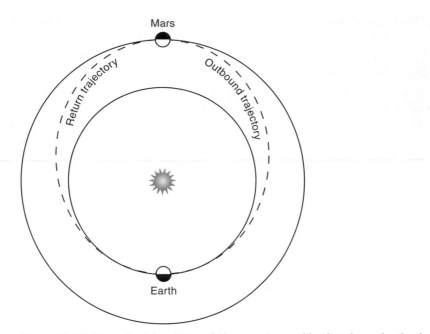

Figure 2.2 Flights to (and from) Mars follow a trajectory like that shown by the dashed line, which uses the least amount of fuel. Earth must be at the point shown at departure, and Mars at the point shown at arrival, and vice versa for the return journey, a situation that occurs at intervals of just over two years. Every seven hundred and eighty days or about twenty-six months (the synodic period of Mars), Mars and Earth are in the same place in their orbits with respect to each other, with the two planets directly opposite each other on opposite sides of the Sun. At that time, the energy required to send a spacecraft directly from Earth to Mars is a minimum. A launch opportunity or 'window' is the interval, centred on the optimum time and typically two to five weeks in duration, in which a spacecraft can be launched and still reach Mars. Recent and upcoming occurrences and missions planned to use them are in October 2007 (*Phoenix*), October 2009 (*Phobos-Grunt*), December 2011(*Mars Science Laboratory*), December 2013 (*MAVEN*), and January 2016 (*ExoMars*).It will eventually, when sample return and human expeditions finally happen, be noteworthy that minimum energy windows for launch from Mars back to Earth occur with the same frequency. Due to the non-circularity of the orbits, especially that of Mars, not all launch windows are equally good, and a larger or smaller booster may be required to propel a given mass to or from Mars, depending on the opportunity used.

days of its flight, however, *Mariner 4* perversely locked on to the bright double star γ Velorum instead, and headed resolutely away from Mars. The mission controllers back on Earth worked out that small particles of dust and debris, brighter than any star, were passing close to the sensor and confusing it. A fresh search for Canopus, initiated by command from ground control, put it back on track for a while, and could be repeated whenever the problem occurred. This did the trick, and (as one contemporary observer put it) the spacecraft fumbled its way to Mars, drawing near on 14 July 1965.

Figure 2.3 *Mariner 4*, the first spacecraft to explore Mars successfully, had a mass of two hundred and sixty kilograms. The octagonal main body is a hundred and twenty-seven centimetres across, with a span of just under seven metres including the solar panels. These supplied three hundred and ten watts of power at Mars.

While *Mariner 4*'s overwhelming important objective was to photograph the surface of Mars and transmit the pictures back to the Earth, it also carried experiments to measure the density and composition of the solar wind,[4] the magnetic field near Mars, and any associated Van Allen-type belts of energetic subatomic particles around the planet. A sensitive way to detect the presence of a small magnetic field is to look for its effect on atoms of helium gas. Such a magnetometer on *Mariner 4* searched without success for a global Martian magnetic field – we now know there is none, except for a very weak remnant field stored in magnetic material in the surface rocks. Its cosmic dust detector was more positive in its findings, using surface penetration detectors and a microphone attached to an aluminium plate to record the momentum, direction, and number of hits by solid particles in space. Very little was known before that time about this potential hazard from the smallest solid components of the Solar System. It turned out to be less of a problem than many had feared.

---

[4] The solar wind is a stream of charged particles that is continuously emitted from the Sun at high speed (hundreds of kilometres per second) and reaches all of the planets, being responsible for producing the aurorae in Earth's upper atmosphere when focussed to the poles by the magnetic field. At Mars, where there is little or no field, the solar wind impinges more directly onto the atmosphere. This could have changed the rate at which gases were stripped away into space, a process that may have contributed significantly to the low density of the present Martian atmosphere (see the discussion in Chapter 5).

Observing the surface of Mars close-up with cameras meant achieving much higher contrast and spatial resolution than the best telescopes in observatories on Earth, and the problems with local atmospheric absorption and turbulence ('seeing') that plague terrestrial astronomers were eliminated altogether. Thus the high-resolution pictures from television cameras mounted on the probes were expected to reveal many exciting details of the features on the surface of Mars for the first time. Anticipation was high as the pictures were transmitted in coded form back to the Earth by radio, and reconstructed at NASA's planetary mission control centre at the Jet Propulsion Laboratory.

What the scientists saw, and their reaction, when the images were received at JPL, was recorded in real time by a media team. Watching this grainy, yellowing monochrome film today, the sense of the excitement of the *Mariner* team's achievement in obtaining the pictures is clearly tempered with disappointment at the austere nature of what they saw (the example in Figure 2.4 is one of the more exciting ones). The journalists reported that the cold, cratered surface and tenuous atmosphere seemed to have more in common with the Moon than the Earth, and bore none of the hoped-for resemblance to the landscapes or lifeforms that were familiar to many from the fictional works of romantic authors like Edgar Rice Burroughs and H. G. Wells,[5] and certainly no trace of Percival Lowell's planetary-scale civil engineering.[6]

Mars was indeed much more heavily cratered than anyone had expected. Most craters in the Solar System formed long ago, when there were many more impacts; they survive on the Moon because volcanism ceased early in that small body's history, limiting the obliteration of craters by flooding with lava, and because of the lack of erosion by atmosphere or water. Their survival in such numbers on Mars meant an ancient surface, largely free of most of the interesting phenomena that have removed nearly all traces of impact craters on the Earth, although they were once as numerous as on the Moon. The leader of the *Mariner 4* imaging team, Robert Leighton, a geologist from Caltech,[7] estimated that they were looking at features on Mars that were two to five billion years old, and very well preserved. There were no signs of life, no canals, and no vegetation, in fact no really Earth-like geological features, such as mountains and valleys or ocean basins, either. The new estimates of mean surface temperature and pressure were

---

[5] The 'Tarzan' author Edgar Rice Burroughs also wrote a series of stories set on a populated Mars, beginning with *Under the Moons of Mars* in 1912, and ending with *John Carter of Mars* in 1964. Wells published *The War of the Worlds*, about the invasion of Earth by Martians, in 1898.

[6] The canal hypothesis had survived in some quarters long after the death of Percival Lowell in 1916. As late as 1962, Earl Slipher, the brother of Vesto who had been Lowell's assistant and successor at the Flagstaff observatory, published a book entitled *The Photographic Story of Mars* in which he wrote: 'every skilled observer who goes to the best available site for his observations has had no great difficulty of seeing and convincing himself of the reality of the canals.'

[7] The California Institute of Technology in Pasadena, a few miles away from the Jet Propulsion Laboratory, which it administers on behalf of NASA.

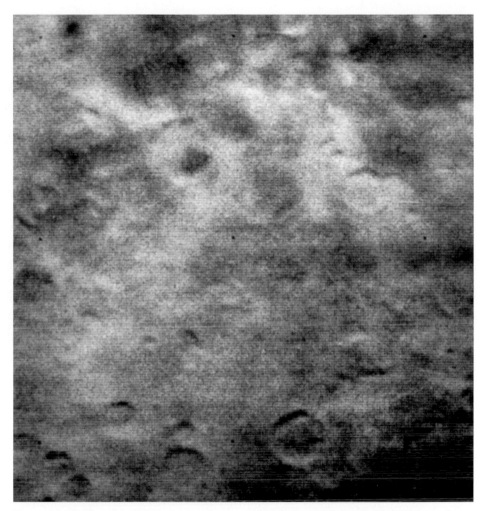

Figure 2.4 *Mariner 4* took this picture of craters on Mars on 15 July 1965, from a range of about fourteen thousand kilometres. It covers an area about three hundred kilometres across, centred at 14° S latitude, 174° W longitude, in a region south of Amazonis Planitia.

so low (at about minus forty degrees centigrade, and half a per cent of Earth, respectively) that the polar caps were now seen as more likely to be frozen carbon dioxide than water ice. The investigators gave as rosy an opinion as they could by concluding that the photos 'neither demonstrate nor preclude the possible existence of life on Mars'.

There were only twenty-two pictures in total from *Mariner 4*, and they covered less than one per cent of the surface area of the planet. The pictures were of poor quality, even by the standards of the fledgling space photography business, probably due to a light leak in the camera. A big improvement was expected from *Mariners 6* and *7* when they arrived at Mars in 1969. These had better cameras, and could take many more pictures, including some from a range as

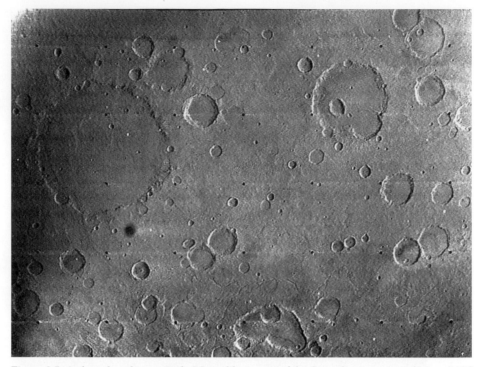

Figure 2.5 A sharp but depressingly Moon-like image of the Deucalionis region of Mars (15° S 340° W), obtained by *Mariner 6* on 31 July 1969. Pictures like this were a factor in killing off Wernher von Braun's vision of a manned Mars mission to follow the successful *Apollo* landings on the Moon.

close as only three and a half thousand kilometres from the surface. All this offered the possibility of fresh insights.

The two identical spacecraft both worked well and, between them, returned one thousand, one hundred and seventy-seven sharp, clear pictures, now covering more than ten per cent of the surface of the planet. As the example in Figure 2.5 shows, they did not do much to dispel the depressing picture left by their predecessor, however. There still seemed to be an awful lot of craters, with very little evidence, direct or indirect, for past or present water on the surface, and certainly none for vegetation or any other kind of life. The attenuation of the radio signals when the two spacecraft passed behind Mars was used to confirm the surface pressure as six and a half millibars in the Sinus Meridiani region and only three and a half over Hellespontica Depressio, an elevated region near the south polar cap that (like Mitchel's mountains) was misnamed by early astronomers. Measurements by an infrared radiometer on the spacecraft of the heat radiated from the surface showed that the temperature of the polar cap was minus a hundred and twenty-three degrees centigrade, definitely corresponding to frozen carbon dioxide. All considered, conditions on Mars were apparently so bleak that the only forms of life worth searching for would be microbes, and finding those would require landing on the surface.

At the time that *Mariners 6* and *7* were painting a gloomy picture of an ancient, cratered, dead world with a cold surface and thin air, their successors, *Mariners 8* and *9*, were being prepared to move to the next stage of exploration by becoming the first artificial satellites of Mars. The detailed survey of the whole of the planet that these could provide was an essential precursor to landing, not least for choosing the best place to land. In any case, most concepts for landed stations require getting into orbit first and descending from there. But was it worth proceeding with these increasingly expensive and risky missions? Perhaps not, if the rest of Mars was a cold, cratered desert like the tenth of the surface already covered by the first three *Mariners*, and the almost airless nature of the planet was not encouraging either.

The scientists rallied. They pointed out that there were still the clouds, storms and polar caps that had fascinated observers for centuries, and that some locations showed signs of interesting geological evolution. The floor of the Hellas basin, for example, was virtually free of craters, and since it must have been bombarded at the same time as the ancient cratered terrain, some more recent process must have smoothed it over. Geologists were also intrigued by chaotic, ridged regions, also nearly craterless, which were not like anything found on the Moon. Professor Leighton told the Press that the first *Mariners* had discovered significant new and unexpected terrain, saying, 'I leave it to you to figure out how many new surprises there are still waiting for us on Mars.' His colleague Norman Horowitz, a biologist, came out and said that versions of terrestrial life could survive under the distinctly non-Earth-like conditions now seen to prevail on Mars. He advocated landing an automated laboratory that could make a 'direct test' for life, thereby adding an irresistible challenge to the renewed momentum already built up.

Placing the next generation of *Mariners* into orbit about Mars would need much more precise navigation than that which sufficed for a fly-by mission. Passing too far away from the planet would result in insufficient gravitational attraction to trap the spacecraft into orbit; too close, and it would crash onto the surface. A current NASA Mars programme manager recently likened the very small margins for error involved to throwing a basketball from New York to Los Angeles and putting it through the hoop without touching the sides. It was also necessary to accommodate the extra weight for the retro-rocket and fuel needed to decelerate the spacecraft on arrival. The science teams were not content with an equatorial orbit, either; if the observations were to cover all latitudes, in particular the interesting, icy polar regions, the spacecraft had to burn its motor longer in order to achieve a high-inclination orbit. With all of the fuel that implied, plus bigger cameras than before and additional scientific instruments, *Mariners 8* and *9* (the latter seen being prepared for launch in Figure 2.6) each weighed nearly a ton at launch, more than *Mariners 6* and *7* combined. Once again, the first launch failed when, on 8 May 1971, *Mariner 8* fell into the Atlantic

Figure 2.6 *Mariner 9* being prepared for launch at Cape Canaveral in May 1971. The solar panels are folded upwards to create a profile that will fit inside the shroud (part of which can be seen to the left), which fits on top of the *Atlas-Centaur* launch vehicle.

Ocean following problems with the *Centaur* upper stage. NASA had reason to be thankful for its policy of redundancy when, three weeks later, its twin launched successfully, and on 14 November 1971 *Mariner 9* became the first man-made moon, not just of Mars, but also of any planet other than the Earth.

Peering down from its orbit one thousand, six hundred and fifty kilometres above the surface,[8] *Mariner 9* found the planet in the throes of a major global dust storm. It was more than a month before the dust cleared enough to reveal the surface but, when it did, the results were spectacular: giant volcanoes, far larger than Everest; sinuous channels resembling dried riverbeds, complete with

---

[8] This is the distance at closest approach. The orbit is elliptical, taking the spacecraft about ten times further away from the surface at its greatest distance. A circular orbit would have been better for mapping, but more expensive in terms of fuel to achieve.

tributaries and deltas; a huge canyon that could dwarf Earth's grandest in Arizona; many more of the vast, relatively uncratered plains previously glimpsed in Hellas; abundant volatiles in the form of clouds and ice, especially at the poles. Considerable structure could now be seen in the polar caps, including the swirl of ridges and valleys seen in Figure 2.7 in the northern cap, and bizarre layered terrain around the southern one that spoke of long-term climatic change. During its one-year lifetime, *Mariner 9*'s camera covered about seventy per cent of the surface at a resolution that approached one kilometre in the best of its more than seven thousand pictures. The Moon-like character glumly suggested by the earlier missions was replaced almost overnight by the concept of Mars as a planet with a water-rich history like the Earth's, much more suitable for the development of life.

The great canyon, now named Valles Marineris after the spacecraft that revealed its true nature, seen in Figure 2.8, is an immense complex of canyons, more than six kilometres deep, a hundred kilometres wide, and several thousand kilometres long, just south of the equator to the east of the giant Tharsis

Figure 2.7 A *Mariner 9* image of the north polar cap of Mars, seen surrounded by clouds shortly after summer solstice when the cap is at its smallest, about a thousand kilometres across, and the underlying topography shows through as an irregular spiral pattern.

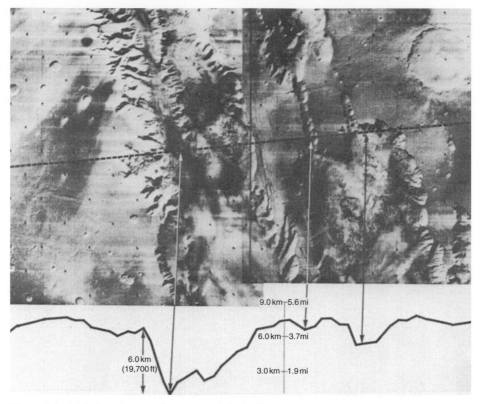

Figure 2.8 A *Mariner 9* view of a section of the Valles Marineris system, covering an area six hundred kilometres across. Pressure measurements along the dotted line by the ultraviolet spectrometer also on *Mariner 9* are shown converted to depths in the profile below. This part of the canyon is about six kilometres deep, compared to less than two kilometres for one of the deepest valleys on Earth, the Grand Canyon in Arizona.

volcanoes.[9] At its western end, this becomes a maze of smaller canyons known as Labyrinthus Noctis, while to the east is found some of the chaotic collapsed terrain. Valles Marineris itself is what Earth-bound geologists call a rift valley, formed when large-scale movements probably associated with the formation of the Tharsis ridge ripped the crust apart, but with considerable evidence that running water was also involved. For instance, the chaotic terrain at the eastern end appears to be the source of a number of dried-up rivers that run into the northern plains, themselves apparently the bed of a former ocean. They were not visible to *Mariner 9*, but higher-resolution observations of the floor of the canyon by later missions have since shown layering that must be sedimentary in origin, and signs of erosion in the canyon walls, including some large landslides, suggesting a great age.

---

[9] Carl Sagan noted that Valles Marineris coincides with one of the canals, called Agathodaemon. in Percival Lowell's maps of Mars. However, he could find few other correlations between 'canals' and real features in the *Mariner 9* images.

In contrast to this immensely deep valley were the enormously high mountains, the largest of which, Olympus Mons, had for a long time been known to astronomers as Nix Olympica (nix meaning snow) because it appears bright when seen through Earth-based telescopes. As long ago as the time of Schiaparelli it was thought to be a high region, as it remained bright when the 'wave of darkening' that is now known to be storm-driven airborne dust obscured all of the features on the rest of the planet. This is the same phenomenon that *Mariner 9* observed on arrival at Mars. When the dust from the 1971 storm subsided, Mariner's cameras took pictures that showed Nix Olympica to be an extinct volcano twenty-five kilometres high and six hundred kilometres across, prompting the name change. Olympus is bigger than anything comparable known elsewhere in the Solar System, since Maxwell on Venus is only about eleven kilometres high, and Everest less than nine kilometres above sea level.

The shallow sinuous valleys tens and hundreds of kilometres long that occur throughout the ancient cratered terrain resemble the networks of streams that drain high ground on Earth. Altimetry measurements have shown that they used to do the same thing on Mars, suggesting that the water may have fallen as rain or snow during a period of moderate climate that lasted long enough to carve the valleys. It was apparent that this must have occurred early in Mars' history, since the valleys are nearly all found on the oldest part of the Martian surface, although more recent water-related features have since been identified by later missions.

Mariner 9 is still orbiting Mars to this day, although it was switched off, its mission complete and its attitude control gas depleted, on 27 October 1972.[10] It had operated for three hundred and forty-nine days, in which time it succeeded in making Mars interesting again by revealing its rich geological history and suggesting a warmer, wetter, more Earth-like past, when conditions on the surface may have been similarly conducive for life. This poses the exciting question of whether the apparent existence of conditions suitable for life on Mars means life actually arose there at about the same time it was appearing on Earth. The only way to answer this key question would be to look for bacteria in the Martian soil, and for possible traces of earlier life forms as fossils of some kind. NASA had been working since 1967 on a mission called *Viking* that could seek to provide the answer; this now forged ahead with renewed vigour.

---

[10] The end date of a mission is determined by any of a number of factors. Ideally, the goals of the mission are deemed to have been attained after a fixed length of time and the controllers simply switch the spacecraft off by command from Earth. This period is agreed before the spacecraft launches, and is sometimes extended if enough scientists argue sufficiently persuasively that new discoveries need to be followed up by an extended mission. They have to make a good case, because keeping the ground station and the various support teams going longer than originally planned involves additional cost and may affect other missions. Alternatively, of course, if a major system fails making it pointless to continue, a mission may terminate before its appointed time.

2.3 Soviet Mars missions

Meanwhile, the Russians were also busy. Curiously, while they achieved spectacular successes in landing on the hostile surface of Venus, they were unable to do the same at Mars. The Soviet Union was actually the first nation to launch a spacecraft towards Mars, in November 1962. Contact with *Mars 1* was lost when it when it was at a distance of one hundred and six million kilometres from Earth, and it was never heard from again. This kind of complete, instant failure of what was, up to that point, an apparently healthy spacecraft, is particularly frustrating for the team because the loss of the data relay, along with everything else, allows no possibility of ever knowing what went wrong. A similar fate has befallen quite a number of Mars missions since, including NASA's *Mars Observer* in 1991 and *Polar Lander* in 1999, and ESA's *Beagle 2* in 2003. Speculation about the cause of the accident always follows, some of it (because it is about Mars) often quite bizarre (including alleged sabotage by Martians, or intervention on the part of the FBI, the KGB, or other sinister government agency on Earth, united in their intent to prevent *Observer*'s high-resolution cameras from revealing artefacts of civilisations on the surface[11]).

Two back-up launches, *Mars 1962 A* and *B* (like NASA, the Soviets name their missions by the year of launch until they are successfully en route, and then switch to a sequential numbering system) failed to get away from the Earth. Undaunted, they went on to fly two *Viking*-like orbiter-lander combinations, *Mars 2* and *3*, as early as 1971, using the same launch window used by NASA's *Mariner 9* orbiter and arriving just a couple of weeks later. Both of the Soviet orbiters functioned, somewhat erratically, for eight months, managing to photograph the surface, search for the elusive Martian magnetic field, and make spectroscopic measurements of water and dust in the atmosphere.

The landers were less successful. *Mars 2* entered the Martian atmosphere at too steep an angle, failed to deploy its parachute, and crashed on the surface on 27 November 1971, winning the Pyrrhic honour of being the first man-made object from Earth to reach the surface of Mars. The wreckage lies at 45° S, 302° W, awaiting the gaze of future explorers.[12] *Mars 3* used its parachutes and retro-rockets to land its one thousand, two hundred kilogram mass safely on 2 December 1971 at 45° S, 158° W, during a raging dust storm (the same storm that was observed from aloft by *Mariner 9*, as well as the two Soviet orbiters). The transmission from the surface failed after only ninety seconds, before the four

---

[11] A best-selling book, *Dark Mission: the Secret History of NASA* by Richard Hoagland and Mike Bara includes a number of such stories. See also Hoagland's *Monuments of Mars* for further off-the-wall views. An interesting debate can centre around whether such rampant speculation, sincere or otherwise, helps or hinders serious scientific Mars exploration.

[12] In a recent movie, *The Red Planet* (2000), a team of stranded astronauts goes one better by not only finding the fifty-year-old wreckage of a Russian lander, but also deftly repairing it and using it to fly themselves up into Mars orbit and thence back to Earth. This sort of nonsense definitely does little to help serious scientific Mars exploration.

and a half kilogram rover it carried, which was to have walked on skis on the Martian surface at the end of a fifteen-metre tether, could be deployed. Only a small fragment of the first of many planned pictures was relayed. The mission controllers speculated that the dusty, windy conditions might have produced electrical discharges that damaged the spacecraft electronics.

In 1973, no fewer than four Soviet spacecraft were launched towards the red planet, *Mars 4* and *5* on 21 and 25 June, and *Mars 6* and *7* on 5 and 9 August. In 1973 the launch window was less favourable than it had been in 1971, so that even the enormous *Proton* rocket available to the Soviets could not boost the orbiter-lander-rover combinations to Mars. Instead, the first pair of launches was to place spacecraft in Martian orbit, ready to receive the data relayed from the surface by the landers that would be delivered by the second pair. In the event, *Mars 4* missed the planet due to a failure of the braking rockets to fire, and *Mars 7* had a similar fate due to an electronics failure. *Mars 5*, pictured in Figure 2.9, was partially successful, operating for twenty-two days in a looping twenty-five hour orbit inclined at thirty-five degrees to the Martian equator and approaching within two thousand kilometres of the surface. The remaining lander, *Mars 6*, deployed its parachutes and reached the surface, but sent no signals. It missed its intended landing site by a wide margin and came in over chaotic terrain that included Samara Vallis, a gorge comparable with the Grand Canyon on Earth, which may explain why it apparently crashed on landing.

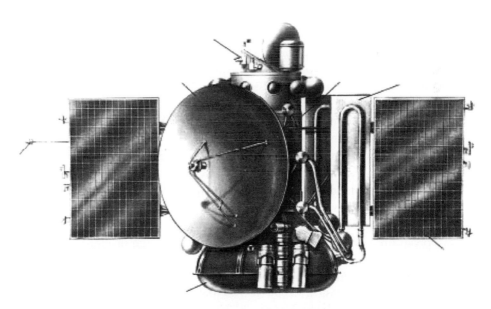

Figure 2.9 The Soviet *Mars 5* spacecraft, which arrived in Mars orbit on 12 February 1974. In addition to two cameras, an infrared radiometer and an ultraviolet photometer, it carried two French experiments, one to study the distribution and intensity of fluxes of solar protons and electrons and another to study solar radio emissions.

A long pause ensued in the Soviet Mars exploration programme, lasting until 1988 when an ambitious double mission to land on the small Martian moon Phobos was attempted, followed eight years later by the massive *Mars 96*, which weighed in at nearly seven tons. The failure of both of these, and the ongoing political changes in the Soviet Union, meant a further hiatus that continues to the present day.

2.4   The *Vikings*

While the 'space race' to get to new worlds before the Russians was the driving force for NASA's lunar exploration programme in the late sixties and early seventies, the reason increasingly given for pressing on to Mars after that was the search for extraterrestrial life. This was probably fortunate, since scientific goals are less prone to changes of fashion or circumstance than political ones, and having a principled motivation would mean that NASA could go on to build a strong Mars exploration programme even after the Russians had long ceased to compete. This they did, in spite of opposition and cynicism in some quarters, most noticeable among members of the press and politicians who wanted to 'spend the money here on Earth' – as if it was spent anywhere else.

The trouble with astrobiology as a goal was that *Mariner 9* had confirmed that the environment on Mars was unlikely to be home to any kind of advanced species. Although a few pundits continued to stress, as one or two do to this day, that advanced civilisations might have developed during the early warm, wet era and since secreted themselves underground, the focus had to be shifted onto bacteria and other microscopic organisms that might have developed along the same lines as on Earth, and then adapted to the changing climate. Since the 1990s it has been known that some terrestrial microbes are indeed extremely hardy, and could survive in Mars-like environments, especially if shielded from the ultraviolet radiation from the Sun, but at the time of *Viking* this was a moot point. The UV component of sunshine can be harmful to life even on the Earth, which is protected by a thick atmosphere containing a layer of UV-absorbing ozone for which there is no equivalent on Mars. However, the *Viking* planners decided to test the theory that microscopic life could survive on Mars just below the surface, if they could land and dig and analyse the soil.

The goal of landing on the surface of the planet and searching for signs of life had already been anticipated in the heady days of the early 1960s in studies of a programme called *Voyager*. This envisioned a large orbiter-lander combination propelled to Mars by a *Saturn 1B-Centaur* combination or even, in later versions, by the huge *Saturn V* being used to take men to the Moon. The possibility of a manned expedition to Mars as a sequel to *Apollo* and *Voyager* was also on the cards, as we saw in Chapter 1.

The Mars *Voyager*, unlike its namesake which flew to the outer planets a decade or so later, was extensively studied and designed, but never built and flown. The reason was basically the perennial one of balancing goals against cost and risk: NASA felt that too little was known about the environment of Mars to risk an expensive landing, even if only robots were facing possible disaster. This view was vindicated later by the problems *Mariner 9* found with the global dust storm that obscured the planet as it arrived and led to the early demise of the only successful landing ever made by the Soviets. In 1967 the Jet Propulsion Laboratory, now in partnership with NASA's Langley Research Center in Virginia, responded with a cheaper version of *Voyager*, which became *Viking*, and which could be launched by the *Titan 3C* booster with a *Centaur* upper stage at a tenth of the cost of a *Saturn V*. This project gained the support of President Lyndon B. Johnson, who said to Congress in January 1968, 'We will not abandon the field of planetary exploration', recommending the 'development of a new spacecraft for launch in 1973 to orbit and land on Mars'.

Landing on Mars for the first time, with delicate equipment requiring a soft touchdown, raised technical issues that got more challenging the closer they were examined (Figure 2.10). Many frustrating delays and cost over-runs later, the *Vikings* finally headed for Mars following launches on 20 August and 9 September 1975. Each consisted of two identical composite spacecraft, with an orbiter and a large surface station that flew, bolted together, into Mars orbit. The lander would then detach and use a combi-nation of parachutes, retro-rockets and a radar altimeter to touch down gently on the surface. Each orbiter had more than four times the mass of *Mariner 9*, with cameras capable of colour imaging and advanced remote sensing instruments. The latter included an infrared radiometer to map the surface temperature and, among other goals, search for anomalously warm regions on the surface that might be the result of residual volcanic activity. A spectrometer called MAWD, for *Mars Atmospheric Water Detector*, was tuned to measure water lines so it could map the humidity across the face of the planet and watch it change with season, representing the ultimate evolution of Slipher, Spinrad and Young's Earth-based experiments.

Viking Lander 1 landed on 20 July 1976 in Chryse Planitia at 22° N latitude, 47° W longitude and operated for more than six years, until 13 November 1982. Its twin landed on Utopia Planitia on 3 September 1976 at 47° N latitude, 225° W longitude, and also exceeded its ninety-day design lifetime, transmitting data including views like that in Plate 5a to Earth until it went silent on 11 April 1980. These landing sites were not the ones that had been chosen in advance: the surveillance that was carried out from orbit before the landers detached and descended showed that the latter were much too rough to risk a landing. The search for new sites, the blander the better, took time, and meant that the first landing did not take place on 4 July 1976 to add to the bicentennial gaiety of the American nation as originally intended. This was a wise precaution, how-ever, because even the plains that looked the smoothest from above turned out to be strewn with some dangerously large boulders, as the first ever picture from the surface of Mars (Figure 2.11) soon revealed.

Figure 2.10 Setting a heavy spacecraft down on Mars is difficult, and not helped by the fact that parachutes do not work very well in the thin atmosphere. In 1976 two *Viking* landers, each with a mass of nearly six hundred kilograms, made a powered descent for the last few hundred metres, using downward-facing retro-rockets (fuelled with specially purified hydrazine to limit the contamination of the Martian surface) to decelerate, and radar to sense the proximity of the surface (inset). Thirty-two years later, the three hundred and fifty kilogram Phoenix used a similar technique to land near the north pole of Mars (main picture).

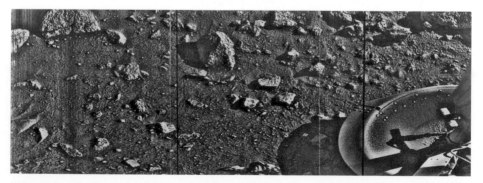

Figure 2.11 The first-ever photograph taken on the surface of Mars, obtained by the camera on *Viking 1* shortly after its landing on Chryse Planitia (the Plain of Gold) on 20 July 1976.

Possibly much against the odds, both landers set down safely. The instruments on the two *Viking* orbiters also functioned extremely well, and had long lifetimes. *Viking 2* terminated operations on 25 July 1978 and *Viking 1* on 17 August 1980, after 706 and 1485 orbits respectively. Technically and scientifically, *Viking* was an enormous success all round.

2.5 The post-*Viking* picture of Mars

After the new data had been assimilated, published and discussed, the scientific community found itself with what seemed to be, if not by any means a complete understanding of Mars, the first reasonably detailed, well-rounded picture of the red planet and its similarities to, and differences from, the Earth. First among the highlights are the images: the views of the surface of Mars at the two *Viking* landing sites revealed the landscape and the local geology on a human scale for the first time. Of course, these were desert sites, chosen for landing because of their blandness, and turning out, in the event, to consist of sandy soil strewn with rocks. Still, the sense of 'being there' at last was very real, particularly when viewing the glorious panoramic pictures, like that in Figure 2.12. The cameras on the orbiters collected a huge number of high-quality photographs of the surface – almost sixteen thousand images from *Viking 1* alone – to compile a record of the global geology, which turned out to be remarkably diverse.

Volcanic constructs, large topographic features like the huge Tharsis bulge, the polar caps, great rift valleys, impact craters of various kinds, vast plains, channels and aeolian (wind-formed) features were all recorded and mapped in detail. Images like Figure 2.13 also produced a convincing body of evidence for channels cut by water and other evidence of massive climate change in the past. Data on the temperature of the surface and of the atmosphere were collected on a similarly large scale, and this, with the water vapour mapping by MAWD, defined the present-day Martian climate in some detail.

The landers were equipped with an arm to scoop up samples of Martian soil, and an automated laboratory to perform wet-chemistry experiments designed to

Figure 2.12 A panoramic view from *Viking 1* of Chryse Planitia on which it landed on 20 July 1976.

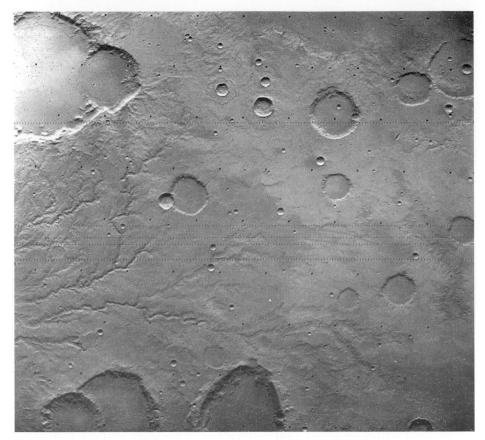

Figure 2.13 A *Viking 1* orbiter image showing Parana Valles (23.8° S, 8.7° W), a region containing channels apparently cut by running water in the past. This area, about two hundred and fifty kilometres across, falls within the Margaritifer Sinus region of Mars, which was targeted in 1956 by Wernher von Braun as the landing site in his plans for a human Mars expedition (see Chapter 1).

reveal evidence of life-related processes. A mass spectrometer[13] made the first detailed analysis of atmospheric composition, including gases present only in tiny amounts, especially the noble gases argon, neon and krypton that are very difficult to detect by optical spectroscopy, and the isotopic ratios of some of these. Argon, for example, has two common isotopes,[14] one of which is produced

---

[13] A mass spectrometer is a device for analysing gases by ingesting a sample and determining the masses of the various molecules present from the trajectory they follow through a magnetic field. It should not be confused with an optical spectrometer (see the description with Figure 1.5), which analyses the light emitted or reflected from an object from an arbitrarily large distance away.

[14] Isotopes are different versions of the same element that differ slightly in mass, due to having additional neutral particles (neutrons) in the nucleus of the atom. Most common elements have several naturally occurring isotopes, for example, oxygen has three, ^{16}O, ^{17}O and ^{18}O, with masses in the proportion 16:17:18, the first of these being the most common.

by the decay of radioactive potassium in rocks and therefore is time dependent in abundance, and one of which is primordial. Neither reacts chemically, so their ratio should depend only on the time elapsed at a known rate of decay, making potassium a kind of geological clock. This also works on Earth and Venus, so even if the interpretation is wrong in absolute terms, differences between the planetary siblings should stand out clearly.

Combined with a gas chromatograph, the mass spectrometer could search for organic molecules in soil samples, right down to the parts per billion level, by heating the soil and analysing the gases given off. No life-related signs were detected by *Viking*, however. Instead of the organically rich material that had been predicted to lie underneath a thin crust of UV-sterilised soil on the surface, the Martian soil produced only water and carbon dioxide when it was heated. The likely reason for this soon became clear when the soil was found to be not only sterile but also highly oxidising down as far as *Viking* could dig, which was about fifteen centimetres. The oxidising agent was identified as hydrogen per-oxide, the same substance that is used to sterilise wounds and surgical instru-ments on Earth. On Mars it forms in small quantities in the atmosphere, again by photochemistry under the influence of solar ultraviolet radiation on atmospheric water vapour, and is then adsorbed onto the soil particles. The surface layer on Mars, at least at the two landing sites, had been penetrated by, and mixed with, the atmosphere more than had been anticipated by the *Viking* planners. In retrospect, they knew that they should have designed a way to dig much deeper if they were to establish the presence or absence of Martian bacteria.

The discovery of the presence of hydrogen peroxide not only postponed the resolution of the life question. It also provoked a re-assessment of the chemistry of Mars' atmosphere, and helped to provide a solution to another long-standing puzzle: why is carbon dioxide the major atmospheric constituent, when the ultraviolet radiation from the Sun should have converted it to carbon monoxide, long ago? The answer appears to be that oxidising reactions involving hydrogen peroxide continuously convert CO back again to CO_2, keeping the level of CO down. According to measurements, it is less than one per cent of the present atmosphere, consistent with detailed models of this theory. A similar process involving water vapour is probably what limits any tendency for an ozone layer like the Earth's to develop on Mars, particularly since traces of ozone are in fact detected at times and places where the water abundance is particularly low.

In addition to the life-detection equipment, the *Viking* landers carried mete-orological instruments to measure temperature, pressure, and wind strength and direction. Shortly after the first landing came the first weather report from the surface of another planet from the meteorology team: 'Light winds from the east in the late afternoon, changing to light winds from the south-west after midnight. Maximum winds were fifteen mph. Temperature ranged from minus 122 °F just after dawn to minus 22 °F. Pressure steady at 7.7 mb.'[15]

---

[15] Minus 122 °F to minus 22 °F (Fahrenheit) is equivalent to minus 85 °C to minus 30 °C (centigrade).

Because the atmosphere of Mars is quite thin, the wind velocities needed to raise dust are much higher than on Earth. Any local gust strong enough to increase the dust loading also thereby increases the absorption of heat by the atmosphere. The consequent temperature and pressure gradient between the dust-laden region and nearby clearer air tends to enhance the local winds. This positive feedback can continue until the winds reach the threshold speed for raising more dust, when the storm can spread rapidly. The *Viking* orbiters observed that the dust particles are sometimes lifted to great heights (fifty kilometres or more) above the surface, and take a long time to settle out, as seen in Figure 2.14, because they are not purged by rainfall as on Earth. Thus, the global dust storms can last for several weeks before running their course and petering out.

With this kind of wide-ranging distribution mechanism stirring the loose material all over the planet, it is to be expected that the Martian soil (regolith) has much the same composition everywhere, and indeed it was found to be the

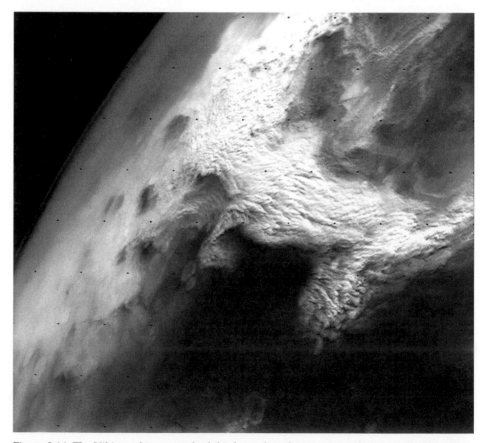

Figure 2.14 The Viking orbiters watched this large disturbance raising dust over the Thaumasia Planum, just south of Valles Marineris on the equator of Mars as it grew into a global dust storm.

same at both *Viking* landing sites. Regions where the rocky surface is exposed, such as high ground and steep slopes, have turned out to be mostly solidified lava of various kinds, dominated by basalts, which is the commonest kind of volcanic rock on the Earth. The dust also seems to be mostly basaltic in composition, presumably eroded from the landforms and mixed around the planet by the high winds, especially during global dust storms.

Before the atmosphere was sampled directly by *Viking* it had been confirmed from orbital spectroscopic and radio-occultation data that the surface pressure was less than one per cent of Earth's, and that carbon dioxide was indeed the major constituent, not diluted by large quantities of inert gas like nitrogen or argon as early speculators like Lowell and von Braun had hoped, although both of those are present in small amounts. Traces of water vapour (H_2O), oxygen (O_2), ozone (O_3) and carbon monoxide (CO) had been measured from orbit, but nitrogen (N_2) and argon (Ar) are difficult to observe with remote spectroscopy because their symmetric structure does not give rise to strong infrared lines. Direct measurements made on air drawn into the sensitive mass spectrometers during *Viking's* entry into the atmosphere and again on the surface after landing showed that ninety-five per cent of the atmosphere is carbon dioxide, while the rest is mostly nitrogen and argon.

The relatively small proportion of nitrogen that the *Viking* landers found was of interest for three important reasons: firstly, of course, nitrogen is the commonest gas in Earth's atmosphere, making up about eighty per cent, and it is important to consider why Mars is so different. Secondly, nitrogen is essential for life once 'fixed', i.e. converted from the relatively inert molecular form into ammonia, nitriles, amino acids and proteins. Finally, because the molecular form is so inert, the isotopic ratios in nitrogen are potent clues to the evolution of the atmosphere, including the history of the surface pressure on Mars.

The low concentration of water vapour in the atmosphere is less surprising, since this is mandated by the low temperatures and is not, on its own, an indication that Mars as a whole is water-poor. Frozen water is abundant, in fact, and close enough to saturation in Mars' atmosphere to form clouds at times. The combination of the *Viking* measurements of surface temperature and humidity could be used in models to make maps of the depth of permafrost below the surface, data that are still in use and being refined by the most recent missions.

It was also known before landing that the winter polar caps contained carbon dioxide ice that condensed in winter and evaporated in summer, producing a variation in the atmospheric pressure with season all over the planet. The pressure sensors on the *Viking* landers showed that the amplitude of the seasonal pressure swing was larger than had been predicted, from less than seven to about nine millibars at the *Viking 1* site, and from just over seven to about ten millibars at the *Viking 2* site (the latter being higher on average because it is at a lower elevation). This corresponds to something like ten trillion tons of carbon dioxide moving in and out of each polar cap every year.

The tracking of the paths followed by the *Viking* orbiters as they circled the planet was precise enough to make maps of the gravity field of Mars that in turn were sensitive enough to detect the perturbations when passing over the five largest volcanoes or, in the opposite sense, over the void that is Valles Marineris. These revealed that Mars, like the Earth, is pear-shaped. However, while on Earth the hemispheric difference is measured in tens of metres, on Mars the southern hemisphere is higher by about ten thousand metres. Something, perhaps the collision of a large asteroid billions of years ago, seems to have blasted away the top part of the outermost solid layer, the crust, over a large part of the northern half of the planet. Even so, Mars' crust must be significantly thicker than the Earth's, since at the equator it supports a huge dome, ten kilometres high and several thousand kilometres in radius. Known as *Tharsis*, this incorporates a row of three giant shield volcanoes, the Tharsis Montes, shown in Plate 17, and, at its north-western edge, the twenty-five kilometres high Olympus Mons, pictured in Plate 18. The loading of the crust by such a giant mass has to lead to some level of cracking and shifting of the substrate bearing the weight, and the *Viking* landers each carried a seismometer to measure the strength of the Martian 'earthquakes' expected to result from this. Although the instruments operated successfully, and picked up the vibrations induced in the structure of the spacecraft by the wind, no sign of any marsquakes was detected. This remains the situation to the present day, probably because sufficiently sensitive seismometers have yet to be placed on the surface or operated for long enough.

2.6 The legacy of *Viking*

The *Viking* missions were the climax of some two centuries of serious research on the nature of Mars as a planet. Sensitive instruments determined the composition of the Martian atmosphere with great precision for the first time, while the weather stations on the landers and the remote sensing instruments on the orbiters added their long records of cloud images, water vapour maps, and temperature and wind measurements to those of *Mariner 9* to give an extended picture of the meteorology on Mars. Along with a basic knowledge of the chemistry and dynamics of the atmosphere came some understanding of the behaviour of the volatiles present as solids and gases, primarily water and carbon dioxide, and of the role played by airborne dust in the present-day climate. Clear hints had emerged of an interesting climatic history; a basic understanding of the structure of the interior had been inferred from global gravity and topography, and of the surface morphology and composition from spectroscopy and from the study of meteorites from Mars.

However, *Viking* failed to find any signs of life, present or past. NASA and the planetary sciences community had a hard lesson to learn: having got everything right, including landing and operating on Mars for the first time, their reward was not the new follow-on missions they had hoped for and expected. Quite the opposite: the advanced versions of *Viking*, with mobility, drills and sample return, being actively considered by mission designers at the Jet Propulsion Laboratory and elsewhere, stayed on the drawing board. The lesson was that, once it is mentioned (or even when it is not), the search for life automatically becomes perceived by and reported in the media as the main motivation for going to Mars. NASA and the *Viking* team had explicitly put the life issue up front, and paid the price by being widely seen as having failed in its primary objective despite its technical success.

Mars itself was once again pictured as being sterile and dull, and NASA shelved all plans for more missions. It was to be fifteen years before the mood changed back and NASA regained the desire to launch spacecraft in the direction of Mars, and twenty years before one arrived successfully. Without the stimulus and the funding to advance their understanding of the planet, the number of scientists working on any aspect of Mars science dwindled as researchers moved on to more profitable challenges.

Gradually, among those who remained, the emphasis on Mars exploration shifted from looking for extant life on a planet now revealed to be cold, dry and probably dead, to understanding the time scale on which the evolution of the planet had wound down to such a sorry state, and what processes had been responsible. The current strong awareness that the exploration of the terrestrial planets has a powerful message for Earth dwellers, in terms not only of their common origins, but also of their past and current evolution, dates from about this time. New research was showing that both Mars and Venus may once have had climates that were much more like the Earth; global change, driven by processes that are essentially the same on all three bodies, had resulted in a scorching greenhouse on Venus and a frozen desert on Mars, with Earth balanced precariously in between. The science of comparative planetology, although not entirely new, certainly moved much closer to centre stage, in part as a consequence of *Viking*. We will return to this theme in later chapters.

Also of course, there remained below the surface – literally and metaphorically – the possibility that if life developed in a warmer, wetter era, it would have left detectible traces of its presence as organic material and even more substantial fossils of some kind. Perhaps surviving descendants on some scale, probably microscopic, that had adapted to the changing climate, lived on in some ecological niche somewhere below the surface. In the immediate aftermath of *Viking*, it was a brave person who would stake their career on proposing or authorising expenditure on a mission capable of finding them, however. Whether or not the planet itself was sterile, a long barren period was about to begin in the exploration of Mars.

Further reading

An official NASA history of the period of spacecraft exploration covered by this chapter, including a blow-by-blow account of the *Viking* project, can be found in:

On Mars: Exploration of the Red Planet 1958–1978, by Edward and Linda Ezell and
    published by NASA in 1984 (NASA SP-4212).

Another excellent account, especially in terms of its coverage of ground-based
observations of Mars from the earliest times to the *Viking* era, is:

The Planet Mars: A History of Observation and Discovery, by William Sheehan,
    University of Arizona Press, 1996.

The origins and early history of the Jet Propulsion Laboratory are recounted in:

JPL and the American Space Program, by Clayton R. Koppes, Yale University Press,
    1982.

A comprehensive summary of post-*Viking* scientific knowledge about Mars,
nearly 1500 pages long and strongly featuring the *Viking* science team members
among the multiple authorship, was published in 1992:

Mars, by H. H. Kieffer, B. M. Jackosky, C. W. Snyder, and M. S. Matthews (Eds.),
    University of Arizona Press, 1992.

Plate 1. Mars through the *Hubble Space Telescope*.

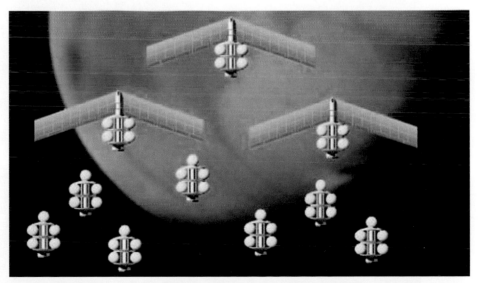

Plate 2. Von Braun's first plan for an expedition to Mars. The flotilla consisted of seven ships carrying ten men each, plus three unmanned cargo vessels, which each incorporated a winged lander that would carry fifty of the seventy crew down to the surface of Mars for a four-hundred-day exploration. (© Bonestell Space Art, used with permission)

Plate 3. *Landing on Mars* (*c*. 1955) by Chesley Bonestell. From *The Exploration of Mars* by Willy Ley and Wernher von Braun, reproduced courtesy of private collection.

Plate 4. Gateway to Mars: the Jet Propulsion Laboratory (JPL), home of NASA's planetary programme, as it looks today.

Plate 5. (a) The view of the surface of Mars from the first-ever landing. *Viking 1* touched down in Chryse Planitia on 20 July 1976. The large rock is about two metres across, and one metre high.

Plate 5. (b) Sunrise on Mars as seen from *Pathfinder*. The scattering effects of dust suspended in the atmosphere can be seen, and estimates of particle size and density have been made from the brightness profile and wavelength dependence of the scattered light.

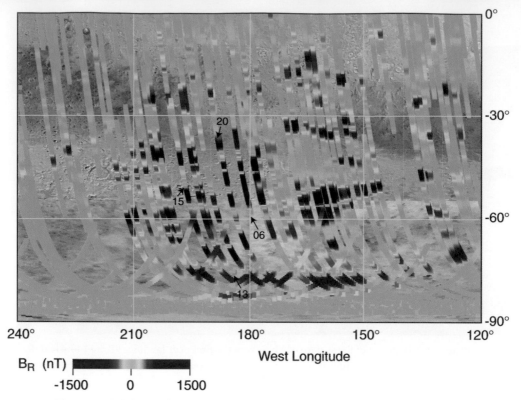

BR (nT) ▬▬▬▬▬▬
-1500   0   1500

Plate 6. A global map of Mars' crustal magnetism, assembled from four years of observations from orbit by the magnetometer on *Mars Global Surveyor*. Mars no longer has an internally generated field; these values are due to the magnetisation it induced in the rocks near the surface before it vanished billions of years ago. In the key to the colours used, B_R stands for the remnant magnetic field, measured in nanoteslas (nT). The highest value seen on Mars today is only about one per cent of the current magnetic field at the surface of Earth, which still has an internal source, but is large enough to suggest that Mars' ancient field may have been stronger than the Earth's.

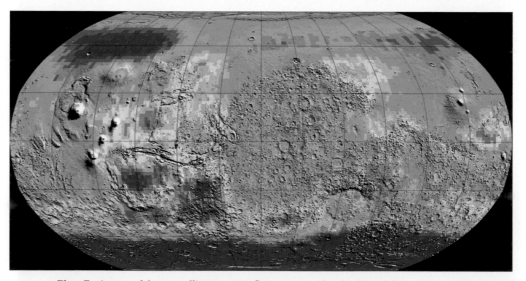

Plate 7. A map of the upwelling neutron flux measured by the *Mars Odyssey Neutron Spectrometer*, with regions of low flux shown in blue. These are interpreted as having high hydrogen concentrations less than one metre below the surface, most likely in the form of water ice, which should be stable near the surface in the polar regions according to theory supported by thermal inertia measurements. The coverage of the north pole was affected at the time of observation by the winter polar cap of carbon dioxide ice; without this, it probably would appear similar to the south pole in these data.

Plate 8. An example of *Mars Express* imagery, showing what has been interpreted as a dust-covered frozen sea with kilometre-sized plates of 'pack ice' on Elysium Planitia near the Martian equator.

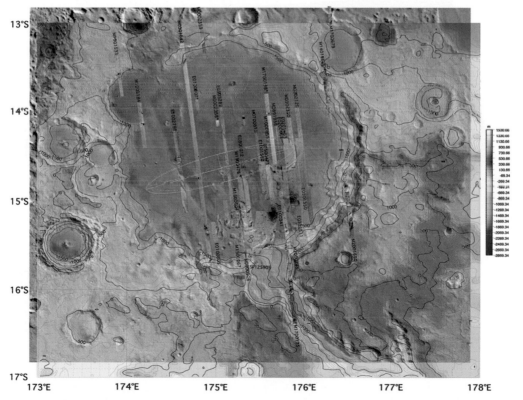

Plate 9. The *Mars Exploration Rover Spirit* landed near the centre of Gusev Crater, a ninety-mile-wide remnant of an asteroid impact just south of Mars' equator. A broad channel system drains into Gusev, evidently carrying liquid water that flooded the crater several billion years ago.

Plate 10. This realistic-looking image of the rover *Spirit* on the flank of Husband Hill, in the Columbia Hills, was produced by combining a picture of the rover with a panorama obtained by its own cameras, using the tracks left by the rover's wheels to provide the correct scaling.

Plate 11. The view from the rover *Opportunity* of the outcrop named Cape St Vincent in Victoria Crater. The detailed layering of this and other features in the crater wall reveals much about the local geological history.

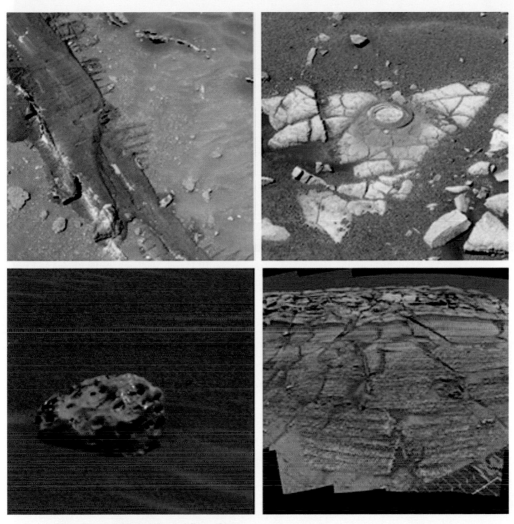

Plate 12. Exploring the surface of Mars.

Top left: *Spirit*'s wheels have churned up light-coloured soil containing soluble salts that indicate the former presence of water.

Top right: On 29 July 2006 *Opportunity* ground this three-millimetre-deep hole using the rock abrasion tool in an outcrop twenty-five metres from the south-west rim of Beagle Crater on the Plains of Meridiani.

Bottom left: *Opportunity* found this basketball-sized iron-nickel meteorite on 6 January 2005. It was probably buried on impact but over time the wind has cleared the soil and dust from around it, leaving the heavy object perching on the surface.

Bottom right: Burns Cliff, on the southern rim of Endurance Crater, is made up of layers formed by deposition of solid material, including haematite 'blueberries', from solution. Apparently the entire area was covered by a shallow sea or lake that went through wet and dry episodes, estimated to have taken place a few hundred million years ago.

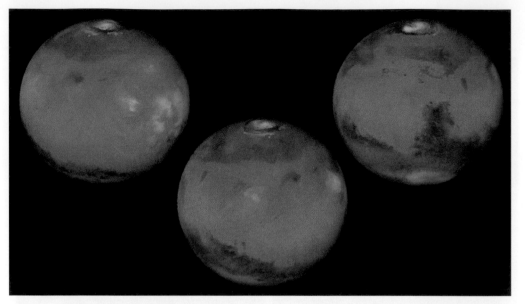

Plate 13. Mars through the *Hubble Space Telescope* on 10 March 1997, when the planet was near its closest to the Earth, and just before summer solstice in the northern hemisphere.

Plate 14. This *HiRISE* image shows part of Nili Fossae, a chasm about six hundred kilometres long and about twenty-five kilometres wide, in a region that is one of those favoured for future exploration by an advanced rover such as *Mars Science Laboratory*. Spectroscopic evidence has shown that Nili Fossae has deposits of several interesting minerals, including clay-like phyllosilicates, which typically form in the presence of water, olivine, pyroxene and small patches of carbonates, and it may be a source of atmospheric methane.

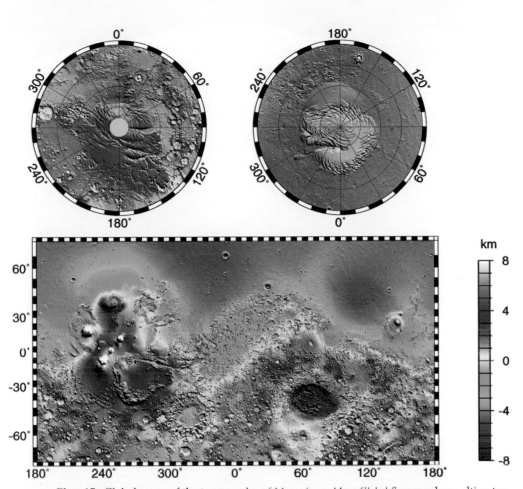

Plate 15. Global maps of the topography of Mars, from *Mars Global Surveyor* laser altimeter
data. The blue to green colours correspond to terrain of below-average elevation, and the red to
white colours to high ground and tall features. The total range from lowest to highest is about
twenty kilometres, with the lowest place being the Hellas basin below and to the right of centre,
and the highest the giant Tharsis volcanoes at upper left.

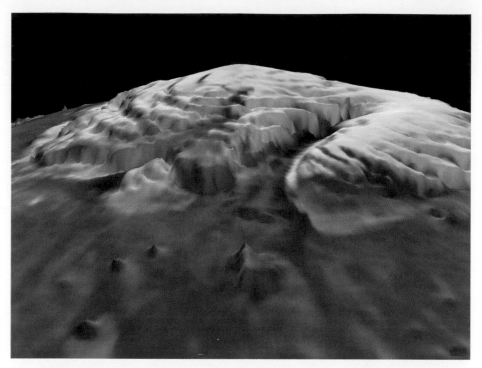

Plate 16. The volume of the water ice cap at the north pole could be estimated accurately for the first time from this three-dimensional picture obtained by the altimeter on *Mars Global Surveyor*. More than two million laser pulse measurements were mapped onto a grid with a spatial resolution of one kilometre and a vertical accuracy of about ten metres to produce the image. The vertical scale is expanded to bring out the detail: in reality the cliffs in the foreground are about one kilometre high, while the whole cap is about a thousand kilometres across.

Plate 17. A simulated view of the three large volcanoes of the Tharsis Montes: Arsia Mons, Pavonis Mons and Ascraeus Mons, with the smaller volcanoes Biblis Patera and Ulysses Patera to the left. (© Kees Veenenbos)

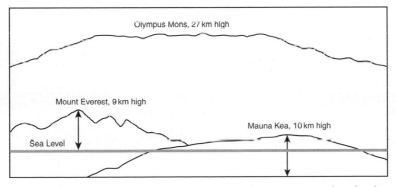

Olympus Mons, 27 km high

Mount Everest, 9 km high

Mauna Kea, 10 km high

Sea Level

Plate 18. Twenty-seven kilometres tall and stretching over more than five hundred kilometres east to west, Olympus Mons is the largest of the Tharsis volcanoes and the highest known mountain in the Solar System. The caldera at the summit, containing at least seven roughly circular craters, is seventy kilometres across. The picture above from *Mars Express* shows a section of the eastern scarp of the volcano, which is six kilometres high in places and shows remnants of lava flows that geologists estimate took place a few hundred million years ago.

Plate 19. Echus Chasma, north of Valles Marineris, is over one hundred kilometres long, ten kilometres wide, and four kilometres deep. It appears to have been carved by running water; if so, the amount and persistence of the flow must have been formidable.

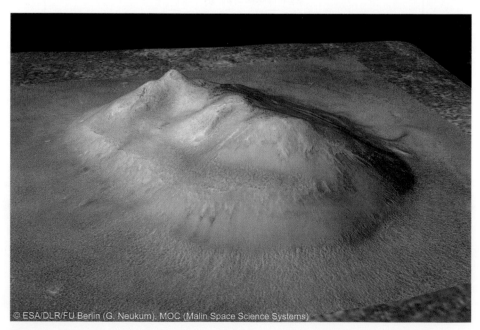

© ESA/DLR/FU Berlin (G. Neukum). MOC (Malin Space Science Systems)

Plate 20. *Mars Express* obtained this view of the 'face' on Mars on 22 July 2006. See also Figure 7.2.

Plate 21. Top: The view photographed by the *Spirit* rover from the peak of Husband Hill. The rocks in the foreground are named Hillary, Tenzing and Whittaker after the Everest pioneers. Below: a close-up of Hillary.

Plate 22. The bleak northern plains of Mars imaged from the *Phoenix* lander on 7 October 2008, one hundred and thirty one days into its mission. The atmospheric temperature is about minus thirty degrees centigrade.

Plate 23. A *Global Surveyor* view of the Mountains of Mitchel near the boundary of the seasonal south polar frost cap of Mars at 70° S. It was first noted in 1846 that this region stays brighter longer than its surroundings in early spring when it comes out of the darkness of the polar winter. Mitchel, an astronomer at the University of Cincinnati, attributed this behaviour to frost or snow on a range of hills or mountains, although the *Mars Orbiter Laser Altimeter* has shown that the area, although rough, is not especially lofty. The probable reason for the persistence of the frost is a combination of south-facing slopes, which shield it from the Sun, and meteorological conditions that circulate cold air from the polar cap.

Plate 24. The south polar region of Mars in winter – an artistic view.

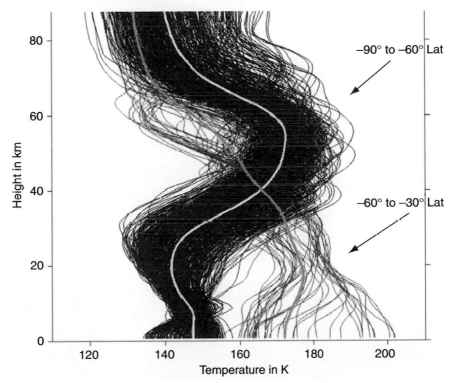

−90° to −60° Lat

−60° to −30° Lat

Plate 25. Mars atmospheric temperature profiles over the south polar region in winter, obtained by the *Mars Climate Sounder* on the *Mars Reconnaissance Orbiter* shortly after arrival at the planet in September 2006. The dark blue profiles lie over the polar cap, where the surface temperature is constrained by the sublimation of carbon dioxide ice to values near its freezing point. The magenta profiles, falling outside the ice cap, have higher and more variable surface temperatures. Note the high temperatures at around fifty kilometres altitude, over the pole; this is produced by compression caused by the circulation of the atmosphere, which at these latitudes is in permanent darkness for the duration of the winter.

Plate 26. Mars rovers: *Sojourner* (top) in the laboratory, and an artist's concept of a *Mars Exploration Rover* (*Spirit* or *Opportunity*) (bottom) in the path of a dust storm.

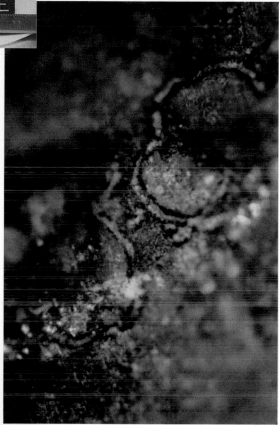

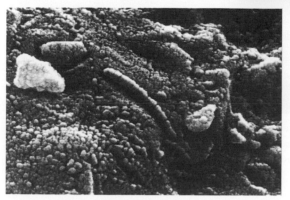

Plate 27. Rock from Mars: (top) the SNC meteorite ALH 84001, (middle) a microscopic view of a thin slice of the object viewed in transmitted light through a polariser, and (bottom) a photomicrograph of the putative nanobacteria found in the orange-coloured, water-deposited carbonate grains inside the meteorite.

Plate 28. *Mars Orbiter Camera* captured this view of a localised dust storm just south of the valley complex known as Labyrinthus Noctis, to the bottom of the picture. Water ice clouds are also seen, especially leeward of the five large volcanoes.

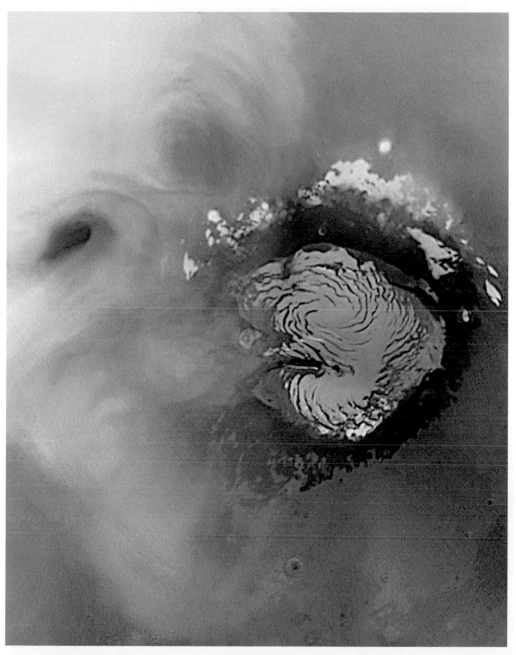

Plate 29. In the north polar region, mid-summer temperature differences between bright areas of year-round ice and dark areas of sand and rock create strong winds that mix the atmosphere and create waves of clouds that swirl around the polar cap. Sometimes they form tight cyclones; other times, they weave an intricate pattern reflecting the turbulence of the circulation of the atmosphere. The clouds viewed in this image are typical for this season on Mars, and show forms often seen on Earth. Waves of clouds are moving from the upper portion of the frame towards the bottom under the influence of circumpolar winds. Cloud speeds vary from day to day, averaging a moderate fifteen kilometres per hour.

Plate 30. *Spirit*'s instrumented arm inspects a rock, dubbed Adirondack, during its thirtieth day on Mars.

Plate 31. A panorama taken by *Spirit* on 23 August 2005, after the rover completed its climb up Husband Hill, revealing a windswept plateau of scattered rocks, sand dunes and exposures of outcrop. The view is toward the north, looking down into the Tennessee Valley.

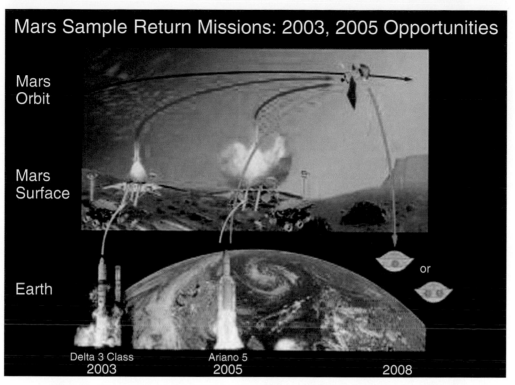

Plate 32. Chart showing NASA's 'Faster, Cheaper, Better' plan for Mars exploration as formulated in 1997, featuring sample return missions in 2003 and 2005.

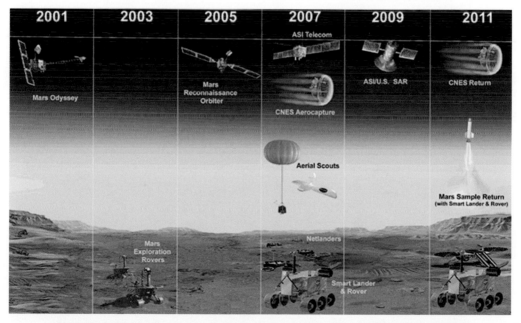

Plate 33. NASA's Mars programme plan as it was in the fall of 2000, with the CNES-supported *NetLander* and sample return missions still planned for 2007 and 2011 respectively.

Plate 34. *Mars Science Laboratory* should be in action on Mars from 2013 on. Much larger than its predecessors, it has a nuclear power unit (the cylindrical device on the back) which supplies about five times as much electricity as the solar cells on *Spirit* and *Opportunity*.

Plate 35. The European *ExoMars* rover, due on Mars in 2016, has similar goals to NASA's *Astrobiology Field Laboratory*, likely to fly a few years later. Both will carry advanced tools for investigating sites that have been shown by earlier missions to have clear life-related signs, such as geothermal activity or emission of methane gas.

Plate 36. The *Deep Drill* mission is the leading concept within NASA to fly late in the decade 2010–20 if no promising astrobiology sites have been identified. The vertical dimension, below the ground, would then be searched for such sites by drilling down tens or even hundreds of metres, to find aquifers and carbon-based material.

Plate 37. ESA has sample return in its plans. This is how the sample container might look during lift-off from Mars for the return to Earth.

Plate 38. NASA sample return studies generally have favoured the use of a rover (not necessarily programmed for self-preservation, as here) to seek the most interesting accessible samples.

Plate 39. Ground-breaking sample return, which has no mobility but just an arm to obtain a 'grab sample', is the simplest concept studied. However, the scientific value of the sample may be limited by the restricted choice available, and this approach is currently out of favour.

Plate 40. A rendezvous in Mars orbit of the ascent vehicle with a larger interplanetary spacecraft is a feature of most mission designs for Mars sample return. In NASA's plans, the manoeuvre will first be rehearsed at the Moon.

Plate 41. A recent concept by NASA for an unmanned reconnaissance aircraft to explore Mars. The *Kitty Hawk* was proposed for deployment on Mars in 2003 to mark on the centenary of the historic first flight of its namesake by the Wright brothers. (Malin Space Science Systems & Mars Airborne Geophysical Explorer Consortium)

Plate 42. An early idea of what a future human base on Mars might look like, with living quarters under a dome, and food growing in cloches outside. Current thinking would place the inhabited part of the base underground, for better thermal stability and for protection against the cosmic ray flux from space. On Earth, the thicker atmosphere and the magnetic field make these precautions less essential. (© Bonestell Space Art, used with permission)

Plate 43. The *International Space Station*: stepping-stone to Mars, or just a distraction?

Plate 44. A *Virgin Galactic* shuttle returns to Earth after a test flight prior to taking passengers into orbit and perhaps eventually to the *Space Station*.

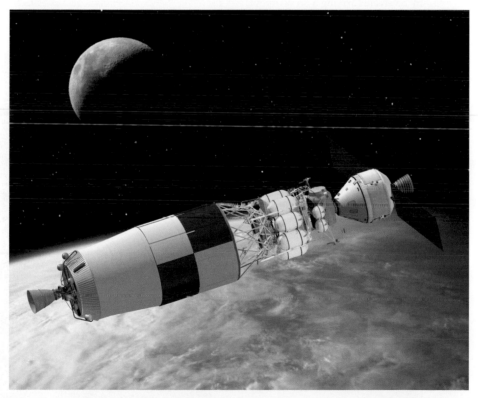

Plate 45. In this NASA concept, two spacecraft capable of accommodating astronauts are attached to a powerful booster stage in Earth orbit. On the right is the *Crew Exploration Vehicle*, which accommodates the crew during flight to and from lunar orbit, while in the centre is a separate landing module. A system like this may transport crews to the Moon in the first instance, and later form the basis for an expedition to Mars.

Plate 46. A possible configuration being considered by NASA for a descent to the lunar surface by a vehicle carrying a crew of four.

Plate 47. Later landings will be at a manned lunar base for which the individual components, including pressurised vehicles, spacesuit maintenance and storage modules, and living and working quarters for the crew, will have been delivered by unmanned cargo flights and assembled robotically before the crew arrives.

Plate 48. After a base has been established, routine supplies could be delivered to the Moon and Mars using relatively simple unmanned carriers launched from Earth and capable of a high-precision, automated landing at the base on the surface of the planet.

Plate 49. Flights to Mars will build on the experience built up by lunar expeditions, but must travel much further and need to carry more mass. One of many possible upgrades is to replace chemical propulsion stages by a thermonuclear-powered rocket that could carry large loads, including crews of eight or more people, to Mars and back. In this NASA concept the blunt, cylindrical vessel carries a built-in heat shield at the front for deceleration using upper atmospheric drag, both on arrival at Mars and on return to Earth, during which the communications dish at top centre would be retracted.

Plate 50. A realistic design for a manned, nuclear powered spacecraft that might ply between Earth, Moon, Mars and beyond in the more distant future. The living quarters are well separated and shielded from the propulsion unit to minimise radiation damage to the crew on a long flight, and long cooling vanes carry away excess heat from the reactor and radiate it to space. The vehicle might be over a kilometre long and have a mass of several hundred tons. (© 2009 David Robinson, www.bambam131.com)

Plate 51. Achieving orbit by aerocapture gives the crew an exciting ride, involving their spacecraft skimming the surface at a height less than that of the highest mountain on Mars, while the spacecraft's heat shield burns away due to friction with the air.

Plate 52. The first human expedition lands on Mars, having used its capability to hover and manoeuvre, possibly in a gusty wind, in order to find a landing site free of large rocks. (© 2009 by David Robinson, www.bambam131.com)

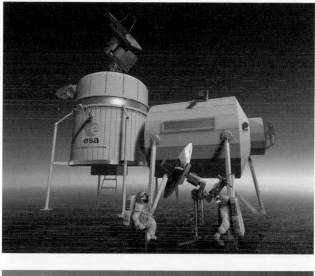

Plate 53. The first manned bases on Mars will probably look something like the European concept in the top picture, featuring a typical Martian landscape with dust obscuring the horizon. Further into the future, more advanced bases like the NASA concept in the lower picture may feature permanent facilities for shelter, food and fuel production, and water extraction. Notice the shuttle touching down on the landing field in the background, and the reconnaissance aircraft flying above.

Plate 54. In the distant future, a base on Phobos looks down on Mars as it undergoes terraforming. Under a thicker, cloudier atmosphere, the huge canyon named Valles Marineris after the spacecraft that discovered it in 1971 is visible near the centre of the illuminated part of the globe. Three hundred years later, it is full of liquid water. (Artwork © David A. Hardy, www.astroart.org)

Chapter 3
After *Viking*: the twenty-year hiatus

3.1   The mood after *Viking*: so high, so low

It is ironic that it was a successful project, Viking, which led to the long hiatus in Mars exploration that marked the end of the first era of space missions to the planet. Many people, especially the scientists and engineers, were thrilled to have explored the surface of Mars close-up and were excited about what they saw as the inevitable next steps. Others, including key NASA administrators, were less sanguine. They had massive problems with the space shuttle, then at a key stage in its development, which were making pressing calls on their budget. They had 'done' Mars, and found it to be dry and lifeless. They knew that the public, and the politicians, found the reality disappointing compared to the expectations stoked by those scientists who believed in life on Mars. Such people were still to be found, intuitively clinging to the principle that such an Earth-like world must have, or have had, life on it despite the dwindling scientific case.

For a time the mission designers at the Jet Propulsion Laboratory and elsewhere pressed on undaunted, driven by the momentum built up by the programme over the previous two decades. While the disciples of the new science of astrobiology, some of them at least, may have been discouraged by the data *Viking* sent back from the surface of Mars, the rest of the scientific community saw plenty of fresh objectives to follow up, and was keen to press on with new missions. During the period from the end of the *Viking* nominal mission on 15 November 1976, ninety days after the landing, until *Pathfinder* touched down on 4 July 1997, more than twenty years later, there was certainly no shortage of proposals and ideas, or of excitement and ambition. The United States, the Soviet Union, and now the Europeans, all studied a number of advanced concepts, some in great detail. Some of them were even built and launched, eventually, but none of those arrived successfully at Mars.

3.2   Phantom missions: proposals, working groups
and Phase A studies

The sequence that starts with an idea for a new space mission, and then leads to new experiments carried out in orbit around Mars or on its surface, is often a long one. More often than not, the outcome of a previous mission provides the inspiration for the next one, with the frustrating outcome of the *Viking* life experiments an outstanding example. The tantalising and obdurate lack of evidence for life that many felt must be there somewhere left the scientific community burning to follow up with mobility, to find more promising locations to probe than just the immediate vicinity of the safest landing site, and for extended capabilities, such as drilling to explore the subsurface. Knowing what they want to do, a group of like-minded scientists may decide to prepare a proposal to make the case for a new mission. This can begin as something quite informal – a presentation at a conference,[1] or a paper published in a journal – but at some stage it has to attract the attention of a space agency if it is ever to fly. There are two main ways to do this, either by approaching the advanced study departments at places such as JPL or ESTEC,[2] who are always looking for new ideas with a view to planning their future programme, or by waiting for the agency itself to issue an Announcement of Opportunity. These come out whenever there are funds for a new project and are sometimes targeted (on Mars, for example) or quite general, like the recent Cosmic Visions exercise in Europe, which attracted proposals for space missions ranging in discipline from cosmology to fundamental physics as well as the inner and outer planets, and the small bodies (comets, asteroids etc.).[3] Often, the agency will set up one or more study groups before inviting proposals, in order to stimulate interest in the community in the direction that the agency's overall plan takes it, and to do

---

[1] For instance, papers can be presented or 'splinter groups' organised to present and discuss a new mission concept at the annual meeting of the Division for Planetary Sciences of the American Astronomical Society or the European Planetary Sciences Congress Other opportunities that occur every year include the meetings of the American Geophysical Union (which meets in Baltimore in the spring, and in San Francisco each December) and the European Geosciences Union (which currently meets around Eastertide in Vienna); both have large planetary sciences sessions.
In addition, the Committee on Space Research, COSPAR, and the International Union of Geodesy and Geophysics hold biannual meetings in different locations around the world. There are several other societies which organise relevant sessions, plus of course dedicated meetings like the quadrennial Mars Conference at Caltech and numerous specialised Mars meetings organised on an ad hoc basis.

[2] The European Space Research and Technology Centre at Noordwijk in the Netherlands, occupying former tulip fields between Amsterdam and Leiden, is ESA's equivalent to the Jet Propulsion Laboratory in the USA.

[3] At the time of writing, a new mission to the outer planets, conducted jointly with NASA, is the front-runner in this particular competition, or possibly sample return from an asteroid. Mars is considered to be catered for by the *Exomars* project due to launch in 2016, and Venus and Mercury are served by current missions, *Venus Express* and *BepiColombo* respectively.

some preliminary assessments of the technology advances needed and their likely cost profile over the years ahead.

Membership of a NASA or ESA sponsored study group gives a team of like-minded scientists, led by an aspiring team leader (usually known as the Principal Investigator), an inside track when it eventually comes to proposing to participate in a mission, for instance by providing an instrument to form part of the payload. Not only does the study member help to choose what sort of mission it will be, he or she gains an intimate understanding of the technological interfaces the instrument must satisfy and the suite of scientific goals to which it must contribute. For these and other reasons, including the prestige granted by recognition of the individual as an expert, appointment to a study group is much prized, even if the outcome is statistically likely to end in disappointment. A secondary, but by no means negligible, benefit of participation in a major (that is, agency-funded) study is the fun one has while doing it. Skilled engineering and programmatic support concerning the capabilities of the various launch vehicles, spacecraft and other technologies available, is on hand to allow a scientist to put flesh on the bones of his ideas, and slowly turn them into, if not yet reality, at least something realistic. The meetings are generally rotated around the home institutions of the members, or different agency centres, often far-flung around the continent or even around the world. The resulting travel,[4] hospitality and long, intimate discussions are good for the development of ideas and careers alike.

The most important goal of the work that goes into a study or working group is of course the scientific endorsement of a mission that may, or more likely may not, lead to actual hardware flying to Mars. Most space agencies follow NASA in planning a mission in *phases*, where at the end of each phase a decision is made whether or not to continue. The decision is based on the scientific, technical, political and (especially) the financial health of the project at that point, as determined by a number of in-depth reviews. The pivotal moment comes when a concept is accepted for Phase A, which represents the beginning of a serious commitment to the project and the expenditure of large sums of money. Figure 3.1 shows how this fits in to the overall sequence. In Phase A, the agency's experts assess the technical requirements, including things like the mass of the spacecraft, the launch vehicle to be used, the time scale, and an estimate of the likely cost. In Phase B, industrial contractors produce a detailed design, with an associated schedule and budget, while the actual hardware is manufactured and launched in Phase C/D, followed by flight to the planet and mission operations there in Phase E. The European Space Agency uses a similar approach, but

---

[4] Travel used to be more fun when it was not so commonplace as it is now, nor so regimented. Until the late 1970s, the Jet Propulsion Laboratory had a private helicopter that took off and landed from the tallest building on the laboratory site, and which would fly staff to and from meetings in the region. For longer journeys, it would fly to Burbank airport and land next to JPL's private plane for the onward journey. If long-haul was required, the helicopter would land on the tarmac outside the departure gate at Los Angeles International Airport. Needless to say, none of this happens today.

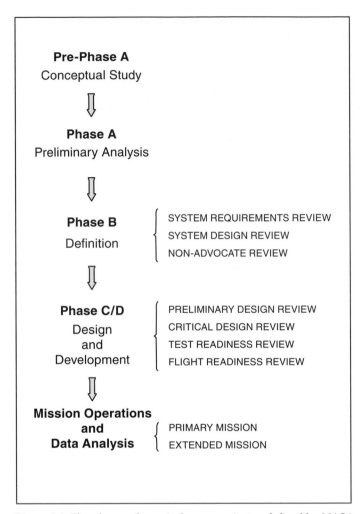

Figure 3.1 The phases of a typical space project as defined by NASA, showing the main reviews and roughly when in the time-line, which flows from the top of the diagram to the bottom, each takes place. In the pre-Phase A, scientists discuss the measurements they need to address key goals, for example, the return of samples from Mars to assess the climate history of a region. (On the walls and notice boards of various NASA centres, and other facilities involved in developing missions, it was popular for a time in the 1970s to show a parody of this diagram that showed the five phases as (1) Excitement and euphoria, (2) Grim dedication, (3) Total panic, (4) Reprimands and redeployment for the innocent, and (5) Rewards and acclaim for those not much involved.)

organised into seven major phases (Phase 0, Mission analysis; Phase A, Feasibility; Phase B, Preliminary definition; Phase C, Detailed definition; Phase D, Production and Testing; Phase E, Utilisation; Phase F, Disposal).

Many mission concepts and designs go through pre-Phase A studies of various lengths and complexities, either as part of the work of an ad hoc group, a standing committee like COMPLEX (the Committee for Planetary Exploration of the US National Academy of Sciences), or just in the space science laboratories of research centres and universities, often forming extensive, even international, consortia. Few of these actually become the subject of a Phase A study, as these are expensive and indicate a high level of commitment to proceeding further. Fewer still go on to be designed and built. Once approved for Phase C/D, it is rare indeed for a mission to be cancelled, since contracts have been signed for large sums of money, much of which is then unrecoverable. When a mission under construction does get cancelled, the reason usually is that its cost has grown enormously beyond that foreseen because of technical problems that were overlooked or underestimated in the preceding studies.[5]

Thus, the history of a period like that which followed *Viking* is full of descriptions of missions that were developed to various stages, some quite advanced, and then shelved. The first success, in the sense of being politically approved, built and launched, did not come until the polar orbiter, *Mars Observer*, which reached the vicinity of Mars in 1993, almost two decades after *Viking*. The failure of *Observer* three days out from Mars meant there would not be a successful Mars mission until *Pathfinder* arrived in July 1997, itself a relict of the phantom mission called *Mars Environmental Survey*.

3.3 The NASA Mars Science Working Group (1976–9)

In the aftermath of *Viking*, NASA did not immediately or deliberately decide it had no plans to return to Mars; indeed, in late 1976, as the still-operating *Viking* spacecraft began their extended mission, the agency directed the Jet Propulsion Laboratory to make plans for a new project to fly in 1984. This was to conduct an even more in-depth study of the planet, as a precursor to a mission to return samples of surface materials and atmospheric gases in 1990 or thereabouts. In its efforts to match technology readiness to scientific ambition and agency budgets, JPL was supported by an eclectic twenty-seven-member body called the Mars Science Working Group, under the chairmanship of Thomas (Tim) Mutch, a geologist from Brown University in New Hampshire, who was the leader of the *Viking Lander* Imaging Team.

The Mars Science Working Group reported in July 1977, with a recommendation that the 1984 mission should consist of two launches of the newly commissioned space shuttle into Earth orbit. From there, a secondary booster would take each of two identical spacecraft on to Mars, where they would be placed into different orbits, one high-inclination, passing over the pole, the other circling nearer to the equator. Once safely installed around Mars, each orbiter would release three hard landers to form a surface network of six stations, followed by a

---

[5] Another aphorism, comprising part humour and part bitter experience, heard around spacecraft and instrument contractors, is that 'ninety per cent of the budget is spent on the first half of the project. The other ninety per cent is spent in the second half.'

soft-landed roving vehicle. The rovers would be capable of operating on the surface for as long as two years, during which time they could trek over three hundred metres per day, steered from the Earth.

With *Viking* still operating and plenty of expensive problems elsewhere in its programme, not least the shuttle itself, NASA found it could not fit the ambitious *Mars '84* mission into its budget for the following year. Since a 1984 launch would not then be practicable, if indeed it ever was, the Mars Science Working Group was asked to reconvene in order to look at later, and perhaps less expensive, options. Tim Mutch captured the excitement and frustrations that come from this sort of exercise when, in October 1978, he wrote to his colleagues on the working group:

The prospect of spending so much time on committee business – somewhere between science and administration but, legitimately, neither – is worrisome. Added to this is the uncertainty of the outcome. I was persuaded to continue not by any assurances that this time the approval of a Mars mission seemed certain (indeed, no such assurances were offered). Instead, I was persuaded that there is widespread agreement among scientists and NASA administrators that there must be some continuing exploration of Mars. Definition of 'some', the primary task of the Mars Science Working Group, is exceedingly difficult.

Shortly afterwards, the JPL Chief Scientist, Arden Albee, took over from Professor Mutch as chairman of the Mars Science Working Group.

The Group considered the most important research objectives left unsolved by *Viking*, and invoked a wide range of mission types to address them. The ones they chose to focus on were given as:

1. A Mars polar orbiter that would conduct a comprehensive global mapping of the geochemical provinces of the planet, its atmosphere, its gravitational field, its magnetic field, and the charged particles in its vicinity.
2. Hard landers and surface penetrators, weighing less than a hundred kilograms each, which would be released from an orbiter to establish a network of stations making meteorological, seismic and perhaps other measurements.
3. Hydrazine-powered aeroplanes weighing about three hundred kilograms that could remain aloft for about a day, and could carry instruments for making magnetic, gravity and geochemical measurements while in flight, take very high resolution pictures, or even deploy more permanent scientific stations on the surface.
4. Gigantic inflated spheres, perhaps as large as a kilometre in diameter, which could carry a variety of scientific instruments, and, propelled by the wind or their own radioisotope-powered propulsion system, might roam for hundreds of kilometres over the surface.
5. A variety of wheeled vehicles that might rove the surface for distances of tens of metres to hundreds of kilometres.
6. A range of Mars sample return techniques.

This list is the bedrock on which the unmanned Mars exploration programme is built, right up to the present day, not just that of NASA in the USA but of all nations with ambitions in space. It merits a more detailed discussion of each of the concepts in turn, starting with the simplest.

The early designs for a Mars polar orbiter to conduct global mapping of the geochemical provinces of the planet (i.e. identifying the minerals and rock types present in different regions) and study the present-day climate, aimed for a very low cost mission that would kick-start the Mars exploration programme again after *Viking*. The result was the *Mars Geochemistry Climatology Orbiter*, which used subsystems from the *Fleet Satellite Communications* Earth orbiting military communications satellite, the *SATCOM* communications satellite, and the *TIROS* weather satellite. Figure 3.2 shows the first of a series of ten *TIROS* satellites

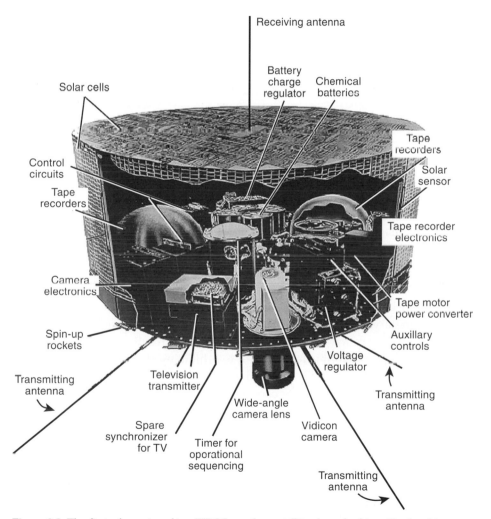

Figure 3.2 The first of a series of ten *TIROS* weather satellites launched into Earth orbit between 1960 and 1965. The low-cost *Mars Geochemistry Climatology Orbiter* concept was to use subsystems from this and from the *Fleet Satellite Communications* Earth orbiting military communications satellite.

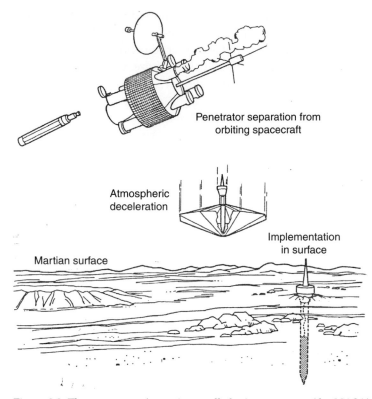

Penetrator separation from
orbiting spacecraft

Atmospheric
deceleration

Implementation
in surface

Martian surface

Figure 3.3 The penetrator (sometimes called a 'penetrometer' by NASA) is fired from orbit to strike the surface at several hundred metres per second, so the forebody forces its way into the regolith while the afterbody separates on impact and remains on the surface to provide power and communications.

launched into Earth orbit between 1960 and 1965. In 1985 the name of the planetary mission was changed to *Mars Observer*, and it finally launched in 1992 after much design evolution and cost growth. After a successful flight to Mars, a catastrophic failure prior to orbital insertion resulted in the loss of all of the long-anticipated scientific data.

The surface penetrator concept illustrated in Figure 3.3 involved dropping an instrumented projectile from orbit to strike the surface at high speed (several hundred metres per second). In most designs studied, the afterbody separates on impact and remains on the surface, connected by a cable to the forebody which forces its way into the regolith. The afterbody contains the communications and power subsystems; everything has to be designed to survive and operate after a shock on impact that can be as high as twenty thousand g.[6] Both parts carry scientific instruments to make meteorological and seismic measurements, and to

---

[6] This means that the deceleration when the device hits the ground is equal to twenty thousand times the acceleration due to gravity at the surface of the Earth.

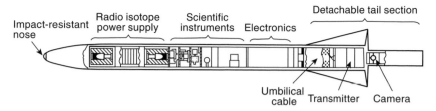

Figure 3.4 This 1980s NASA design for a penetrator was a hundred and forty centimetres in length, nine centimetres in diameter and weighed thirty-one kilograms. Radioisotope thermal generators provide electrical power, since using batteries alone would greatly restrict the power available to the instruments. They also dissipate heat that keeps the battery and electronics at a safe operating temperature in the icy Martian environment.

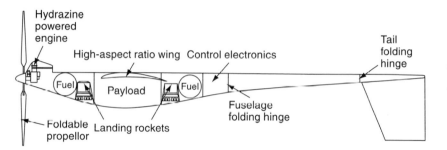

Figure 3.5 This 1978 design for a Mars aeroplane had a wingspan of twenty-one metres, but a mass of only thirty-six kilograms due to its lightweight, glider-like construction. It folded up into a compact size for delivery into the Martian atmosphere, and could deploy in mid air.

analyse the atmosphere and soil, as shown in Figure 3.4. Each penetrator is sufficiently small (typically around one metre long and ten kilograms in mass) so that a dozen or so of them can be deployed in a network that covers most of the planet and many different types of terrain.

The Mars aeroplane is an enduring concept that has been extensively studied for decades as an efficient way of surveying the surface of Mars using high-resolution, stereoscopic imaging, but is yet to fly in any form. The version in Figure 3.5, from a Jet Propulsion Laboratory study in 1978, was considered as a possible follow-on to *Viking*. Equipped with radar and variable-thrust rockets as well as a fifteen horsepower hydrazine engine driving a two-metre propeller, it was capable of soft-landing forty kilograms of payload on the surface. The plan called for twelve aircraft to be deployed simultaneously, carrying survey equipment, surface stations and mini-rovers.

Inflatable Mars rovers were first proposed in 1977 by a team of French scientists, led by Jacques Blamont of the University of Paris, and since then various versions have been designed and Earth-bound prototypes have been built and tested. The most successful of these was made up of a number of individual airbags that could be inflated with ambient Martian air after landing

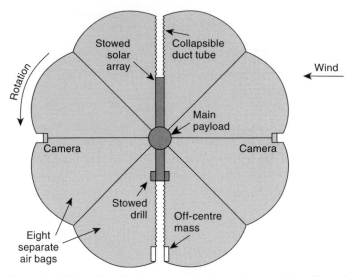

Figure 3.6 This inflatable Mars rover design, nicknamed *Tumbleweed*, was an impressive ten metres in diameter but had a mass of only forty kilograms. It was intended to be inflated during descent from orbit to reduce speed through drag and to protect the payload at its centre during final impact, making it a very inexpensive way to put a rover on Mars. The main drawback was it was difficult to steer.

using a system of pumps and valves, to make a six-metre high ball that rolls along, blown by the wind. In some versions, the segments could be deflated and inflated in a sequence that produces a slow rolling motion, for use when relatively short, directed travel is required.[7] Deflating several adjacent bags simultaneously has the effect of arresting all motion when an objective has been reached. Deflation also lowers the payload, carried in the axle, down onto the surface where instruments, including a drill, can be deployed. In the original *Mars Boule* concept, the power for the instruments and the pumps was provided by a radioisotope source, and the entire vehicle weighed two hundred and fifty kilograms. Later NASA versions, like the *Tumbleweed* shown in Figure 3.6, were much lighter (forty kilograms) and the fabric so strong that the ball could be inflated before descent onto Mars, becoming its own parachute and using the airbags to absorb the final impact. The designers claimed that the device could climb a forty-five degree hill if propelled by a strong wind of thirty metres per second or more.

Vehicles with wheels, tracks or legs can traverse the surface of Mars for distances of tens of metres to hundreds of kilometres, depending on their design and purpose. In the 1980s, concepts ranging from a simple '*Viking* on wheels' to

---

[7] A full-scale prototype was demonstrated at a number of planetary science conferences in the 1970s. Unexpectedly encountering this huge, hissing, lumbering monster on the lawn outside the meeting room could be a terrifying experience.

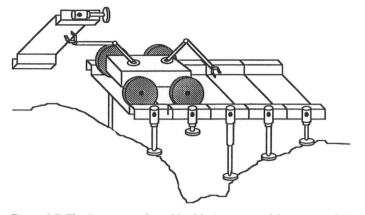

Figure 3.7 The 'incremental road builder' was one of the more ambitious of a large number of mobility concepts that were considered before the relatively simple six-wheeled rover became the preferred robotic approach to traversing the surface of Mars.

the implausibly elaborate 'incremental road builder' in Figure 3.7 were considered, along with vehicles of intermediate complexity that formed the basis of the designs for the actual missions that would go on to place *Sojourner*, *Spirit*, *Opportunity*, and soon *Mars Science Laboratory* and *ExoMars*, on the surface of Mars.

Sample return has many advantages over the *Viking* technique of analysing the sample remotely with instruments on Mars. Firstly, much more sensitive and sophisticated techniques can be used so that, for example, the absolute age of the rocks and detailed chemical composition and mineralogy of the surface material can be determined very accurately. Secondly, experiments can be repeated, iterative experiments based on new results are possible, and new, more powerful, not yet perfected techniques can be used as they are developed. Finally, a broad range of scientific disciplines and a larger section of the international scientific community can become involved. The Mars sample return 'option tree' shown in Figure 3.8 was prepared by a study group looking at *Viking* follow-on missions, which reported in 1978. The group came down against a simple 'grab-sampling' concept, on the grounds that a disappointing result was possible, whereas the incremental cost for intelligent and effective sampling was (they optimistically claimed) relatively small, requiring (1) local mobility, (2) site characterisation from orbit and with precursor missions such as aircraft, (3) at least three carefully chosen sampling sites. They recommended a programme with two launches in 1988 (a global mapping orbiter and a sampling mini-rover), four launches in 1990 (two sampling landers with mini-rovers, a six-penetrator network, and a communications satellite) and three launches in 1992 (another site with sampling and a larger rover, plus four Mars airplanes with sample collecting mini-rovers), all at a cost of nearly two and a half billion dollars (in 1979 prices).

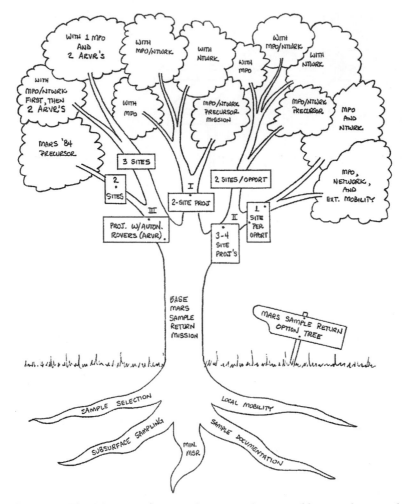

Figure 3.8 This Mars sample return 'option tree', prepared by a study group looking at *Viking* follow-on missions in 1978, illustrates the wide, verging on bewildering, range of options they considered.

It is interesting to also quote the scientific objectives seen at this time as the motivation for going ahead with some or all of this ambitious selection of techniques. The following were the 'top ten' key areas still to be investigated and understood in 1979, early in the hiatus but after most of the *Viking* results were in hand, according to the *Viking Orbiter* Project Scientist, Conway Snyder:

1. the internal structure, dynamics and physical state of the planet;
2. the chemical composition and mineralogy of surface and near-surface materials on a regional and global scale;
3. the chemical composition, mineralogy and absolute ages of rocks and soil for the principal geological provinces on Mars;
4. the interaction of the atmosphere with the regolith;

5. the chemical composition, distribution and transport of volatile compounds that relate to the formation and chemical evolution of the atmosphere;
6. the dynamics of the atmosphere on a global scale;
7. the planetary magnetic field and its interaction with the upper atmosphere, solar radiation and the solar wind;
8. the processes that have produced the landforms on the planet;
9. the extent of organic chemical and biological evolution on Mars;
10. the history of the planet and [of] these evolutionary processes.

The biggest problem is getting the trade-offs right. Acquiring the most valuable samples involves covering a wide range of usually difficult terrains and examining the material carefully on Mars before deciding to bring it back. Doing all of this properly increases the cost and complexity of the mission a lot, and reduces the reliability at the same time. On the other hand, simple missions may be cheaper and more likely to meet their modest goals, but they attract less enthusiastic support, not least within the study group members, who as Mars scientists find it hard to resist the most ambitious targets. For what was the first but would not be the last time, a sample return mission proposal overreached itself scientifically and proved politically non-viable. By the time Dr Snyder wrote, in August 1979,[8] it was clear that none of the Mars Science Working Group's plans was high enough on NASA's list of priorities to go ahead any time soon. Dr Snyder concluded diplomatically, 'it is now clear that the decade of the 1980s will be spent in analysing and understanding the data provided by *Viking* and in planning for missions in the 1990s.' So it was to prove.

3.4 *Kepler, Mars Aeronomy Orbiter* and *Mars Geoscience/Climatology Orbiter*

In Europe, after a slow and shaky start, a space agency representing most of the nations forming the continent had come of age, encouraged by the successful interception of Halley's Comet by the *Giotto* mission[9] in 1976, and began to develop an increasingly ambitious programme. A few dynamic individuals,[10] some of whom had had the chance to be involved in American missions, encouraged the European Space Agency (ESA) to look towards the planets, and especially Mars. The first fruit of this was a proposal to develop a spin-stabilised[11] orbiter called *Kepler* (Figure 3.9), to be placed in a highly eccentric orbit ranging

---

[8] Conway W. Snyder, The planet Mars as seen at the end of the Viking mission, *Journal of Geophysical Research*, **84**, 8487–8519, 1979.

[9] The *Giotto* spacecraft was built by British Aerospace at the former Bristol Aeroplane Company site at Filton, a major milestone in the history of the plant where Bristol Fighter biplanes had been produced during the First World War.

[10] Led by Professor Ulf von Zahn of the University of Bonn, see Appendix C.

[11] Spacecraft have to have a stable alignment so they can point their cameras and other instruments at the target and keep their communications antenna and solar cells pointed at the Earth and the Sun,

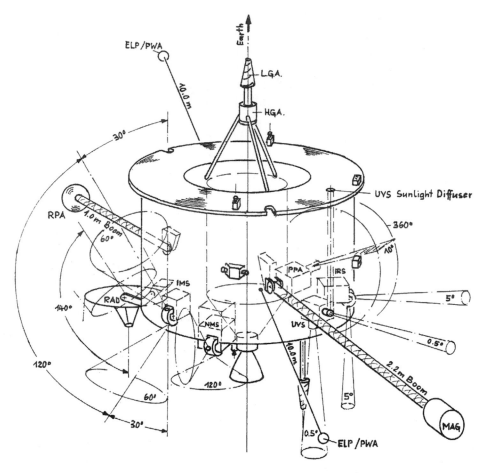

Figure 3.9 A sketch by the European study team of their proposed *Kepler* Mars orbiter, showing the positioning of the scientific instruments, and their fields-of-view where appropriate. The spacecraft was designed to be spin-stabilised, with the central axis pointing towards the Earth at all times to keep the communications dish on top of the drum-shaped main structure aligned. From Mars, the Sun is always in approximately the same direction as Earth, so the photoelectric array to provide power can be arranged in a flat annulus surrounding the communications antenna. The nozzle for the rocket motor to be used for on-orbit manoeuvres can be seen at the bottom of the spacecraft.

from a hundred and fifty kilometres minimum to ten thousand kilometres maximum altitude above the surface of the planet. This low-cost project would focus on Martian aeronomy, a general term for the upper atmosphere, ionosphere and near-space environment of a planet.

---

respectively. This can be achieved by a slow, steady rate of spin (five revolutions per minute for *Kepler*) around an axis through the centre of mass, like a top, or by using gyros and small gas jets. The latter, called three-axis stabilisation, makes instrument pointing more versatile but is more expensive to achieve and adds mass to the spacecraft.

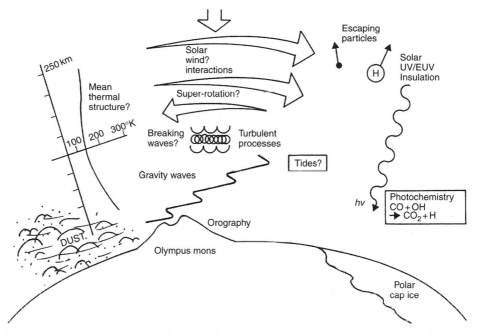

Figure 3.10 In 1986, NASA studied *Mars Aeronomy Orbiter*, a mission very similar to the European *Kepler* in terms of objectives and spin-stabilised spacecraft design. This sketch from the study team's report shows the principal processes in the upper atmosphere of Mars that were to be studied; it could apply equally to either the American or the European mission.

Five years on, the project had twice been the subject of a Phase A study by ESA, with the German Messerschmitt aerospace company[12] as the main industrial contractor. Each time, these detailed evaluations showed that a successful mission was possible within ESA's budget, effectively qualifying the mission as a candidate to be a 'new start' at the next opportunity in the agency's programme. Despite a protracted effort by interested scientists, supported by those who felt that it was a sensible candidate for the fledgling European planetary programme, being multidisciplinary and not too ambitious technically or fiscally, *Kepler* faced severe competition for limited resources and never achieved flight status. However, many in the larger planetary community in the USA also supported the concept and had a similar project in mind, which they proceeded to develop further when *Kepler* dropped out of contention. This was *Mars Aeronomy Orbiter*, with the goals illustrated in Figure 3.10 Although it had a good scientific following and was very low in cost compared to its competitors, NASA thought the *Aeronomy Orbiter* dull and lacking in ambition as the successor to a successful

---

[12] By this time part of a consortium called MBB (Messerschmitt–Boelkow–Blohm).

landing. In the end, it somehow fell through the cracks between the conservative and the cavalier proponents of Mars exploration at the time.[13]

A better compromise was found with the *Mars Geoscience/Climatology Orbiter*, another low-cost mission designed to address the requirement for more detailed studies of the Martian environment which strategy groups like COMPLEX had stressed was an essential precursor to landing and sample return from Mars. The working title for the project was chosen to emphasise that this was not a search for life, but an investigation into the surface composition and topography, the gravitational and intrinsic magnetic fields, and the seasonal behaviour of volatiles, dust and the atmosphere. The concept eventually won approval for flight status as *Mars Observer*, leading to a launch in September 1992.

3.5 The US–European Joint Working Group on Planetary Exploration (1982–4)

When ESA began to take *Kepler* seriously, American scientists soon realised that they had a new ally in their search for new flight opportunities to Mars. The Chairman of the Space Science Committee of the European Science Foundation, Johannes Geiss, a professor at the University of Berne in Switzerland, worked with the chairman of the Committee on Planetary and Lunar Exploration (COMPLEX) of the Space Science Board of the US National Academy of Sciences, Eugene Levy, to set up the US–European Joint Working Group on Planetary Exploration, which had its first meeting in June 1982 in Paris.

The task of the Joint Working Group was to define a set of mission options that could be carried out jointly by ESA and NASA, and then to take steps to see at least some of them implemented. The agreed strategy was to consider three missions, representing low, medium and high cost options, in each of the three broad areas of planetary science: (1) terrestrial planets, (2) outer planets, and (3) small bodies (comets and asteroids). A member of the Joint Working Group was elected chairman of each area and was able to call on other members, and co-opt experts from both sides of the Atlantic, to compose a working group to define its three proposed missions in detail. These were relayed back to the Working Group, which released its final report in 1986.[14]

The Terrestrial Planets Working Group decided to focus on Mars for all three of its missions.[15] The inexpensive option was the *Mars Dual Orbiter* mission

---

[13] It is interesting to note that the mission recently announced as approved by NASA and in preparation for launch in late 2013, known as *MAVEN* (for *Mars Atmosphere and Volatile EvolutioN*), is essentially the *Mars Aeronomy Orbiter*, or *Kepler*, resurrected twenty years on. See the description in Chapter 8.

[14] *United States and Western Europe Cooperation in Planetary Exploration*. National Academy Press, Washington, DC, 1986. See also Appendix C.

[15] Part of the reason for not choosing Venus was that NASA had just approved a *Venus Radar Mapper* mission (later named *Magellan*), while Mars exploration would remain in the doldrums if it could not get approval to start the *Geochemistry Climatology Orbiter* project.

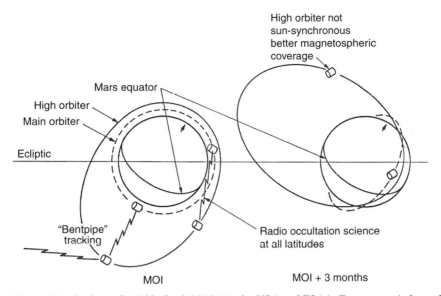

Figure 3.11 In the early 1980s, both NASA in the USA and ESA in Europe carried out detailed technical studies of a Mars polar orbiter that would conduct a comprehensive global mapping of the geochemical provinces of the planet, its atmosphere, its gravitational field, its magnetic field, and the charged particles in its vicinity. Both had difficulty in obtaining the broad support necessary to proceed to a flight project, and in 1983 they joined forces to produce a proposal called *Mars Dual Orbiter*, in which the two spacecraft would be sent separately, but at the same time, in order to operate simultaneously and work together on commonly agreed science goals.

shown in Figure 3.11, which involved American and European spacecraft orbiting Mars at the same time and working together to deliver joint science. The intermediate-price mission was a *Mars Surface Network* of twelve small stations, employing penetrators to form a grid of meteorological and seismic stations across the planet. Finally, the high-priced *Mars Surface Rover* mission was to consist of 'two, or possibly three' identical rovers, placed at widely separated locations of different surface types, and a telecommunications satellite to maintain contact with them and relay commands and data back to Earth.

3.5.1 Mars Dual Orbiter

The background to this concept combined the first European Mars mission to be seriously studied, *Kepler*, and the NASA *Mars Geochemistry Climatology Orbiter*, both of which had already been developed and designed as individual projects. The Joint Working Group pointed out that flying both together would offer more scientific return than the sum of the two parts, and launching the two components separately kept the interface between the agencies simple. It also avoided the possibility that the whole mission could be lost due to an error or misjudgement (or cancellation) by either partner. The NASA orbiter would go into a low altitude circular orbit, while ESA's spin-stabilised spacecraft occupied a high, eccentric orbit. Among the additional benefits that accrue from using two space-craft simultaneously is the ability to provide a radio link between the two spacecraft, as well as between each spacecraft and Earth. This provides high vertical resolution data on the Martian atmosphere during occultations, while improving the downlink to the Earth and so giving much better coverage of the planet by both spacecraft.

3.5.2 Mars Surface Network

The concept of a network of penetrators is attractive because it offers the possi-bility of a large number of instrumented stations on Mars, covering the whole planet and a variety of terrains, at a relatively low price. It offers easy access to regions that are difficult to reach any other way, such as the icy polar caps, the tops of the giant volcanoes or the bottom of the deep rift valleys. Penetrators also probe the subsurface of Mars, where ice and possibly even biological material is to be found, without the complexity of drilling. The main disadvantage is the difficulty involved in producing scientific instruments and subsystems that can survive the very high shock that results from striking the surface at a speed of several hundred metres per second, although the technology to do this is actually fairly readily available, thanks to years of military development. A subtler disadvantage is the need to use radioactive thermal generators[16] in order to supply power and keep the instruments warm enough to operate successfully. Most of us do not feel comfortable about showering nuclear-tipped missiles down onto another planet, even non-explosive ones delivered with benign intentions.

Still, the scientific potential of a penetrator network is great. The Joint Working Group listed the main objectives as follows:

1. The global distribution of igneous (volcanic) and sedimentary (water-deposited) rock under the weathered layer (i.e. the soil built up from dust carried and widely deposited by winds).
2. The distribution of subsurface volatiles, principally water ice. Most of the water that once flowed on the surface of Mars is frozen a few metres and more below the surface, probably within striking distance for a penetrator probe.
3. The interior structure of Mars, using a network of seismic measurements. These work best when the seismometers are in intimate contact with the regolith, i.e. driven into it some considerable depth, rather than lying on the surface.

---

[16] Radioisotope thermal generators (RTGs) contain pellets of radioactive material, usually plutonium, which produce heat to generate electricity using thermocouples. They are used to charge a spacecraft's batteries whenever the preferred method, photovoltaic solar cells, cannot be used, for instance far from the Sun (as with the *Cassini* mission to Saturn) or (as here) underground.

4. The magnetic properties of the surface material, and measurements of residual fields (induced when Mars had a global magnetic field, in the distant past).
5. Meteorology and general circulation of the atmosphere, investigated with twelve weather stations distributed around the planet.
6. The heat flow in the crust (assuming the interior of Mars is still cooling) and its variability with location on the planet.
7. Surface imaging, to study the local geological environment in many different locations.

3.5.3 Mars Surface Rover

In order to put at least two rovers on Mars within the budget for a large mission (which had been set at one billion US dollars) the Joint Working Group recognised that they would have to be relatively simple devices driven from the Earth, and not the intelligent, autonomous robots that were still some way off in terms of development. There were two obvious candidates, the American six-wheeled rover studied as part of the abandoned *Mars '84* mission, or the inflatable ball type developed in Europe. The latter had a huge cost advantage, in that it could relatively easily survive a hard or semi-hard landing, whereas the NASA rover required a *Viking*-style soft landing using retro-rockets controlled by surface-detecting radar.[17] Either way, these would be large rovers (three hundred and fifty kilograms) that could carry much larger and more sophisticated scientific payloads than penetrators (around a hundred kilograms, versus about ten kilograms), and would make possible the isotopic and chemical analysis of carefully selected samples of rock, soil, atmosphere and ice.

3.6 The Joint Committee on US–European Cooperation in Space Science (1995–7)

Of the three Mars missions studied by the Joint Working Group, the dual orbiter and network missions never flew, and, although the rover mission description had a great deal in common with the *Mars Exploration Rovers* that launched in July 2003, that was a purely American mission.[18]

---

[17] The Joint Working Group's Terrestrial Planets team foresaw the use of inflatable 'landing dampers' not unlike those that would be used to deliver six-wheeled rovers as part of the *Pathfinder* and *Mars Exploration Rover* missions more than a decade later, but concluded that this approach could not be recommended since 'no study or costing has been performed on such systems'.

[18] The Joint Working Group scored a bull's-eye with *Saturn Orbiter–Titan Probe*, the 'expensive' recommendation of its Outer Planets Study Group, led by the late Fred Scarf. As *Cassini–Huygens*, this was eventually launched in 1997, and consisted of an American Saturn orbiter spacecraft (albeit not in the event a duplicate of the *Galileo* Jupiter orbiter, as the report had recommended) with a European Titan probe.

Nevertheless, the Working Group was felt to have been a success, at least diplomatically, and in terms of getting ideas on the table that would eventually get implemented, one way or another. In fact, the group was reconstituted more than ten years later as component of a Joint Committee on US–European Cooperation in Space Science, which had similar goals to the earlier initiative but across a broader range of disciplines. In its 1997 report, the Planetary Science section reviewed the earlier study and its outcomes, successful and not, and again re-affirmed Mars as the target of choice for a new joint project. Sample return remained the 'holy grail' for the scientists, but the surface network was the most strongly endorsed of the options considered for immediate implementation, as being excellent science, timely and inexpensive, and relatively straightforward to divide between the two agencies. The Committee did not initiate any new technical studies, but pointed out that the necessary work had already been completed under the banners of *Mars Rover–Sample Return*, *Marsnet*, *MESUR* and various spin-off studies, leaving networks on Mars and the return of samples both ready to move straight into a joint Phase A, in which the engineering tasks would be divided between the two partners.

The report of the Joint Committee was well received, and led to a considerable amount of further study and diplomatic activity, but ultimately no flights to Mars that can be directly attributed to its work. The report noted: 'Cooperation in the exploration of Mars is perhaps the most discussed of any large international project, and the one with the highest public profile. Yet it stubbornly refuses to come into being.' This is still true, at least the second part, more than ten years later. Networks and sample return are still very much on the agenda in 2009, however, and we will encounter them again in Chapter 7 when we look forward to the missions that will take place in the next few decades of the twenty-first century.

3.6.1 *Mars Rover–Sample Return*

After *Viking*, it seemed that a mobile lander combined with sample return had all the right attributes to become the big follow-on mission that everyone that had contributed to the first successful landing expected. NASA managers were not so keen, however, and by the time they somewhat grudgingly authorised a pre-Phase A in 1987, it had long been clear to the realists that samples would be returned to Earth later rather than sooner. This was actually an advantage in some ways, since such a complex and expensive mission could use a lot of development time, but the scientists wanted to get started. The prospect of actually having carefully selected samples of the Mars environment in the laboratories on Earth was irresistible, but it had to be done properly. This might mean acquiring samples from relatively inaccessible places, and it certainly meant a careful investigation of the setting they came from. At the very least, it would be important to make sure the samples were really valuable and not just soil made of the windblown dust that covers the whole planet. So, mobility was a key component, and since the advances in robotics implied in producing a competent rover would have countless potential applications elsewhere on Earth, this was a technological development almost everyone could support in spite of its high cost.

The study group certainly felt that, this time, they were working on a mission that would fly, both literally and metaphorically, as a direct consequence of their definition work. They stated the goals of the mission as:

1. To reconstruct the geological, climatological and biological history of Mars and determine the nature of its near-surface materials.
2. To obtain key environmental information and test key technologies necessary to maximise the safety and effectiveness of eventual human exploration, thus cannily and succinctly emphasising the all-inclusive nature of the science, and highlighting the technology advances and the contribution to eventual manned exploration.

Much effort was expended on the recurring question of how to choose the best landing site, such that it would have maximum scientific interest while still being a safe place to set down a heavy and fairly delicate vehicle. Among several factors that made the payload vulnerable was of course the fact that it had to land loaded with explosive fuel, in order to be equipped to take off again with the samples. The criteria adopted for choosing the landing site were:

1. A variety of material types, sedimentary as well as igneous, should be available for sampling; they should include as many key types as possible (what they meant by this is defined in Figure 3.12).
2. The site should be well characterised in advance so the rover operators know what they are looking for; this does not exclude responding to discoveries as well.
3. Some of the accessible materials should be examples of 'typical' Mars surface components found over large areas on the planet.

In order to cover all of the requirements, the rover would have to collect and return between five and ten kilograms of material, made up of soil samples and cores, over a hundred rock fragments, and several separate samples of atmospheric gas. All had to be labelled, so they could be distinguished when they were unpacked back on Earth.

The final choice of landing site would require a high-resolution survey from orbit. At that time, there was no other mission in prospect that could provide this so the mission that delivered the rover and sample return capability would need also to deploy an orbiter. This would not only verify the safety of the site, and its suitability for collecting samples, but also support the rover in choosing the best path for its fifty-kilometre trek, and help to understand the broad context of the local terrain it crossed. Imaging with a resolution better than one metre (comparable to the camera on the 2006 *Mars Reconnaissance Orbiter*) would be

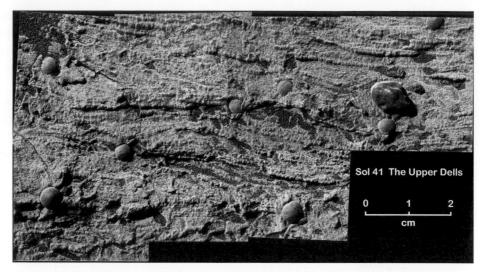

Figure 3.12 Choosing the materials for sample return. The *Mars Rover–Sample Return* team made the following list of materials that they would like to see returned from Mars, and explained why. While recognising that a single rover could probably not access all of these, it nevertheless recommended that the site selection should be made to try to take in as many as possible.

- Unweathered igneous rocks with a wide range of ages. These come from deep inside the planet and provide information on its overall composition. By comparing samples of different ages the history of the evolution of the planet is revealed. Weathering – long exposure to water or to the atmosphere, or to shock impacts and outer space as well in the case of the Martian meteorites found on Earth – destroys much of this evidence.
- Fragments ('breccias') from large impact basins. These are likely to have originated tens of kilometres below the surface, representing samples of the very early crust, or possibly even from the mantle that lies below the crust.
- Water-laid sediments, which should contain evidence of the climate history and also of any past biological activity.
- Drift material, including soil; these are probably the main component of the surface planet-wide.
- Salts, including carbonates, sulphates and nitrates. These are key to assessing the volatile inventory of the planet, and understanding changes in the atmospheric density and composition over time.
- Layered sediments, recording climate change and thermophysical history related to changes in Mars' orbit, volcanic activity and other factors.
- Atmospheric gases, to obtain key isotopic ratios and abundances of trace constituents, all of which are hard to obtain accurately by remote spectroscopy.

required, and supported by spectral characterisation of the rock types and their distribution around the site.

Much effort was also expended on deciding what should be the range of the rover on the surface, in order to get the best trade-off between variety of samples on one hand, and mass, cost and risk on the other. The safety requirement meant that the landing was likely to take place on a flat area like a plain where the surroundings were similar for a considerable distance – not near a polar cap or the edge of an exposed cliff, for example (although the landing by *Spirit* would later show that it is possible to be fortunate in this regard). In 1987, the conclusion was that, in order to guarantee that more than one type of geological unit was sampled, the rover should be able to travel several tens of kilometres away from the lander, and of course it had to return to it with the samples without losing its way.

The vertical dimension is also important; samples obtained at several depths down to at least two metres were considered essential in order to provide a profile through the zone where the surface is modified by contact with the atmosphere and by exposure to the radiation environment at the surface. This requirement was recently re-iterated by the team planning *ExoMars*, leading to the addition of a two-metre long drill to the rover being readied for launch in 2016.

Finally, time; how long should the mission last? This is relatively easy, since it is set by the time taken to travel the distance specified, and also by launch windows for arrival and return.

The rover would have to be equipped not only for navigation and sample acquisition, but also with sensors that could tell an interesting sample from one that was boring or repetitive. The range of possibilities included alpha, proton, neutron, X-ray and infrared spectrometers, and gas analysers attached to ovens for heating samples to make them give off gas for analysis. Many of these devices did eventually fly on landers and rovers and they and their results will be described later. In what was in retrospect probably a fatal decision, the team decided that the rover–sample return mission also presented an irresistible opportunity for 'add-on' science, and it designed in the deployment of meteorological stations, a seismic network, and electromagnetic sounding for penetrating the surface to look for strata, including water or ice, down to a depth of several kilometres. A similar capability for subsurface sounding over a wider range was added to the orbiter, anticipating the ground-penetrating radars to be flown on *Mars Express* and *Mars Reconnaissance Orbiter* some twenty years later.

Once the required capabilities had been spelt out by the scientists and in-house engineers and mission planning specialists, NASA put out contracts to the aerospace industry for detailed designs. The result, following parallel Phase A studies by two large companies with strong track records in space, TRW[19] and Martin Marietta,[20] is shown in Figure 3.13. Two large *Titan-Centaur* launches would be required, the first planned for the 1996 opportunity to deliver the imaging and communications relay satellite and the sample return vehicle into Mars orbit. The latter would simply wait there, while the former was busy

---

[19] TRW was Thompson Ramo Wooldridge Inc. before 1965. Since 2002 they have been part of Northrop Grumman Corporation.

[20] Now Lockheed Martin, following a merger in 1995.

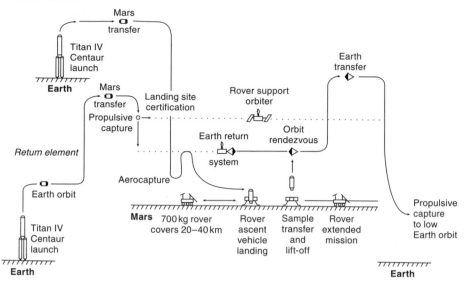

Figure 3.13 The planned sequence of events that emerged from the NASA *Mars Rover–Sample Return* mission study in 1989. The two launches were to take place in successive launch windows, the first in 1996 and the second in 1998, to allow the imaging/communications rover support orbiter time to map the planet at high resolution and find a good landing site. The samples were to have been back on Earth in 2000 or thereabouts.

imaging the surface to support the selection of an optimum landing site. The second launch, two years later, would deliver the rover and the sample-carrying ascent vehicle to the surface. The latter would also wait patiently while the rover carried out its mission of finding and collecting samples. These would then be transferred to the ascent vehicle – itself a non-trivial task as can be appreciated by a quick look at Figure 3.14 – which would take off into orbit, there to rendezvous with the return vehicle. At the appropriate time for a minimum-energy transfer to Earth, the rockets would fire to begin the journey. They would be used again seven months later upon return to Earth to achieve a low orbit where the capsule with its precious cargo could be collected by a space shuttle flight.

The complexity of the multi-launch, multi-vehicle approach required, and the generous approach to the objectives adopted by the science definition team, made for a very capable mission with wide appeal (a 'Christmas tree' in the derogatory language of those within the community who believed that the most austere and economical approach always had the best chance of success). However, it came at a high cost, and when the accountants on the team added up the estimates for the multiple spacecraft, its scientific instruments, launch and mission operations, the grand total was well over three billion dollars. Probably well over ten billion in today's money, this total was well beyond what NASA saw as its means at the time, and the concept was swiftly shelved. Recognising

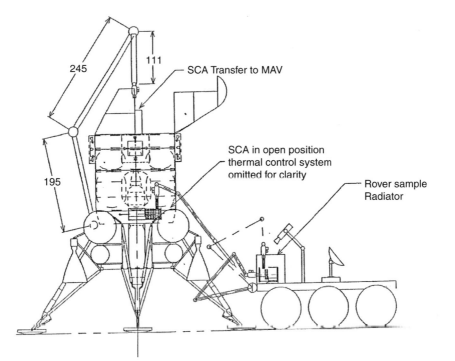

Figure 3.14 A simplified drawing of the procedure whereby the rover hands over its cache of samples to the ascent vehicle, as envisioned by the *Mars Rover–Sample Return* mission study in 1989.

that they had been overconfident, the team considered revisiting all of its work and scrimping it to produce a 'baseline' version, but was discouraged by the reduced science and increased risk that invariably appeared when savings were made, even at a level that was probably still too expensive.

3.6.2 Network Missions: *MarsNet* and *Mars Environmental Survey*

Having failed to initiate sample return, not for the first time or the last, JPL and the community it supported decided to look at the other mission concept-in-waiting, the global network, recognising that this could be done without serious compromise for much less than the cost of sample return. At this time, the early 1990s, a study of such a mission was also going on in Europe, although to keep the cost within bounds, ESA's *MarsNet* involved just four landers.

The *MarsNet* landers were each identically equipped for meteorology, seismology, atmospheric structure, geology, geochemistry and mineralogy measurements. In order to use solar cells and avoid the need for radioisotope power sources, which were not available in Europe, the landing sites were constrained to the latitude range from 45° N to 45° S. The ideal distribution for seismic measurements would place three of the landers on a triangle about one thousand to three thousand kilometres apart, close to a likely seismic source such as the massive uplands of Tharsis or Elysium, with the fourth station located on the opposite side of the planet. Such an arrangement is reasonably optimised for meteorology and geochemistry also, although perhaps with some widening of the triangle to maximise the latitude coverage and provide a variety of geological units and ages.

With *Mars Environment Survey* (*MESUR*), which was studied from 1991 to 1994, NASA aimed to have at least sixteen stations covering the whole planet. The payload designed by the study team included imagers, an atmospheric structure experiment, a weather station, a spectrometer and a seismometer. In order to operate at high latitudes, even in the polar night, power was to be provided by radioisotope thermal generators, which could also be used to keep the payload warm in freezing conditions. The semi-soft landing system would involve entry into Mars' atmosphere directly from its Earth–Mars trajectory and deceleration using a large heat shield, parachutes, and finally airbags deployed seconds before landing.[21] All of this led to a considerable mass – around a hundred and sixty kilograms – for each *MESUR* station, compared to only sixty kilograms for *MarsNet*. A communications orbiter would also be required to keep in touch with the widely distributed stations on the surface and relay their data back to Earth.

The *MESUR* team devoted a great deal of effort to choosing sixteen landing sites. In their report, they specified that three stations would make a seismic triangle near each other on the north rim of Valles Marineris, and another three, the same arrangement at the foot of the giant extinct volcano Olympus Mons. Stations optimised for meteorology would include one in Chryse Planitia, where *Viking 1* accumulated data from 1976 to 1983, and one east of Solis Planum, a region of known dust storm activity, and one in Aonia Terra, south-west of the vast Argyre basin. Two would land near the north and south poles, two would be on the opposite side of the globe to the seismic triangles, to enable the size of Mars' core to be determined, and the rest would complete the global coverage.

To spread out the cost, and allow the use of relatively small *Delta* launch vehicles, the network and the communications orbiter were to be put in place in three stages, during the launch windows in 1999, 2001 and 2003. Primarily to reduce the risk associated with the unproven landing system, the network would be preceded by *MESUR Pathfinder*, a single-spacecraft mission devoted entirely to testing the technology, which includes a system of 'petals' to force each lander into an upright position. Despite eventually relaxing the requirement for radio-isotope power, when the study was completed and the cost totted up it came out at much too high a price for the political climate at the time, and *MESUR* was shelved. Only the *Pathfinder* survived the cuts.

---

[21] Direct entry results in hitting the top of the atmosphere at about seven kilometres per second, about half the speed of entry from orbit like *Viking*.

3.7 *Mars Together*

At about the same time that the Joint Committee was deliberating US–European joint exploration of Mars, another set of phantom missions for the joint exploration of Mars was being studied by the United States with another partner, Russia. The concept, *Mars Together*, was stimulated by the end of the cold war, and fuelled by a feeling, widely held at the time, for a while, that governments in the two nations would welcome an exciting, peaceful, high-tech substitute for the arms race. The initial idea was to develop innovative ion propulsion systems using large solar arrays or nuclear reactors capable of generating up to thirty to forty thousand watts. These mighty thrusters would be demonstrated as part of a project to place a large spacecraft into Martian orbit, where a side-looking radar would undertake systematic digital mapping of the surface. Before that, a more modest prototype spacecraft with a mass of a hundred and fifty kilograms, a solar panel area of thirty square metres, and ion engines with a thrust of three kilowatts would orbit the Earth with a changing orbital altitude as the ion engine thrust continuously for several hundred hours.

Along with this ambitious scheme, slightly more realistic plans were discussed by the joint Russian/American *Mars Together* team for a collaborative effort to place a *Marsokhod* rover on the surface in 2001, and to follow this with a sample return mission in the mid 2000s. What actually happened was that a Russian Co-Principal Investigator (Co-PI) and Russian hardware were incorporated into the *Pressure Modulated Infrared Radiometer* experiment to be flown on the *Mars Surveyor* 1998 orbiter. The individual selected, Vassili Moroz of the Space Research Institute in Moscow, was an amiable pro-western infrared instrument expert already well known and respected by the US–UK PMIRR team, so the Co-PI arrangement went swimmingly. Less successful was the Russian hardware, which the Americans considered substandard and never actually used. Plans to include an additional Russian instrument on the *Mars Surveyor* 1998 lander also never materialised, fortunately for all concerned as it turned out.

3.8 Failed missions

An even worse fate, if that is possible, for scientists who have seen their favourite missions cancelled after they have been through long planning and lobbying exercises, is to have them built, only to fail during flight before any data are returned. A remarkable feature of the new era of Mars exploration that followed the post-*Viking* slump is the high number of failures associated specifically with missions to Mars; proportionately much higher than for missions to the Moon, Venus or indeed any other Solar System body. In the early, pioneering days of interplanetary flight, of course, problems were to be expected. Both the USA and the USSR had taken many, repeated failures in their strides, and responded by pressing on with increased determination. Before there was any success at all, three spacecraft had failed, two from the USSR and one from the USA. Then, with *Mariner 4* in 1964, the Americans tasted success at Mars, only to have the planet itself prove disappointing, as we have seen.

A list of all of the missions by any nation to Mars, indicating their success or failure, appears in Appendix B. There it can be seen that missions have failed, not only in the early pioneering days of interplanetary flight, but quite recently as well. In fact, the Russians, have never had a really successful mission to Mars, despite nineteen attempts and some results from the *Phobos 2* orbiter in 1988, and a series of very successful missions to Venus. Two-thirds of all of the Mars missions launched by all nations (twenty-nine out of forty-four) have failed; the USA has the best record, with twenty-one missions and fourteen successes.

No fewer than five spacecraft were lost in the six years from 1999 to 2004, including a Japanese orbiter called *Nozomi* and the European *Beagle 2* lander. Losses, of course, are inevitable in such a daring venture into the unknown, particularly since each mission must necessarily be more ambitious than the last in order to seem worth doing at all. While hardware failure is just barely acceptable, because sometimes unavoidable, politicians stick short of risking lives to go to Mars. While the explorers of the sixteenth to the nineteenth centuries could not wait to risk their lives in remote corners of the Earth in return for knowledge and glory and perhaps even riches beyond compare, and were encouraged to do so by those who languished at home and awaited the spoils, at Mars, so far, only robots have risked their lives.

There is also a big human cost when robots fail. The scientist who devotes a big piece of his career to developing an experiment to fly on a space mission risks a great deal if it does not deliver the results needed to write papers and illustrate talks at international meetings. In extremis, as happened in Russia when their space programme collapsed during the turmoil that led to the demise of the Soviet Union, many of the best space scientists and engineers end up without an income or a career at all, unless they seek jobs in other countries. Quite a few of the Russians with international reputations moved to the United States, where they found an outlet for their energy and ideas with NASA or in universities with space research groups. Most of the fresh technological challenges were not, in the 1980s, in sample return and human precursor missions to once-again-boring Mars, but in ambitious missions to the glamorous gas giant planets of the outer Solar System. *Voyager* was on its grand tour of Jupiter, Saturn, Uranus and Neptune, *Galileo* was being built to orbit Jupiter and make close encounters with its family of satellites, and *Cassini–Huygens* was in its early planning phase for launch to Saturn and its cloudy moon, Titan.

Mars would not go away, however. It might be barren, but it is still a nearby, Earth-like world, laden with clues about where our home planet came from and how it evolved. And what about ancient Mars, with its riverbeds and coastlines? Was there life then? Had any of it hung on in remote sites, or at depths not accessed by *Viking*? What could be learned from the global change that very obviously took place on Mars about the sorts of changes in Earth's environment that were starting to capture the headlines? Motivations like these could still affect key decisions and events, in government, in Hollywood, and in the public consciousness.

3.8.1 The Soviet revival: *Phobos*

It was actually the Soviet Union who, after a fifteen-year gap, first demonstrated a renewed commitment to building hardware to go to Mars, with an ambitious mission to land probes on the surface of the larger of Mars' two moons, Phobos. It was also hoped to hover over the surface and sample its composition with a laser, as pictured in Figure 3.15. Two identical spacecraft were launched on 7 July 1988 and 12 July 1988. The stated goals were to:

1. place the *Phobos 1* and *2* spacecraft in a Martian orbit;
2. approach the Martian satellite Phobos;
3. land two automatic stations on Phobos' surface;
4. conduct a remote distance study of Phobos soil chemistry from a fly-by trajectory;
5. conduct an extensive study of Mars from its satellite orbit.

The first craft, *Phobos 1*, failed to deploy properly at Mars due to a computer programming error. *Phobos 2* did enter Mars orbit in early 1989, but the signals from it ceased one week before the scheduled landing on the tiny moon.

The final fling of the Soviet, by then Russian, Mars programme was *Mars-96*, a large orbiter with an on-orbit dry mass of over three tons, which presumably would have become *Mars 8* had it achieved its objectives. Attaining these would involve not just operations in orbit, but two penetrators and two 'small stations' that were to be delivered to the surface of Mars. Unfortunately for what would have been a very glamorous mission, the launch vehicle failed and the payload fell back to Earth.

Mars-96, despite its failure to get to Mars, had a considerable heritage, not all of it negative. The penetrators were a bold attempt to access the subsurface of Mars without drills. Like the basic idea that NASA had already studied but considered too risky, a sharp-nosed torpedo-shaped projectile was to be dropped from orbit to impact the surface at high speed and penetrate deep into it. The tail of the device would separate on impact and remain on the surface, connected to the instruments in the burrowing forebody by a long wire. The part on the surface then relayed the data from the instruments in the nose, which could reach depths of around ten metres in relatively soft soil, back to the orbiter overhead and thence to Earth. Of course, survival was less certain if the penetrator struck solid rock, and in any event the instruments had to be designed to survive a large shock on impact at around a hundred metres per second. Such instruments do exist, and measurements of the temperature and composition (including, with luck, water ice) at depth could certainly have been achieved if the devices had made it to Mars and were fortunate in their points of impact.

Figure 3.15 The Soviet *Phobos 2* spacecraft shown exploring the surface of the eponymous thirteen-kilometre long moon with a laser beam, an experiment that it did not survive to perform.

What the indigenous Martians, if there were any, would think about these projectiles is horrific to contemplate, particularly since each missile contained a nuclear device, intended to provide power to the payload.

The small stations were no less ambitious. Carrying seven instruments each, these were to be jettisoned five days before the spacecraft reached Mars orbit, following a separate trajectory around the planet and coming to ground in a location (intended to be in the Amazonia-Arcadia region, around 35° N and 165° W) that would be visible to the orbiter instruments overhead as they circled Mars for the first time. Both the penetrators and the stations were to enter Mars' atmosphere at nearly five kilometres per second, and use aeroshields to shed most of the heat before they landed. The stations also had parachutes to reduce the landing speed to a survivable twenty metres per second, and airbags to absorb the ensuing two hundred g shock. The thirty-seven kilogram package would then have unfolded petals to force it into an upright stance where it could deploy its twelve kilograms of scientific instruments. In addition to meteorological and surface composition measurements, these included a magnetometer and a seismometer to probe the interior of the planet.

The payloads of the *Mars-96* orbiter and surface stations included a number of scientific instruments that were being carried to Mars on behalf of European investigators, and when the launch failed to get beyond Earth orbit on 16 November 1996, the desire to re-fly those experiments became the force behind the first European Mars mission, *Mars Express*. This was not as ambitious, or as large, as the *Mars-96* mission, but ESA did want a lander of some kind on board, and the Russian experience provided heritage, if not hardware, for that.

3.8.2 *Mars Observer*

The *Mars Geoscience/Climatology Orbiter* was intended to get NASA back on track for Mars sample return and eventual human missions at a modest cost by completing the survey of the planet begun by *Viking*. In addition to reconnaissance, there was to be a focus on the science of global change on Mars, including mysteries such as understanding the importance of water, and identifying and quantifying its sources and sinks, in the present and past climates of Mars. NASA listed the objectives as:

1. determine the global elemental and mineralogical character of the surface material;
2. define globally the topography and gravitational field;
3. establish the nature of the Martian magnetic field;
4. determine the temporal and spatial distribution, abundance, sources and sinks of volatiles and dust over a seasonal cycle;
5. explore the structure and circulation of the atmosphere.

Attempts were made to keep the cost down by basing the spacecraft on existing designs, including the venerable and reliable *TIROS* weather satellite, manufactured by RCA and used by the US Defense Meteorological Satellite Program. The instruments were selected competitively to reflect the current interests in what survived of the Mars scientific community, and to encourage some new blood to enter the field. The total cost, using these 'existing' hardware and designs, was initially estimated at a very modest level of less than a hundred million dollars.

During the development, the mass and complexity of the spacecraft and its payload grew, to the concern of new NASA Administrator Daniel Goldin, who saw the run-out cost increase during the last few months before launch to more than eight hundred million dollars.

Renamed *Mars Observer* in 1985, the spacecraft was launched in 1992, two years later than originally planned, heading for a near-circular polar orbit and carrying a payload of seven instruments:

1. *Mars Orbiter Laser Altimeter* to transmit and receive infrared laser pulses and measure the time of flight to determine the range from the spacecraft to the Martian surface to construct a topographic map of Mars for studies in geophysics, geology and atmospheric circulation.
2. *Gamma-ray spectrometer* to measure the abundance of elements (uranium, thorium, potassium, iron and silicon, for example) on the surface of Mars.
3. *Thermal emission spectrometer* to map the mineral content of surface rocks, frosts and the composition of clouds.
4. *Mars Observer Camera* to take low-resolution images of Mars on a daily basis for studies of the climate, and medium- and high-resolution images of selected areas to study surface geology and interactions between the surface and the atmosphere.
6. *Pressure-Modulator Infrared Radiometer*[22] to measure profiles of temperature, water vapour, dust and condensates in the atmosphere, as they change with latitude, longitude and season.
7. *Radio Science Investigation*, using the effect of the atmosphere on the spacecraft radio beam to measure the temperature profile as it varies with altitude, and tracking data to measure the gravity field of Mars.
8. *Magnetometer*, to determine the nature of the magnetic field of Mars, and its interactions with the solar wind.

None of these experiments obtained any useful data at Mars, but we will encounter them again in happier circumstances in a later chapter. Just three days before *Observer* was due to fire its engines to reduce speed and enter orbit about Mars, the fuel system had to be pressurised, an operation that involved firing two pyrotechnic devices.[23] These were intended to open the valves that would allow high-pressure gaseous helium to pressurise the nitrogen tetroxide oxidiser tank and the monomethyl hydrazine fuel tank. This in turn required turning off the transmitter that communicated with the Earth, because of concerns that the

---

[22] The 'pressure modulator' part refers to a special technique that uses pressure-cycled cells full of the gas of interest (here water vapour and carbon dioxide) to identify and isolate the infrared radiation emitted from that gas in the atmosphere.

[23] Pyrotechnic devices, small capsules of electrically ignited explosive, have been used on many spacecraft, including the *Apollo* manned flights to the Moon. They are a compact, inexpensive, and usually reliable way of actuating 'one-off' releases of valves or stowed equipment like solar panels. Their main disadvantage is the shock produced by the explosion, although this is usually small and no more than a minor inconvenience.

shock from the pyros might damage the travelling wave tube, an integral part of the transmitter, if it was on at the time.

The communications shutdown was timed to last ten minutes; in fact, the spacecraft was never heard from again. The failure review board, with no direct evidence to go on, surmised that the components of the rocket fuel mixed prematurely in the fuel transfer line and exploded, bursting the line and sending the orbiter into a rapid spin, which the on-board stabilising systems could not control. Even if the transmitter had turned on, the communications antenna on the spacecraft was no longer pointing towards the Earth, and the solar cell panels were no longer pointed towards the Sun. Without power, the spacecraft batteries would run down in a few hours and the spacecraft would fall silent forever. Without the rocket burn, it would miss Mars and go on to describe an orbit around the Sun. It is very likely that this is what actually happened.

NASA had quite recently lost the space shuttle *Challenger*, and had now given the watching world the impression that it was no longer able to accomplish even relatively simple tasks like orbiting a nearby planet. The review board concluded that the reliable Earth satellite components on which *Observer* was based had either been changed so much they no longer had any heritage, or else had not been adequately qualified for the different conditions under which they were to be used, in particular the long cruise to Mars.

Only weeks after the loss, NASA and the Jet Propulsion Laboratory began to discuss a bid to recover from the failure by sending a new mission to Mars. The agency needed a success: elsewhere in its programme, shuttle launches had become less frequent and more costly than originally advertised to the Congress, and there had been alarmingly rapid growth in the cost of the *International Space Station*. Originally promised for seventeen billion dollars, as of 2004 the projected cost was over thirty-two billion; in 2007, ESA estimated the total run-out would be around one hundred billion. Newly arrived from a career in the aerospace industry, Administrator Goldin had no responsibility for causing these problems, but he had to solve them, and also find the resources to move on to new projects. This he resolved to do using two of the latest industrial management techniques, Chaos Management and Faster, Better, Cheaper, to reshape NASA's culture.

Chapter 4
The modern era

4.1 Faster, Better, Cheaper

Chaos Management is described in a best-selling book, *Thriving on Chaos*, by the management guru Tom Peters.[1] The basic idea is that large organisations like NASA (and indeed major universities, although Peters focussed his analysis mainly on American companies) have become hidebound by employing expensive hierarchies that are actually less effective than unstructured organisations, because individuals within them are much more likely to thrive when stimulated by a quasi-random flow of ideas and opportunities. The savings that could be achieved from this approach were central to the Faster, Better, Cheaper (often shortened to FBC) initiative the NASA Administrator launched in 1992. Daniel Goldin aimed to increase the number of missions within an essentially fixed budget by shortening development times, and increase the scientific return as well, in part by putting more work in private industry. It was felt that this would encourage engineers and scientists to be bold with regard to risk taking, especially if government management and oversight was minimised.[2]

Robotic missions to the planets represented a much better opportunity for developing and demonstrating this new philosophical framework within NASA than did the manned programme, where deliberately increasing risk was not acceptable. Instead, the better practices to be imposed on the planetary programme would hopefully diffuse into the much more expensive world of space shuttle and space station by example. The choice of Mars, in particular, as an early focus for Faster, Better, Cheaper was made obvious by the recent *Mars*

---

[1] Published by Alfred A. Knopf, New York, in 1987.

[2] On one of his visits to JPL, during an address to the assembled scientists and engineers who worked at the Laboratory, Mr Goldin produced from his jacket pocket a small piece of hardware that he said was a prototype of a device that would replace a much larger, heavier and more complicated subsystem routinely used in spacecraft. The employee responsible was praised by name in front of his colleagues, who were exhorted to make similar advances.

Observer experience, where a mission that was conceived cheaply had grown by a large factor, and then failed anyway.

More individual responsibility and risk-taking might not only have avoided the problems that led to the cost growth and delays, but might also have avoided the loss of the mission by increasing individual responsibility at the expense of printed procedures. The official report on the loss of *Mars Observer* included slightly cynical comments from the technical side of the project that the decision to shut down the transmitter, rather than re-engineer it for guaranteed survival of the shocks involved in opening the fuel valves at a cost of six hundred thousand dollars, was exemplary FBC strategy. The more positive thinkers felt that if the costs had been kept down across the board, more launch attempts could have been made within the available budget, and surely not all of them would have failed.

However, the same could be said of missions to the outer planets, where the cost of exploring the Saturn system with the *Cassini* mission (described on at least one occasion as *'Battlestar Galactica'* by Mr Goldin) had grown until it exceeded three billion dollars. However, trips to Mars typically involve distances of 'only' three hundred million kilometres, requiring six months of travel time, compared to more than three billion kilometres and seven years for *Cassini* to get to Saturn. Also, the relatively frequent Earth to Mars launching opportunities (every twenty-six months), make it possible to visit the planet repeatedly for follow-up studies. Researchers obviously attach great importance to the interval of time between their posing of a question or hypothesis about the planet, designing a mission to address it, the radioing back of data related to the answer, and the launch of a subsequent follow-up mission. This interval between question, answer and follow-up is about six to seven years for Mars. In the span of a career, a scientist can participate in as many as four cycles. Contrast this with the interval for the outer planets, which is typically fifteen to twenty years per cycle. Unless they are exceptionally fortunate, an individual scientist would do well to be involved in just one outer planet mission during his or her professional lifetime. With launches planned for each window, so about every two years, robotic missions to Mars would provide excellent opportunities for the intense application of the Faster, Better, Cheaper philosophy. The high repeat rate would make risk of failure more acceptable and render each mission more affordable. Once the cultural revolution was complete, cheaper missions might approach or exceed the success rates achieved in the preceding, more profligate era.

Venus could also be an attractive candidate for the FBC approach, offering even more frequent launch opportunities and shorter journey times compared to Mars. The *Magellan* orbiter was successfully completing its mission around the time of the loss of *Mars Observer*, its synthetic aperture radar revealing the surface of Venus as a living planet with intriguing river beds and 'snow'-capped mountains alongside apparently active giant volcanoes. If the question of extraterrestrial life could be ruled out of consideration, Venus is arguably more Earth-like and interesting than Mars. However, the life issue trumps all else, and the dramatic surface of Venus is at a searing temperature that seemed clearly to rule out all possibilities of the survival or genesis of any conceivable organisms. With Venus life confined to the wings of the Solar System exploration stage (although some people think that there could be microbes in the Venusian clouds, where the temperatures and pressures are quite Earth-like) even the Russians had abandoned Venus and embarked on a mission called *Mars-96*, an attempt to succeed where the two very sophisticated *Phobos* missions had failed in 1988.

4.2 An unexpected bonus: meteorites from Mars

If there had ever been any doubt, the thing that really catapulted Mars into the lead as favourite once more in the planetary exploration stakes was the claimed evidence for life in one of the 'SNC' family of rocky meteorites. These are found on Earth but contain conclusive evidence that they originated on Mars. The rock in question, ALH 84001, weighs just less than two kilograms, and gets its name from the fact that it was the first meteorite to be found in the Alan Hills region of Antarctica in the year 1984. It was recovered by a dedicated meteorite-hunting expedition from the American National Science Foundation.

Large numbers of rocky bodies of all sizes drift through space, and many collide with the Earth each year.[3] Only a fraction of those survive the descent through the atmosphere without burning up completely, but of those that do a small number can be found on the surface. Of the thousands of meteorites that have been recovered, so far thirty-four have been shown to have originated on Mars, and are known as SNC (for Shergottite, Nahklite and Chassignite) meteorites because the earliest falls were at the eponymous locations, in India, Egypt and France, respectively. The first find was at Chassigny, in the Haute-Marne district, as long ago as October 1815 (a few days before the battle of Waterloo), but of course it was not recognised as being from Mars until much later. The Shergottites fell in 1865 and the Nahklites in 1911; more recently, SNCs have been collected in Nigeria (1962), Indiana (1931), Brazil (1958), Libya (1998), Los Angeles (1999), Oman (2000) and Morocco (2001).

They all consist of igneous rocks, formed by the solidification of volcanic magma, but they belong to different mineralogical subclassifications. Some fit into established definitions of volcanic minerals found on the Earth, others don't, and they are all distinct in composition from the basaltic rocks found at the *Mars Pathfinder* landing site, which resemble the common terrestrial mineral known as andesite. The Shergotty meteorite belongs to a class of basaltic minerals not previously known on the Earth; other SNCs have been found elsewhere that are

---

[3] Broadly speaking the number swept up by the Earth is inversely proportional to the size of the meteor, so while vast numbers of dust or sand-sized objects fall onto the planet each year, dangerously large impacts are fortunately very rare. Around thirty thousand intermediate-sized rocks known to be meteorites have been collected altogether, including the thirty-odd from Mars. Many others must remain undiscovered, at the bottom of the ocean for instance.

similar to each other and are now collectively known as shergottites. The chassig-nites are similar to terrestrial dunites, a dense mixture of olivine (magnesium iron silicate) and chromite (chromium iron oxide). Dunite is named after a site on Dun mountain at the northern end of the South Island of New Zealand, which was named in turn for its dun, reddish colour. The nahklites are like terrestrial clino-pyroxenites, a kind of silicate formed at high temperatures and pressures.

How does anyone know these rocks are from Mars? Firstly, they have rela-tively young ages, as determined by isotopic analysis. Most of the meteorites found on the Earth are debris left over from the formation of the planets and have the same age as the Solar System itself, around 4.5 billion years. The SNCs that were first discovered had crystallised from volcanic lava only about 1.3 billion years ago, and some of those found more recently (but not ALH 84001, which is very much older) are as little as 185 million years old. They must therefore have originated on a planet that had active volcanism at that time, which, in addition to Mars, could really only mean Venus or the Earth.

The rocks had definitely spent long periods in space – it is actually possible to date the time spent by analysis of cosmic ray tracks in their interiors – and must have been blasted off the parent by a major impact. The energetics of this process are much more favourable for Mars than for Venus. In addition, the SNCs contain hydrated clay-like material that is unlikely to be from hot, dry Venus.

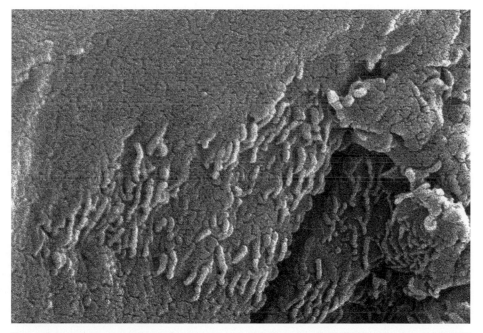

Figure 4.1 A photograph through a microscope of the interior of the SNC meteorite ALH 84001 showing putative 'nanobacteria' less than one micrometre (approximately the wavelength of visible light) in width. These may resemble earthly maggots, but the latter are more than a thousand times bigger.

The crucial indicator is, however, that their oxygen isotope compositions are distinctly different from those of Earth rocks. When the atmospheric composition on Mars was measured by the Viking landers it was found that the relative abundances of nitrogen (N), argon (Ar), krypton (Kr) and xenon (Xe) are identical to those found in bubbles of air trapped inside the meteorites. In particular, both have high concentrations of the less common ^{15}N isotope of nitrogen, ^{40}Ar isotope of argon, ^{129}Xe isotope of xenon and ^{2}H isotope of hydrogen (deuterium), which are rare on the Earth.

ALH 84001 had been sliced open for analysis by a group of scientists from Johnson Space Center in Texas, and by 1996 they were sufficiently convinced that the microstructures and chemical deposits that they found inside the meteorite had biological implications. Their announcement reinforced the impression that pictures like Figure 4.1 give to most people at first sight. On 7 August 1996, as *Global Surveyor* and *Pathfinder* were in the final phases of testing and being readied for launch, rumours began to circulate to the effect that NASA would soon announce something big about Mars. After briefing the President and Vice President, Mr Goldin called a special news conference in the main auditorium of NASA Headquarters in Washington, DC, at which he described the claims of indications of past, very primitive life on another planet. Figure 4.2 shows him speaking on CNN about the finding.

The public's appetite for news about Mars had remained high despite the gloomy verdict from *Viking*, and many people retained the Earth-centric view of Mars that had originally driven popular interest sky-high in the days of Lowell and the canals.

Figure 4.2 NASA press conference, 6 August 1996. Afterwards, Mr Goldin told the Discovery channel: 'After that press conference, I wasn't tired; I wasn't nervous. I was excited. So much adrenalin flowed, that I excused myself … went to my office and I shut the door and I sat there – blinds closed for a half hour – contemplating the impact that this could have on who we think we are…'

Figure 4.3 The press caricatured its own response to the news that NASA had new evidence of possible life on Mars.

That the media responded with unbridled passion, as lampooned in Figure 4.3, was not surprising therefore; more noteworthy was that their zeal was shared by the NASA Administrator himself, who was clearly beside himself with excitement. Within days of the announcement, NASA and the Mars science community were hard at work preparing the search for more evidence of life on Mars. Teams led by the Jet Propulsion Laboratory developed a new strategic plan, and undertook engineering studies of the missions and their costs using the principles of FBC at every opportunity. Soon, the science community, government and industry were all enthusing about the idea of a fleet of less expensive, more frequent missions that collectively achieved much more than the old school of infrequent, over-elaborate one-offs. Inevitably, as they turned from planning to building hardware quickly, cheaply and with unproven technology, the talk among the lower-level managers responsible for implementing the *Surveyor* programme, as it was called for a time in another resurrection of a name from earlier days, turned to the heightened probability of mission failure. Would it really be as acceptable as the strategy demanded?

4.3 The *Mars Surveyor* and *Discovery* programmes

The first task of the optimistic new era was to put together a recovery plan for the objectives that had been set for the ill-fated *Mars Observer*, using the new philosophy. Three small, low-cost orbiters were planned to carry the rebuilt instruments back to Mars, one in each launch window at twenty-six month intervals. The first of the three, *Mars Global Surveyor*, saved money by using many of the pieces of spare hardware that were left over from building *Mars Observer*. It was launched in 1996, carrying five of the original scientific experiments (the camera, laser altimeter, thermal emission spectrometer, magnetometer and radio science). The *Mars Climate Orbiter* which followed it would be even smaller and cheaper than *Global Surveyor*, and carry a rebuilt infrared atmospheric sounder similar to that on *Mars Observer*, plus a colour camera. Finally, *Mars Odyssey* would be launched in 2001, carrying the last leftover instrument, the gamma-ray spectrometer, plus some new instruments to be selected later.

Another programme that fitted well with FBC thinking was *Discovery*, initiated in 1989 as a line of missions that could be managed by academic or research laboratories at a low cost, and in a relatively short time, to address focussed scientific goals. The rules were a fast development time of less than thirty-six months and a cost of less than three hundred million dollars. The first *Discovery* mission to be selected was *Mars Pathfinder*, destined to be the first American lander since *Viking*. Ambitiously, considering its modest cost, it was planned to equip *Pathfinder* with a small roving vehicle that could explore the immediate vicinity of the lander on the end of a tether.

A lot of the cost of exploring Mars is, of course, in getting there. For Faster, Better, Cheaper to work, the new small spacecraft under development would need affordable, smaller launch vehicles, and sophisticated miniature instruments capable of obtaining the same sort of information gained by much larger devices in the laboratory. In other words, both American industry, which builds the rockets and sells them to NASA, and the laboratories in institutes and universities that develop the scientific payload instruments, needed to make investments in new technology that the programme itself was not designed to pay for. Large investments in products that are to undercut existing stock in price are not popular in commercial organisations and the response within the aerospace sector to NASA interest in a new cheap but capable 'Med-Lite' launcher was not wholehearted. The scientists and engineers developing instruments, on the other hand, could see that cheaper missions had many potential advantages. They would have many more opportunities to get their concepts into space, in contrast to the seventies and eighties when they became accustomed to waiting ten or more years for the chance to even propose their best ideas. Even then they could not be assured of a mission for their own team, or even within their own particular subdiscipline, since the scarcity of opportunities meant even keener competition with their peers in the community for the privilege of participation.

Now, there would be an average of about one mission per year, a real cornucopia compared to the *status quo ante*. When the December 1996 launch window opened, there were two Mars missions on the launch pad: *Global Surveyor*, the first of the *Observer* recovery missions, and the first *Discovery* mission, *Pathfinder*.

4.4   *Pathfinder* and *Sojourner*

As we have seen, *Pathfinder* was so called because it was originally intended as a test-bed for the technology needed for the *Mars Environmental Survey* (*MESUR*), which was to set up a network of stations that would be active simultaneously all over the planet. In the event, only the prototype went to Mars, with the now generalised objective of evaluating an inexpensive landing system and initiating the strategy of characterising the Martian surface environment in preparation for future exploration, using its mini-rover, *Sojourner*, pictured in Figure 4.4.

The *Pathfinder* mission in its entirety cost less than the unsuccessful block-buster movie, *Waterworld*, which was on release in cinemas at about the same time the flight hardware was being assembled. Part of the savings was to come by having no scientific instruments on board, on the grounds that this was a

Figure 4.4 The *Sojourner* rover is the size of a milk crate and has a mass of 10.5 kilograms. The top surface is covered with solar cells, which provide the electrical power for movement and control, and for the scientific instruments.

technology mission, but this idea was robustly resisted by the research community. Eventually a compromise was found, in which a modest science payload was contributed by European collaborators at no cost to NASA. The Max Planck Institute in Germany offered an instrument to determine the elemental composition of Martian rocks, and the Finnish Meteorological Institute built a package to measure the local weather at the landing site. Not for the first time, costs were to be contained by flying a mission without a camera to see what is going on and by eschewing the traditional eye-catching pictures for the media. Such misguided thrift has never prevailed, and soon it was decided that a panoramic colour camera paid for by NASA would be added as well, to observe the landscape at the landing site and watch the rover at work. Taken together, this produced a package that was deemed acceptable to most points of view, as described in Figure 4.5. At two hundred and eighty million dollars, *Pathfinder* was still relatively cheap, less than a tenth the cost of *Viking*, for example, which cost three and a half billion after applying inflation to 1997 values.

There were still some scientists who felt the mission was lightweight and therefore at best wasteful and at worst counterproductive, and at the same time there were engineers and managers who felt that *Pathfinder* had cut too many corners and risked failure to an unacceptable degree. Certainly, the size of the budget compared not only to *Viking* but also to most lander/rover mission concepts studied previously meant that many of the usual practices of space hardware development could not be followed. NASA insisted that a small launcher be used, forcing the spacecraft to be smaller and lighter than would have been typical. Saving weight over standard systems costs money and development time, making the job of staying on budget even harder. Faster, Better, Cheaper was facing a major challenge at the first hurdle.

In the event, however, the Jet Propulsion Laboratory took up Mr Goldin's torch and set up a small team that was encouraged by the rules of the *Discovery* programme to operate outside of the traditional engineering and management structure of the Laboratory. A team of young, enthusiastic engineers worked tirelessly and unconventionally, testing everything repeatedly to eliminate hardware and software problems that might otherwise have been caught by the more systematic and costly documentation and review practices that were standard at JPL. Would it work? The talk over lunch among the workforce included terrifying mental pictures of the planned sequence of events, which included steering the spacecraft directly into the atmosphere without going into orbit around the planet, plummeting into the surface at a velocity of nearly twenty metres per second, bouncing high in its protective enclosure of airbags, and finally, after several minutes, rolling to a stop. Then the lander had to unfold like the petals of a flower, forcing itself into the upright position essential for taking pictures and deploying the mini rover. A lot of things had to go just right, a long way away, in an environment full of surprises.

In the flag-flying spirit that is so characteristically American, landing day for *Pathfinder* was planned for US Independence Day, 4 July 1997. Unlike *Viking*, it

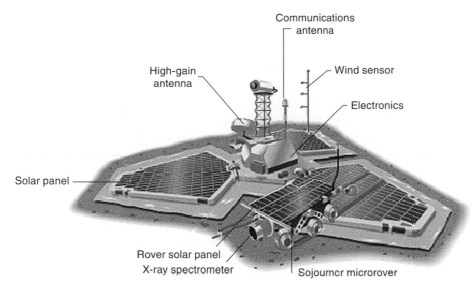

Figure 4.5 **Scientific goals of *Mars Pathfinder***

Pathfinder's original objectives were the testing of new landing and roving techniques, but by the time it flew five scientific goals had been added. NASA described these as follows:

1. Surface morphology and geology at metre scale
    The panoramic stereo camera observations examined geological processes and surface–atmosphere interactions by imaging the landscape, slopes and the distribution of rocks, and the actions of frost, wind erosion, and dust and sand deposition.
2. Petrology and geochemistry of surface materials
    The *Alpha Proton X-Ray Spectrometer* (APXS) detected back-scattered alpha particles, protons and X-rays to determine the elemental chemistry of any surface material against which it could be placed. Visible to near-infrared spectral filters on the camera provided additional information on the composition of the rocks and soils.
3. Magnetic properties and soil mechanics of the surface
    Permanent magnets distributed at various points around the spacecraft collected magnetic minerals from the airborne dust. APXS measurements of this material determined its mineral composition. Wheel-track images were used to study the mechanics of the soil surrounding the landing site.
4. Atmospheric structure and diurnal and seasonal meteorological variations
    The temperature and density of the atmosphere were measured during entry, descent and landing. On the surface, pressure, temperature, wind speed and atmospheric opacity were obtained on a daily basis, with thermocouples mounted on a metre-high mast to obtain temperature profiles with height in the atmospheric boundary layer. Sky and solar spectral observations monitored windborne particle size, shape and distribution with altitude, and the abundance of water vapour.
5. Rotational and orbital dynamics of Mars
    Two-way X-band and Doppler tracking of the *Pathfinder* lander measured its movement relative to the ground station, with an accuracy of a few metres. The orientation and precession rate of the pole with respect to the ecliptic were calculated from this and compared to measurements made with the *Viking* landers twenty years earlier to obtain the moment of inertia, which is a function of the distribution of mass in the Martian interior.

worked this time: the atmosphere at the Jet Propulsion Laboratory, on a public holiday with the media and a number of visiting VIPs in attendance, was electric. Just after breakfast, California time, *Pathfinder* scorched through the Martian atmosphere to make its semi-soft landing using parachutes, rockets and airbags. It came to rest near the mouth of a giant outflow channel, the scene of catastrophic flooding in the distant past, in the Ares Vallis region of Mars at 19.33° N, 33.55° W. Looking around after arrival, it saw and photographed the now-familiar rock-strewn landscape with low hills in the distance, shown in Figure 4.6.

The diminutive six-wheeled *Sojourner* rover was deployed onto Mars' surface on 6 July 1996, two days after landing, and it operated with great panache worthy of its namesake (the former slave and civil rights activist) until 27 September 1996. *Sojourner* was controlled from the Earth, using both rover and lander images for guidance, and had very little built-in autonomy. The downlink delay of about ten minutes, and the time required to examine the images and plan the next move, meant that the little rover moved in a drunken sequence of short, straight lines, all within about ten metres of the lander, until its primary mission objective of analysing the nearby rock and soil was completed.

Pathfinder successfully validated airbags as a practical landing system for Mars exploration (for a small payload, at least) and achieved surface mobility for the first time on another planet. NASA astronauts had driven buggies on the lunar surface in the late 1960s; Russia had deployed its robotic *Lunakhod* in the early 1970s; and now with *Pathfinder* scientists greatly extended the value of their scientific instruments by gaining access to diverse rock-types and different geological units on Mars. While the *Viking* mission arm had a reach of two metres at most, *Sojourner's* reach was tens of metres, and offered the promise that future rovers could carry instruments for much larger distances.

While the images of the landscape pleased the public, the science community's enthusiasm for *Pathfinder's* scientific data was tempered by the modest capabilities of its instruments and by the apparent absence of diversity of composition among the rocks present near the landing site. This was disappointing because the landing site had been selected because it was thought that rocks there had been deposited and shaped by catastrophic floods long ago. Consequently, the site was expected to be a collecting point for a variety of rock types transported by the waters of huge floods carried downhill from adjacent mountains. Within the short distance that *Sojourner* was able to move away from the lander, a maximum of about twenty metres, the rock types appeared to have similar composition, based on measurements of their constituent elements by the *Alpha Proton X-Ray Spectrometer* on the rover. It was interesting, however, that the rocks were not basalts, the commonest type of igneous rock, but instead had a high silicon content, putting them in the class known as andesites, which could have been formed by a number of processes, one of which is the action of water on volcanic rock. The various rock types found on Mars are summarised under Figure 4.7.

The meteorology instrument package on the stationary lander monitored the local weather conditions. An important bonus came when it detected a dust

Figure 4.6 *Pathfinder* took glorious panoramas of the surface (above), while the *Sojourner* mini-rover approached and examined nearby rocks in detail (below).

Figure 4.7 **Mars geology for non-geologists**
The findings about Mars, its surface, and climate history, as derived from studies of the composition of rocks, dust, strata etc., tend to be expressed in terms drawn from comparisons to terrestrial geology. The following is a short glossary for those not familiar with the most common terms that are used by geologists when talking about the surface of Mars.

Acidic:	generally means having high silica content in this context
Andesite:	similar to basalt but darker and containing more than fifty per cent silica
Basalt:	the commonest igneous rock on Earth, rich in iron and magnesium
Endogenic:	material that has been extruded onto the surface, as from a volcano
Feldspar:	a silicate with the formula $(Na,K,Ca)AlSi_3O_6$, where the metals in parentheses vary in their relative amounts
Granite:	a plutonic igneous rock consisting chiefly of quartz and feldspar
Igneous:	rock formed by the solidification of magma (molten rock or lava) from a volcanic source
Olivine:	a silicate with the formula $(Fe,Mg)SiO_4$, where the metals in parentheses vary in their relative amounts
Oxides:	iron oxides on Mars include haematite (Fe_2O_3) and goethite $(FeOOH)$
Palagonite:	a form of phyllosilicate often produced in underwater volcanic activity
Phyllosilicates:	a soft, flaky hydrated aluminium-rich silicate produced from weathered igneous rocks
Plutonic:	igneous rock formed from magma originating at great depths and pressures
Pyroxine:	a silicate with the formula $(Ca,Fe,Mg,Al)Si_2O_6$ where the metals in parentheses can vary in their relative amounts
Quartz:	a hard, glassy mineral made mostly of silica
Serpentine:	a phyllosilicate produced in hydrothermal springs
Salts:	sulphates (common on Mars), carbonates (uncommon on Mars), and nitrates (rare on Mars)
Sedimentary:	rock or clay that has formed by precipitation from water. Common examples include chalk, limestone, sandstone, and shale.
Silica:	silicon dioxide, SiO_2, the main ingredient of quartz, glass and sand
Silicates:	compounds of a metal with silicon and oxygen, e.g. feldspar
Weathered:	minerals that have been altered by exposure to the atmosphere or to water, e.g. phyllosilicates, salts, oxides

devil, a kind of mini-tornado also found on Earth, which passed directly over the lander. These have turned out to be very common on Mars, and are now thought to be the answer to how the Martian atmosphere maintains its high concentration of airborne dust at nearly all times, despite the fact that ordinary winds large enough to raise dust are relatively rare, except in major storms.

The radio communication system on the lander also contributed to the scientific mission by enabling a precise determination of the rotation rate of Mars, and by allowing the position of the lander to be fixed with a precision of a hundred metres. Interpretation of the Doppler content of the radio signals and their variability due to the 'wobble' of the planet, and through intercomparisons with similar measurements made ten years earlier by *Viking*, fitted a model of Mars that has a central metallic core of between one thousand three hundred and two thousand four hundred kilometres in radius.

4.5 *Global Surveyor*: mapping the Martian surface, atmosphere and interior

Mars Global Surveyor, depicted in Figure 4.8, was the first of the three low cost replacements for *Mars Observer*. It was launched on 7 November 1996, a month before *Mars Pathfinder* set off in the same launch window. Because of the requirement to approach on a trajectory suitable for orbit insertion, *Mars Global Surveyor* had a much longer flight time, arriving three months after *Pathfinder* on 11 September 1997. *Global Surveyor*'s mass of just over a ton included an advanced camera called MOC, for *Mars Orbiter* (originally *Mars Observer*) *Camera*, capable of a resolution of one metre at the surface. With this, it was hoped to identify interesting and safe landing sites for future rovers and landers, as well as to study surface features, examples of which are shown in Figures 4.9 to 4.11.

With *Mars Global Surveyor* it was now possible not only to obtain two-dimensional images of mountains and valleys on the surface, but also to measure the associated height variations accurately for the first time. Previously, the topography had to be estimated by taking pairs of stereographic images, or by painstaking measurements of the shadows cast by tall features. The *Mars Orbital Laser Altimeter* on board the orbiter measured the time for a laser-generated pulse of light to reach the surface and return to the spacecraft, thus obtaining the distance. The position of the spacecraft is known with great precision from its orbit, allowing the relative height of the features below to be derived, from which a topographic map of Mars can be built up. These include some remarkable profiles of individual features, including canyons several miles deep like that in Figure 4.12, the contours of the polar caps, and the Mountains of Mitchel, now seen to be disappointingly rather flat.

The *Mars Global Surveyor* payload also included the *Thermal Emission Spectrometer*, a Michelson interferometer used to study the composition of rock, soil, ice, atmospheric dust and clouds from measurements of their spectra at long

Figure 4.8 *Mars Global Surveyor* arrived at Mars on 12 September 1997 and was placed in a highly elliptical orbit. Over the next eighteen months, this was gradually reduced to a close circular orbit by 'aerobraking', which uses the effect of the atmospheric drag encountered during the closest approach to Mars to slow the spacecraft down and bring it closer to the planet.

infrared wavelengths, and made gravity and magnetic field measurements: we discuss these in the next chapter.

Global Surveyor operated successfully for well beyond its design lifetime, finally expiring on 5 November 2006, two days short of the tenth anniversary of its arrival at Mars. It may be coincidental that much of the spacecraft and all of the instruments pre-dated the work ethic of Faster, Better, Cheaper. The space-craft was built at Lockheed Martin Aerospace in Denver, Colorado, after

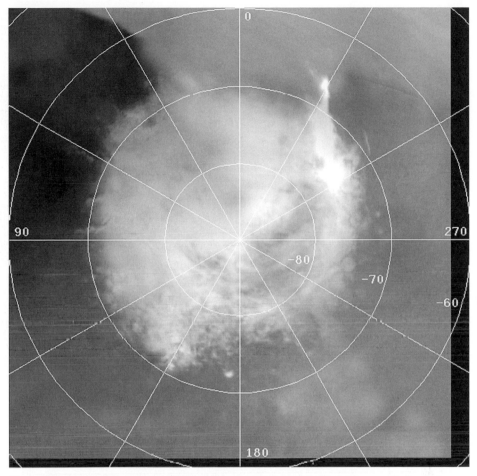

Figure 4.9 A composite image of the south polar region in Martian spring, with the carbon dioxide frost deposit in retreat, obtained by the camera on *Mars Global Surveyor*. The bright, icy Mountains of Mitchel (see Figure 1.3) appear prominently between one and two o'clock on the edge of the cap, as does the dark, dusty 'cryptic region' at nine to eleven o'clock outside the cap edge. (NASA/JPL/Malin Space Science Systems)

Lockheed[4] purchased from RCA the facility that built *Mars Observer*, thereby inheriting the new mission along with the spare hardware left over from the old one. The instruments for *Mars Global Surveyor* were 'built to print' from *Observer* designs, to the extent possible given the decade of changes in technology that had occurred since they were originally designed. However, *Mars Global Surveyor* would carry only part of the *Observer* payload – as planned, the later missions, *Mars Climate Orbiter* and *Mars Odyssey*, would carry the remainder. They, and the mini-*Viking* mission to the polar regions called *Mars Polar Lander* that was being

---

[4] The Loughead Aircraft Manufacturing Company was formed in 1912 by an émigré Scot, who tired of having his name mispronounced by his new countrymen, and switched to phonetic spelling.

Figure 4.10 This *Mars Observer Camera* image of an outcrop in the wall of Terby Crater shows exposed layers in a cliff more than a kilometre high. It is hard to be sure what produced the layers – successive flows of lava, or deposition in water or by wind.

prepared for launch in the same window as *Climate Orbiter*, would be the real test of the power and wisdom of Mr Goldin's vision for frequent, low-cost access to Mars.

4.6 Faster, Cheaper: *Climate Orbiter* and *Polar Lander*

While *Global Surveyor*, *Pathfinder* and *Sojourner* were still delivering data to Earth from their instruments and scoring millions of web hits from a fascinated

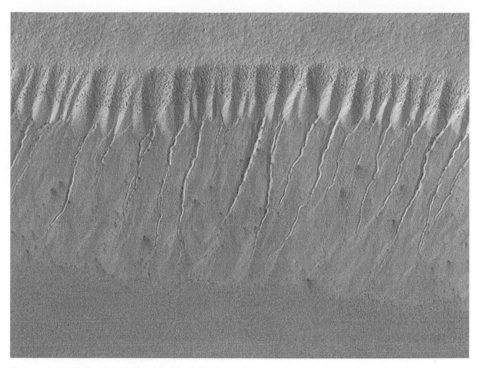

Figure 4.11 A close-up view by *Mars Observer Camera* of a cliff about a kilometre high on the edge of Nirgal Vallis, showing flow features that suggest that a liquid, possibly brine (salt-laden water), ran out of the subsoil and down the slope in the geologically recent past. The cliff is part of an extended 'dry riverbed' feature that had been photographed nearly twenty-five years earlier by *Viking*.

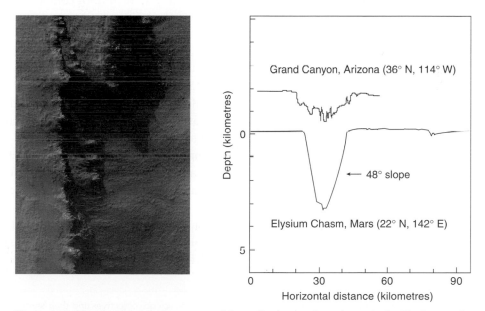

Figure 4.12 The photograph shows part of the walls along a deep chasm in the Elysium region of Mars. On the right is a depth profile across the chasm obtained by the laser altimeter on *Global Surveyor*, with a profile of the Grand Canyon on Earth to the same scale for comparison. The Martian feature is not only deep but also remarkably steep-sided, suggesting that it was cut quickly and relatively recently, otherwise the walls should have collapsed.

audience world-wide, NASA had already started work on the *Mars Climate Orbiter* and *Mars Polar Lander* missions for separate launches in the 1998 window. The orbiter carried the atmospheric sounder and the wide-field camera, two of the three remaining *Mars Observer* instruments that had been guaranteed a re-flight. The young researchers who had invested no less than seven years of their early careers building and testing the original instruments, for no reward in terms of the data they sought on the Martian atmosphere, weather and climate, felt lucky to have the chance to try again. However, their luck was capricious. In September 1999, the *Mars Climate Orbiter* suffered a navigation error and got too close to Mars when trying to orbit, and the drag from the atmosphere pulled it down to crash on the surface. The atmospheric science was lost for the second time.[5]

An official failure board was convened to probe the spacecraft design, the skills of the engineers and managers running the mission, and the overall management approach. It soon surfaced that the problem was of the worst possible kind. Human error had crept in, with imperial and metric units becoming confused when monitoring the trajectory of the spacecraft. The difference was small enough not to be obvious, but large enough to doom the mission. In fact, as every scientist knows, such errors occur all of the time; the difference here was that it had not been picked up, because the usual ponderous paperwork and meticulous checking that underpinned JPL's reputation for quality assurance had been shed in the spirit of FBC.[6] The problem was recognised before the crash occurred, but it was then too late to avoid it. The media, of course, found this risible, and saw it as a fine opportunity to indulge in schadenfreude on a grand scale.

Of more immediate concern to JPL was the fact that the *Mars Polar Lander*, also carrying two surface-science penetrators as a technology demonstration mission called *Deep Space 2*, was approaching Mars at nearly six kilometres per second and would be there in less than ninety days. *Polar Lander* was managed by the same team at the Jet Propulsion Laboratory and built by the same aerospace contractor, Lockheed Martin Aeronautics in Denver, as the ill-fated *Climate Orbiter*. A team of experienced space engineers from NASA and industry was hastily convened to evaluate the mission and spacecraft design and development for *Polar Lander* in an attempt to detect and eliminate any weaknesses or oversights that might threaten its success. They scrutinised the design from top to bottom, interviewed the managers and engineers at Lockheed and JPL, and reviewed the test records, searching for anything that could conceivably go wrong. Individual engineers were quizzed on their designs and the teams that

---

[5] See Chapter 6 for more discussion of the atmospheric science objectives and their significance, and how they were finally attained.

[6] 'Our inability to recognize and correct this simple error has had major implications,' said Dr Edward Stone, director of the Jet Propulsion Laboratory. 'We have underway a thorough investigation to understand this issue.'

assembled the lander were questioned about the procedures that they used. Software engineers were asked to discuss the software they had written. In the final days before the lander began its descent to Mars, the review team concluded that, despite past experience with Mars missions, and the wounds left by *Climate Orbiter*, the probability of success with the new lander was high.

On 3 December 1999, *Mars Polar Lander* entered the atmosphere and began its descent sequence. It was never heard from again. There was no indication from the spacecraft of what went wrong – no communication with Earth was planned during descent, since this would have cost money the project did not have. The camera on *Mars Global Surveyor*, already in orbit, was called upon to try to capture a picture of the lander, presumably wrecked on the surface, but without success. The resolution was marginal and the location highly uncertain, so this was a long shot, but everything that could be thought of was done in an attempt to figure out what had happened. The number of credible possibilities was troubling: perhaps the lander in its cocoon skimmed off the atmosphere, landing way off target; perhaps the heat shield failed or the parachute did not open; possibly the landing rockets failed to fire or the legs did not deploy; maybe it landed on a large rock or crater wall. Most poignantly, the lander could be safely on the surface but with a broken radio, unable to call home. Even today, the only evidence of what went wrong is conjecture based upon laboratory testing of spare parts and engineering models of the spacecraft, back at the Jet Propulsion Laboratory on Earth. This revealed that a fault in the on-board software could have responded to the command to extend the legs for landing by also shutting down the retro-rockets designed to produce a soft landing. The effect would have been to allow the three-hundred kilogram probe to free fall from about a hundred metres above the ground, leaving it to crash into the surface at too high a speed for survival.

The two *Deep Space 2* penetrators, named after the historic polar explorers *Scott* and *Amundsen*,[7] were delivered successfully, but also failed to call home. This had to be regarded as a separate failure, although perhaps one that was easier to bear since the most important objective was to see if a three-kilogram probe hitting the surface at four hundred miles per hour could survive the resulting 30 000g deceleration and penetrate a metre or so into the polar layered terrain at which it was targeted. There, had it survived, it would analyse the soil and in particular its water content, staying alive for around two days before battery failure terminated the mission. Unfortunately, complete silence from both probes meant that any technical evaluation of what had gone wrong was impossible. The review board speculated that the transmitter on the probes was under-designed and may not have survived the impact, or that their batteries, which had been charged before launch almost a year earlier and not checked since then, might not have retained sufficient power. Of course, the probes may

---

[7] Roald Amundsen reached the South Pole of the Earth in 1911, and Robert Scott in 1912.

simply have hit solid rock rather than the anticipated frozen mud or snow at the edge of the polar cap, or struck at too shallow an angle and disintegrated.

In the aftermath of the orbiter, lander and penetrator losses a critical look was taken at Faster, Better, Cheaper and the extent to which it may have contributed to the failures. It emerged that staffing of the navigation team for the *Climate Orbiter* was much less than that on other planetary missions, and that ground testing of the *Polar Lander* was cut short due to budget limitations. Mr Goldin publicly and privately took personal responsibility, but continued to insist that money 'was not the magic ingredient' – he preferred instead to add 'smarter' as a fourth element of the mantra.

In the years that followed, space missions continue to be held to small budgets, but FBC was no longer mentioned as the reigning philosophy. The policy was criticised by the NASA Inspector General, who said, in her report entitled *Faster, Better, Cheaper: Policy, Strategic Planning and Human Resource Alignment*, 'NASA considered [faster, better, cheaper] to be a philosophy that did not need to be formalised into written agency policies or guidance, [but] without [these], NASA cannot effectively communicate [the concept] to program and project managers and contractors, which could negatively affect mission success, weaken accountability for results, and lead to increased cost and delays.' The policy was abandoned when Sean O'Keefe became the tenth NASA Administrator in December 2001.

4.7 Mars Odyssey

Even as the board of enquiry into the loss of the two 1999 missions made its report, the hardware for the next phase in NASA's Mars Exploration programme was nearing completion at the Lockheed Martin plant in Denver, Colorado, ready for launch in 2001. The *Mars Surveyor* 2001 project consisted of two launches, one of an orbiter carrying the last two rebuilt *Mars Observer* instruments and the other of a lander equipped with a rover. However, because the spacecraft had been built according to the same fast, cheap strategy used for *Climate Orbiter* and *Polar Lander*, NASA had to consider to what extent it might be risking another failure. In the end, it was decided to fly the orbiter without the lander, since the only reason the *Climate Orbiter* had failed was human error, and that could be avoided with improved practices without spending much extra money. The lander was put into storage[8] until it could be brought into line with the more cautious strategy that had replaced FBC, including finding a way to maintain a communications link during the descent through the Martian atmosphere. With this, even if the landing failed, at least the mission managers would know what went wrong and could make amends before the next attempt.

---

[8] The 2001 lander hardware was later used as the basis for the *Phoenix* mission, which made the first successful landing in the icy polar regions in 2008, thus redeeming the failure of *Mars Polar Lander*.

Going in 'blind' was easier and cheaper, but then if the mission failed to call home there was no way to know whether it had perished during entry, descent or landing, or even if it had missed the planet altogether.

NASA management also resolved that future descents to the surface would only be authorised after the landing site had been surveyed for potential hazards by better advance surveillance from orbit. In addition, they wanted the lander itself to have the capability to examine the landing site from close range, even as it approached it, and to take evasive action if any new risks were detected. These include the possibility of landing on rocks, crevices, slopes and quicksand, all things that are difficult to detect from orbit. The lander also had to be able to compensate for any high surface winds that it encountered, since localised gusts could occur regardless of what the meteorologists forecasted based on atmospheric data acquired from orbit. To achieve this, some sophisticated and expensive systems would be added to future landers, including imaging radar, image analysis and command software capable of recognising hazards and reacting instantaneously, and a propulsion system that could control the motion of the lander horizontally as well as vertically during the final seconds of descent. Cost was no longer going to be NASA's primary consideration.

Steps were also to be taken to reduce the uncertainty in where the landing will actually occur. For a relatively uncontrolled descent trajectory like that of *Viking* or *Pathfinder* this was large, an ellipse about five hundred kilometres long in the direction of flight, and a hundred kilometres across. No amount of careful reconnaissance is likely to find an area that large on Mars which is free of obstacles. Precision landing improves the odds substantially, and is obviously a good thing for other reasons – targeting scientifically interesting sites, and practicing for manned landings at a future base or the delivery of supplies to a team already on the surface.

The agency now gave high priority to a new mission, *Mars Reconnaissance Orbiter* (earlier called *Site Reconnaissance Orbiter*), which could map Mars with a camera able to detect objects smaller than one metre in diameter. This would require a telescope on board the spacecraft with an aperture and a primary mirror size that is comparable with the instruments used in some of the smaller ground-based observatories. While plans for *Reconnaissance Orbiter* were in the pipeline, the *Surveyor* 2001 orbiter, now renamed *Mars Odyssey*, lifted off on a *Delta II* rocket from Cape Canaveral on 7 April 2001 and reached Mars six and a half months later. The spacecraft carried three experiments: an infrared thermal imaging system for global mapping of the different types of minerals on the surface; a gamma-ray spectrometer combined with neutron detectors to measure the abundances of the elements making up the rocks and ices in the upper layers; and a package to measure the intensity of cosmic rays and solar particle radiation so their potential effects on future human explorers could be assessed.

An early result, of great significance if not entirely unexpected, was the detection of large amounts of subsurface hydrogen, almost certainly in the form of water, probably frozen, from its gamma-ray emission[9] and its effect on the flux of neutrons[10] produced by cosmic ray bombardment of Mars. When the neutron flux data are superimposed on a map of the planet, they show the correlation of low flux, indicating high hydrogen content, with regions of expected stability of sub-surface ice. Sensing down to a depth of about one metre, the data are consistent with a two-layer model in which the upper layer is relatively dry, with a per cent or two of water probably chemically bound in the regolith, while the lower layer is so wet (a third water by weight) that the only reasonable interpretation is a composition of about sixty per cent by volume of ice.

4.8   Europe and Japan look to Mars

Nozomi (Japanese for 'hope' and known before launch as *Planet-B*) was a Mars orbiting aeronomy (upper atmosphere science) mission developed by the Japanese Institute of Space and Astronautical Science (ISAS) with the intention to study the Martian upper atmosphere and its interaction with the solar wind. The spacecraft was launched on 4 July 1998, and originally scheduled to arrive at Mars in October 1999. The failure of a valve in the propulsion system during the planned Earth swing-by on 20 December 1998 caused it to miss its transfer opportunity to Mars, so a new orbit was developed to include two more Earth swing-bys, and the spacecraft finally reached the vicinity of Mars in January of 2004. Once there, however, the probe failed to enter orbit and the mission was abandoned. ISAS has since turned its attention from Mars to Venus, with a mission called *Venus Climate Orbiter* (*Planet-C*) planned for 2010.

While American and Japanese plans for exploring Mars were at a low ebb in 1999, the European Space Agency, ESA, plunged ahead with a bold plan to recover from the disastrous loss of European instruments on the ill-fated Russian mission *Mars-96* by planning a Mars mission of its own. Spares existed for all the instruments that were lost, but another spacecraft would be needed. ESA had very little experience with planetary missions: its strong heritage of ground-based astronomy, solar and magnetospheric physics had prevented the investment necessary in new fields like planetary exploration on which less hide-bound societies like the USA were bent. There had, however, been a successful European mission, *Giotto*, which sent a British-built spacecraft to Halley's comet in 1976, and a second, more ambitious comet mission, *Rosetta*, was under construction. (The early emphasis on comet missions reflected another long-standing vested interest in Europe, namely meteoritic research.)

Mars Express had to be developed on what by NASA standards was a very modest budget. The parallels with NASA's discredited FBC policy might have worried ESA following the loss of the nearly contemporaneous *Mars Climate*

---

[9] Gamma rays consist of very energetic electromagnetic radiation, like X-rays but with an even shorter wavelength of around one millionth of the wavelength of visible light.

[10] Neutrons are subatomic particles that have about the same mass as a proton but no electric charge.

Orbiter, but it did not. After an internal ESA review of the *Mars Express* design, the space agency was confident that it had not made NASA's mistake of cutting resources too close to the bone. If Europe's aerospace industry was to learn how to build affordable missions, ESA needed to press ahead, and it did. *Mars Express* was perceived such a success, in terms of budget and schedule control as well as science, that ESA opted to follow it with *Venus Express* on a similarly rapid and economical strategy.

4.9   Mars Express

The *Mars Express* mission was not simply an orbiter with the reconstituted *Mars-96* payload. ESA decided it had room for a small lander to add extra lustre to the mission; proposals for this were invited from the world community. From these four leading options emerged:

1. an American hard lander/surface penetrator, basically a spare from the failed *Deep Space 2* mission;
2. a Russian-provided lander based on their *Phobos* and *Mars-96* design experience;
3. an all-new European lander based on the *MarsNet* study; and
4. a British capsule designed to look for biosignatures in subsurface soil, called *Beagle 2* to invoke the spirit of Charles Darwin.

ESA's committee chose the last of these options and the combination of *Beagle 2* riding piggyback on *Mars Express* was launched on 2 June 2003, precisely on the schedule that had been laid down five years earlier. The launch vehicle was a Russian *Soyuz-Fregat* two-stage rocket, available on a commercial basis from STARSEM, a commercial company with European partners. The successful lift-off from the cosmodrome at Baikonur in in Kazakhstan[11] was testimony to the efficiency of this arrangement, at relatively low cost, as was the repeat perform-ance three years later that sent a similar spacecraft, *Venus Express*, to the Earth's other planetary neighbour.

One of the goals of *Mars Express*, pictured in Figure 4.13, was to carry out global stereo imaging with its *High-Resolution Stereo Colour Imager*, which pro-duces pictures of selected areas capable of resolving objects two metres across, nested inside larger images at ten metres per pixel resolution. This resolution was not as good as the camera then being built by NASA for its *Reconnaissance Orbiter*, but comparable to that of the camera on the previous American mission, *Mars Global Surveyor*. The spectrometer known as *OMEGA*[12] was designed to map the

---

[11] Actually, the cosmodrome is four hundred kilometres away from the original town of Baikonur. The name is a reminder of cold war days, when the Soviet Union sought to confuse the enemy, and presumably its own citizens too, by this sort of sleight of hand. Recently, things have been put to rights by reassigning the name to a town that is near the launching site.

[12] OMEGA stands for *Observatoire pour la Minéralogie, l'Eau, les Glaces, et l'Activité* – proving that awkwardly contrived acronyms are not the exclusive preserve of Anglophones.

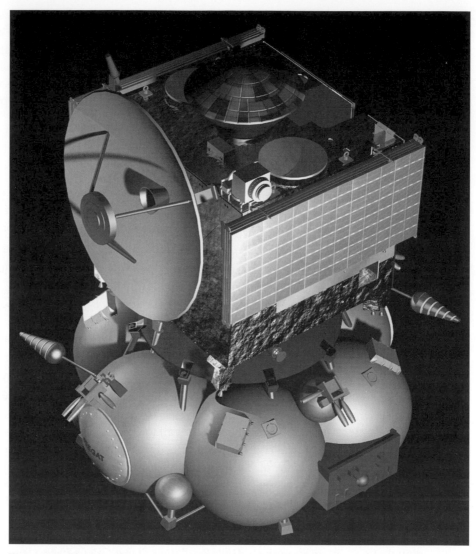

Figure 4.13 The *Mars Express* spacecraft, with *Beagle 2* visible on top, mounted on the *Fregat* upper stage of the *Soyuz* launch vehicle. The spheres contain the fuel for the successive burns that took the spacecraft first into Earth orbit, and then on to Mars.

distribution of minerals by infrared spectroscopy; the subsurface-sounding radar altimeter *MARSIS*[13] to sound the subsurface structure down to the permafrost layer; and a package (*ASPERA*[14]) to measure energetic particles to study the interaction of the atmosphere with the interplanetary medium. Two more spectrometers, the infrared *Planetary Fourier Spectrometer* and the ultraviolet and

---

[13] *Mars Advanced Radar for Subsurface and Ionosphere Sounding.*
[14] *Analyser of Space Plasmas and Energetic Atoms.*

infrared atmospheric occultation spectrometer *SPICAM*,[15] plus the radio science experiment, were dedicated to global atmospheric circulation and composition studies. The spacecraft and all of its instruments have performed well and some of the data are presented and discussed in later chapters.

Representatives of ESA said boldly in the publicity for this, their first Mars mission, that they wanted to address the issue of astrobiology by looking for indications of conditions favourable to the existence of life, present or past, and in particular to study the inventories of liquid, solid or gaseous water on Mars. In pursuit of this theme, the camera takes pictures of ancient riverbeds while the *OMEGA* spectrometer looks for minerals with OH (hydroxyl) radicals formed in the presence of water, and the radar for subsurface ice. The high-resolution spectrometers observe water vapour and other species in the atmosphere, and the *ASPERA* and radio science teams study neutral atom escape from the atmosphere, in particular that of oxygen originating from water and carbon dioxide. In all of these, and more, *Mars Express* has been largely successful, although nearly five years into mission operations the life issue remains opaque. However, most recently, the *Planetary Fourier Spectrometer* team has detected methane (chemical formula CH_4) in the Martian atmosphere, a finding possibly of great exobiological significance. Earth's atmosphere contains large amounts of methane, much of it produced by the decay of organic material and other biological processes, giving rise to its popular name of marsh gas. Spectroscopists using telescopes on Earth also claimed detections of methane at around the same time, but did not support tentative claims that Mars Express might also have detected other 'biomarkers' in the form of traces of formaldehyde (H_2CO) and possibly even ammonia (NH_3). These detections are right at the limit of the performance of the current best instruments, however, and while they provoke an interesting debate, the only certainty is that they define objectives for future experiments.

4.10 *Beagle 2*

The decision to put a lander on *Mars Express* depended on the ability of ESA's technical team to design an orbiter spacecraft that could carry the extra mass of a lander after all of the rebuilt *Mars-96* instruments were on board. The initial estimate was that around a hundred and twenty kilograms would be available, small compared to *Viking* (six hundred kilograms) but considerably more than the thirty-seven kilogram surface stations on *Mars-96*. ESA issued an announcement of opportunity to find a country, for political reasons preferably within Europe, willing to provide and pay for a suitable spacecraft.[16]

---

[15] *Spectroscopy for Investigation of Characteristics of the Atmosphere of Mars.*

[16] Unlike NASA, which represents a single country, ESA has to juggle a budget that comes from eighteen member states. Each member pays a subscription, which covers an agreed 'mandatory' programme of missions, but can also choose to provide funds, alone or with some of the other states, for additional, 'optional' missions.

Figure 4.14 A model of *Beagle 2* on the surface of Bedfordshire with solar panels deployed, showing the deployment of the instruments on the articulated arm at centre and the mole for subsurface burrowing at lower left. All rights reserved by *Beagle 2*.

The response was both predictable and surprising at the same time. The Russians came forward quickly with a spare *Mars-96* surface station and announced that it could readily be adapted to fit on *Mars Express*. The Americans found they could produce a copy of the failed *Deep Space 2* penetrator, which weighed in at less than three kilograms, and try again with that. The surprising part was that the most interesting proposal came from the British. To be more accurate, the proposal came from a Briton, Colin Pillinger, a professor at the Open University in Milton Keynes, England. Largely setting aside the question of whether he could get UK government support and funding, Pillinger put together a talented and enthusiastic team and wrote a proposal to ESA entitled *Beagle 2: A Lander for Mars*. The lander was shaped like a clamshell, which would open on the surface and deploy its instruments and solar cells, as demonstrated by the model in Figure 4.14.

Drawing on Pillinger's expertise at analysing meteoritic samples in the laboratory, for which he had earned a prestigious Fellowship of the Royal Society, *Beagle 2* was following in the footsteps of Darwin to tackle the mystery of life head-on, but now on Mars. It would use a suite of imaging instruments, organic and inorganic chemical analysis, robotic sampling devices and meteorological sensors, and be equipped with a sophisticated robotic sampling arm, which carried a corer/grinder to reach the interesting part of the rock under its weathered exterior. The *Payload Adjustable Workbench (PAW)*[17] could manipulate different types of

---

[17] See previous footnote. *Beagle 2* also had an *Anthropomorphic Robotic Manipulator (ARM)*.

tools and retrieve samples to be analysed by the geochemical instruments mounted on the lander platform. There was also a 'mole', a torpedo-shaped probe driven in small steps by an electromagnetically pulsed internal mass, which produced a hammering action that would cause the probe to move slowly forward over the surface.[18] On encountering a rock, or if deployed vertically, the mole had the potential to burrow into the soil, if it was soft enough, hopefully reaching depths unaffected by the sterilising effect of solar-UV radiation. Once there, it would ingest a few grams of material before being winched by its power cable back into *Beagle 2*, where the sample would be analysed.

ESA liked the *Beagle 2* proposal – it was from a member state, it was exciting and innovative, and unlike its two rivals, it had no history of failing before (since it had no history at all). The Europeans did not have NASA's post-*Viking* reticence over stating outright that they were going to search for life, indeed, they realised that they had a chance to leap ahead of the world's leading space agency by finding it first. ESA, too, put the question of how to pay for the lander on hold and invited Pillinger's team to join the mission.

Then, a cruel blow struck, one that may well have changed the course of Mars exploration history as much as any of the other numerous strokes of fate that have been a regular feature of this grand venture. The ESA mission planners had made a mistake – a more thorough analysis showed that there was little or no spare mass to accommodate the lander on the orbiter. Feeling that the mission would lack impact without a landed element, ESA took another look and decided that they could save some weight by carrying less fuel for orbit insertion, together with a few other economies, but no matter how they strained the maximum available would still be only sixty kilograms, only a little more than half of what they had previously offered and which *Beagle 2* required.

While the selection committee debated what to do, like return to the relatively low-mass Russian or American options for example, to meet legal requirements they put out a revised announcement of opportunity that declared the reduced weight available. Pillinger's team was undaunted and worked around the clock, and in a very short time ESA was surprised to take delivery of a revised proposal, *Beagle 2: A Sixty Kilogram Lander for Mars*. This was accepted, the hardware was delivered on time and – almost – within this mass limit, with most of its scientific capability intact, but its structure had got flimsier and its parachute was smaller and thinner. It remains very hard not to speculate on the effect of the mass savings on its chances of successful deployment.

The official goals of *Beagle 2*, summarised under Figure 4.15, involved studies of the geology, geochemistry, meteorology and astrobiology of the landing site. It was the search for life that once again caught the popular imagination, as *Beagle 2* aimed to look for the presence of water in the soil, rocks and the atmosphere, to search for the presence of methane (CH_4) indicative of extant life,

---

[18] The scientists in Cologne, Germany, who led the mole team decided to compete in the 'most cringe-inducing acronym' stakes by naming the device *PLUTO*, for *Planetary Underground Tool*.

Figure 4.15 *Beagle 2* **objectives**
– Search for criteria relating to past life on Mars including:
  presence of water
  existence of carbonate minerals
  occurrence of organic residues
  complexity and structure of organic material
  isotopic fractionation between organic and inorganic phases.
– Seek trace atmospheric species indicative of extant life.
– Measure the detailed atmospheric composition to establish the geological history of the
  planet and to document the processes involved in seasonal climatic changes or diurnal cycling.
– Investigate the oxidative state of the Martian surface, rock interiors and beneath boulders.
– Examine the geological nature of the rocks, their chemistry, mineralogy, petrology and age.
– Characterise the geomorphology of the landing site.
– Appraise the environmental conditions including temperature, pressure, wind speed, UV
  flux, oxidation potential, dust environment etc.

and the ratio of the lighter ^{12}C isotope compared to the heavier ^{13}C, which varies with biological activity and could indicate a record of extinct life.

Although confidence in *Beagle 2* was never total, even within Europe, and certainly not among the British authorities who had been so effectively out-manoeuvred, the potential payoff of a successful European landing on Mars was so great that all concerned agreed to press ahead. However, the British government, through its Particle Physics and Astronomy Research Council, had already decided that it was not willing to pay the cost, estimated in the region of twenty-five million pounds sterling (about forty million euros and roughly the same in US dollars). Colin Pillinger persisted, encouraged by support from private and commercial sponsors that were poised to pick up at least part of the tab. However, ESA soon concluded that *Beagle 2* needed an injection of agency funds if the lander was going to make it to the launch pad.

Before writing the cheque, ESA insisted on the convening of an independent group of space engineering experts to review the viability of the whole lander component of the mission. The obvious place to find suitable, unbiased experts was NASA's Jet Propulsion Laboratory, and the team was led by John Casani, a massively accomplished individual who had been Project Manager for the *Voyager* spacecraft that visited Jupiter, Saturn, Uranus and Neptune, and for *Galileo*, the highly successful orbiter of Jupiter. The Casani group's assessment of *Beagle 2* was positive; yes, it would be risky, but the accomplishments already achieved by Pillinger's team, whose skills were evident and whose commitment bordered on the fanatical, were truly remarkable. They concluded that *Beagle 2* had a better than even chance of making it.

Based on their own assessments, as well as the Casani team's review, ESA committed twenty million euros of its central funding to the project. Soon after, the UK responded to the surge in public and political interest in *Beagle 2* with a contribution of five million pounds. Pillinger's efforts to raise money from the private sector had been partially successful – although he has never released the details of *Beagle 2's* private funding – but in the final push in the race to meet the *Mars Express* schedule and the launch date in Kazakhstan, neither the time nor money was available to complete all of the testing and redesign that would have helped, although not assured, the success of the hastily lightened *Beagle 2*.

The lander module was designed to descend through the Martian atmosphere in about five minutes and land with an impact velocity of around forty metres per second within an uncertainty ellipse of one hundred by twenty kilometres inside the Isidis Planitia area (10.6° N, 270° W). Once on the surface and its systems deployed, the lander lifetime was to be a few months, its data relayed to Earth through the *Mars Express* orbiter. *Beagle 2* was released successfully from *Mars Express* but never made contact with the team waiting eagerly on Earth to hear from it after its expected landing on Christmas Day, 2004.[19] It must be somewhere on Mars, but the location is unknown and so is the cause of the failure. All that is known for certain is that it was *not* the reason most often cited, 'unusually thin' atmospheric conditions. This fallacy originated in a conference report on one of the *Mars Express* orbiter instruments (*SPICAM*) that was in error because of a software bug in the data analysis software. The fault, and the report, were both soon corrected, but not before the press had latched on to what seemed to make a nicely rounded story. All of the available data in fact show quite normal conditions on Mars at the time of the landing. *Beagle 2's* fate is likely to

---

[19] Like NASA's *Polar Lander*, *Beagle 2* was not designed to communicate during descent, since this would have required extra hardware, mass and complexity.

remain a mystery, at least until the wreckage is inspected by members of a future expedition to the Martian surface.

By attempting *Beagle 2*, there is no doubt that the charismatic Pillinger and his hard-working team, which included not just academic labs but the cream of British industry, often working without the assurance of payment, advanced the goals of not just Mars exploration, but exploration in general. The impression, or illusion, that Faster, Better, Cheaper offered a short-cut emboldened all space-faring nations and, in this case, an individual and his team. The downside to the whole effort is that Britain and Europe show no visible signs of trying again, at least not in this way. The heritage of *Beagle 2* is a vast and spectacular, but bureaucratic and costly, programmatic concept called *Aurora*, which not only seeks evidence for life on Mars but has the eventual aim of putting it there itself in the form of European astronauts. The human missions would be preceded by robotic missions that, at least superficially, look a great deal like those NASA has been considering for the same time frame. The first of these, *ExoMars*, is a large rover scheduled for launch in 2016. While NASA, with all its experience, struggles with the complexities of getting the *Mars Science Laboratory* rover safely down on Mars in 2012, ESA has to figure out how to go directly from the failed landing of a small hard-landed probe to the solution of a much greater challenge.

4.11 The *Mars Exploration Rovers*: *Spirit* and *Opportunity*

At the beginning of the new millennium, with data flowing in from *Global Surveyor*, and preparations underway for the launch in 2001 of *Mars Odyssey*, NASA reviewed its science strategy once again. The *Surveyor* data confirmed that Mars had a complicated and interesting climate history, and that large amounts of water seem to have been involved, but the nature and timeline of the interactions between liquid water, the atmosphere and the surface were still obscure. After *Odyssey*, most of the scientists were keen to proceed as quickly as possible to the return of samples, but there was considerable uncertainty about where to land. The best hope was that the *Reconnaissance Orbiter* would arrive successfully in 2006 and pinpoint at least one site of ancient persistent water, or better still, a place where there is current geothermal activity. Without that information, the nightmare scenario loomed that large amounts of time, money and effort would be invested, only to return samples of rock and soil composed of minerals similar to SNC meteorite material, of which there is plenty on Earth already.

The observed diverse nature of the Martian terrain, and the various theories about internal activity and sources and sinks of ancient and modern water, all pointed to the likely existence of specific locations that held buried treasure in the form of biosignatures. Unless mission planners could identify one of these, land a spacecraft safely there with high precision, and properly access the interesting material by drilling, digging or whatever else it took, a sample return mission could not be certain of contributing to the key questions of climate change and the development of life on Mars. The high cost of mounting a sample return mission required a large investment of money, time and faith; it was, and remains, vital to minimise the risk of a null result, or one that would be perceived as negative or boring by mainstream opinion. However, the knowledge needed to provide any sort of guarantee against such an occurrence has remained tantalisingly out of reach.

For this reason the contemplation of sample return has always led to a collective cold sweat among decision makers, who on each occasion have ultimately resolved to wait and be prudent. In any case, in 2003, any kind of sample return mission was too expensive to be an option for launch within the available budget. A rover developed from the *Mars Pathfinder/Sojourner* design, using the 2001 lander payload (then in storage), did seem to be feasible, however, provided an affordable approach to entry, descent and landing could be found that was acceptable to NASA's newly acquired aversion to risk. This policy also meant that any landed mission, even a rover without sample return, really should wait until the super high resolution camera on the *Reconnaissance Orbiter* had checked out the intended landing site. JPL proposed to resolve the dilemma with a *Pathfinder*-derived rover it called *Mars Exploration Rover* that was larger and more capable than *Sojourner*, but still small enough to be landed safely by *Pathfinder*-style airbags, at a modest cost estimated at approximately three hundred million dollars. This system of landing had been demonstrated to work successfully for a relatively lightweight lander, and had been brought in at a cost of two hundred and fifty million dollars. *Mars Exploration Rover*, it was argued, could contribute some missing pieces to the water/climate jigsaw in advance of sample return, at any of a dozen sites that were already known to be interesting from available pictures and data. It was also in everyone's mind at NASA that the success of a new rover would be good for morale.

NASA agreed to proceed only if two identical *Mars Exploration Rovers* were built and flown in 2003, offering redundancy as a further risk avoidance measure. In so doing, they were returning to the strategy employed in the early stages of planetary exploration, when this approach was standard. The agency also recognised that the incremental cost of copying an existing design is small compared with its development budget, and that there are scientific gains if both spacecraft succeed. Nevertheless, the cost rose to $850 million.[20] NASA confirmed that *Mars Reconnaissance Orbiter* would follow in 2005. The strategy they had embraced said that the joint high-resolution data from the rovers on the surface and the orbiter aloft would combine to lead to a better choice of landing site for sample return from the scientific as well as the safety perspective, especially if the new instruments on both precursor missions could find the missing aqueous minerals.

---

[20] To send two rovers rather than one requires a second launch vehicle. The cost of a *Delta II* launch, like that used for each *Mars Exploration Rover*, is roughly fifty million dollars.

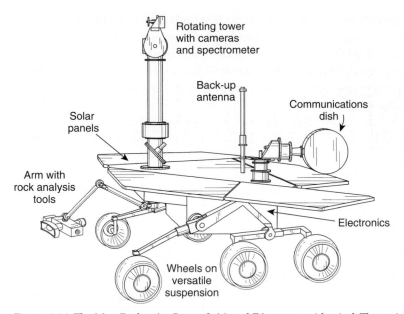

Figure 4.16 The *Mars Exploration Rovers Spirit* and *Discovery* are identical. The various instruments and other components are shown here and in Figure 4.17, and are described in the text.

The Mars scientific community, which in the current era meets regularly in a forum called Mars Exploration Program Analysis Group (MEPAG) chartered by NASA, was generally pleased with the outcome of a rather hurried planning process. The *Exploration Rovers* were touted as robot geologists, the mechanical equivalent of a scientist walking the surface in search of answers about the history of water on Mars. Following two separate launches on 10 June and 7 July 2003, the two rovers were delivered to separate sites on Mars in January 2004. They were given names, *Spirit* and *Opportunity*, by the traditional means of a competition involving the public.

The primary scientific goal of each *Mars Exploration Rover* was to search for and characterise surface materials that might hold clues to water activity on Mars, so they were targeted to sites where there was evidence of liquid water in the past. The science instruments they carried, in the locations shown in Figure 4.16, included a camera mounted high to provide panoramic, stereoscopic views of the local terrain. The *Miniature Thermal Emission Spectrometer* used infrared spectroscopy both to identify interesting rocks and soils for closer examination, and to survey the mineralogy of the whole region wherever it went to find evidence of the processes that had been at work when the terrain was formed. It also looks skyward to provide temperature profiles of the Martian atmosphere. The *Mössbauer Spectrometer* and the *Alpha-Particle X-Ray Spectrometer* are both instruments that work at close range, to determine the minerals and elements present in individual rocks at specific locations, including on cliff faces where the rover could get close enough, and in the soil.

Figure 4.17 The *Mars Exploration Rover* instrument deployment arm and its tools. This painting shows the instrument deployment arm about to place the spinning rock abrasion tool against an interesting boulder. After the weathered coating has been abraded away, the microscopic imager and the *Alpha-Particle X-ray Spectrometer* mounted on the same arm can be brought to bear on the exposed surface.

Magnets on the rovers collected dust particles from the air for analysis. They were also used to determine the ratio of magnetic to non-magnetic particles in the soil, and in rocks that were ground up by the *Rock Abrasion Tool*, a drill-like grinding device that the team likened to a geologist's rock hammer, used to expose the insides of rocks, remove dusty and weathered surfaces, and expose fresh material for examination by the instruments. By a similar analogy, the *Microscopic Imager* is likened to a geologist's hand-held magnifying lens, used for obtaining close-up, high-resolution images of rock and soil textures. Figure 4.17 shows the robotic arm, which is capable of movement in much the same way as a human arm, with an elbow and wrist, and can place instruments directly up against any object of interest that can be reached.

After a fraught development programme, both rovers had a near-perfect flight to Mars, followed by a successful descent and landing. *Spirit* plummeted onto the floor of Gusev Crater, a dried-up lakebed with soil-covered, rock-strewn cratered plains, on 4 January 2004. Its airbags deflated, the cocoon opened, and *Spirit* set off jauntily to examine some of the rocks near its landing site. It quickly found some that were coated and, if porous, penetrated, by material that looked as if it had been carried and deposited by water. This was an excellent start.

Opportunity followed *Spirit* to the surface of Mars on 25 January 2004. It landed in Meridiani Planum, an area where orbital remote sensing data had suggested the presence of haematite, the chief ore of iron, a mineral that forms in liquid water. On arrival, the rover did not have far to look – the haematite was liberally strewn about the surface as small spherules, the size and colour of blueberries. More of these were found embedded in layers of rock in outcrops and crater walls, a sure sign that the blueberries were water-formed, and the obviously sedimentary nature of the rocks made this certain. The layered rocks in the cliffs had been formed by the deposition of sediments, as deep pools or seas of water evaporated. Some of them had weathered away over the ages to free the harder conglomerations of haematite that composed the loose blueberries now lying around the dusty plains.

In one wall of the small crater named Endurance in which it landed, *Opportunity* found a formation, now named Burns Cliff, in which the layered deposits could be easily approached. The strata consist of layers of bedded basaltic sand cemented by large quantities of magnesium, iron and calcium sulphates and chlorides, soluble salts that must have been deposited by water. The spaces between the blocks and fractures in the bedrock are filled with windblown dust, dark basaltic sand and silt, and material eroded from the blocks. The detailed examination showed evidence of both upwelling and downwelling water, suggesting that the floods had receded and returned several times. The analysis of the salt mineralogy and the absence of carbonates both indicated that the solution involved was strongly acidic, possibly loaded with sulphurous material from volcanic eruptions that were going on at the time. As a bonus while heading for other craters where more strata could be examined, *Opportunity* acquired the photograph in Plate 12, which shows a large iron meteorite sitting on the surface of Mars.

Meanwhile, *Spirit* moved on towards some nearby high ground that NASA named the Columbia Hills, after the doomed space shuttle of that name. The soil over which *Spirit* travelled for several kilometres was, disappointingly, not water-deposited material but made of basaltic material of volcanic origin. Evidently this had been eroded from igneous rock formations, possibly far away, and the resulting dust mixed and distributed by the wind. Closer examination of the fines showed that some soluble salts distributed by flowing water were mixed in with the basaltic dust. When *Spirit* reached high ground, it found layered granular deposits, probably of volcanic ash or material ejected by impact events, which had also been modified by the flow of aqueous fluids.

The *Mars Exploration Rovers* vastly exceeded the expectations of those who had planned their science mission by finding indisputable evidence for the action of liquid water on the surface, in large quantities and over extended periods of time. They also monitored the dust opacity of the atmosphere with their cameras, including watching dust devils raising fresh supplies of dust into the air as they

travelled past. The rovers gratified those responsible for their engineering implementation, not only by both landing safely but also by exceeding their design
lifetimes by huge factors. Both successes paved the way for the planning of still
larger and more sophisticated rovers to search for signs of past or present life and
to uncover the resources that might be available for human exploration.

4.12 *Mars Reconnaissance Orbiter*

Plans to follow up on the successes of the *Exploration Rovers* all necessarily
involved landing much heavier vehicles. Sample return missions, in particular,
would have to carry the fuel to take off again and return to Earth. Landing using
airbags, although successfully demonstrated three times in a row, becomes more
and more difficult, and less and less reliable, as the landed mass increases.
Analysis showed that the large rover under development as the immediate successor to the *Exploration Rovers*, now named *Mars Science Laboratory*, would need to
make a controlled, powered landing. This was not just to handle its mass, five time
that of an *Exploration Rover*, but also to be able to avoid to surface irregularities and
to compensate for winds, and to choose with much greater precision exactly where
to touch down. All of these factors would be even more important for coordinating
the various spacecraft expected to be involved in obtaining and returning samples
to Earth, as would the advance intelligence to be obtained from the planned high-
resolution reconnaissance mission. A detailed search from orbit would find the
most scientifically interesting places for future exploration and sample return,
before valuable rovers and sample return vehicles were committed.

Mars Reconnaissance Orbiter arrived in orbit around Mars in March 2006
while the small rovers were still energetically pursuing their goals on the surface.
High-resolution imaging of the surface from orbit poses two demanding
requirements – a camera with large optics, and a high data rate link back to the
Earth. The latter means in turn a large antenna and a powerful transmitter using
large amounts of power. These factors dictated the large size of the spacecraft,
and its mass of over two tons, giving it the distinction of being the biggest vehicle
to date ever to fly to Mars, as the comparison in Figure 4.18 shows. It would have
been even larger were it not for the planned use of aerobraking to move from a
capture orbit to the close, circular orbit needed for mapping.

The aerobraking technique is described in Figure 4.19. When a spacecraft
approaches Mars on a flight from Earth, it is travelling at a speed of several
kilometres per second relative to Mars. If no action is taken, the spacecraft will
normally either continue on past Mars, or if targeted directly at the planet it will
crash into it. Both have been known to happen to intended orbiters, if the orbit
insertion procedure goes wrong due to either targeting or timing errors, or
failure of the propulsion system.

The traditional method for achieving orbit has been to fire a small retro-rocket
in the direction of travel, to slow the spacecraft down. If a precise amount of

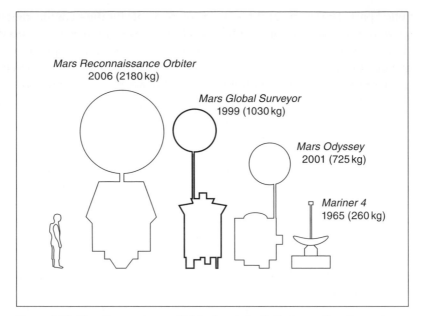

Figure 4.18 *Mars Reconnaissance Orbiter* is substantially larger than its predecessors, and weighed over two tons at launch. The dates refer to when each spacecraft arrived at Mars.

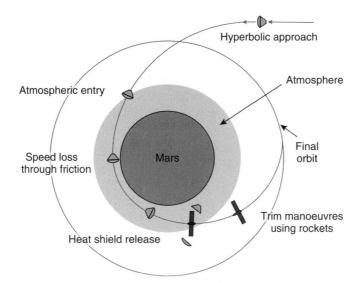

Figure 4.19 Aerobraking and aerocapture are both techniques that steer a spacecraft into the Martian atmosphere in order to use the resulting drag to reduce its speed and change its trajectory. Aerobraking uses atmospheric drag to lower an existing orbit, whereas aerocapture, as shown here, involves the much riskier manoeuvre in which the initial orbit is acquired by aiming the incoming spacecraft so it passes deep into the atmosphere to obtain a carefully calculated degree of atmospheric drag. It then emerges moving slowly enough to go into orbit, but not so slowly that it crashes onto the planet.

thrust is applied for the right amount of time, the spacecraft's position and velocity will be matched to a stable orbit around Mars and it will become a satellite of the planet. If the minimum amount of fuel is used, the orbit will be elongated in the original direction of travel, whereas for optimal science operations a circular orbit is usually needed, in which the spacecraft maintains a roughly constant height above the surface of the planet. There are two ways to change an eccentric (elliptical) orbit into a circular one: by further 'burns' of the retro-rocket, and by aerobraking.

Aerobraking works by using small burns to steer the spacecraft through the outer part of Mars' atmosphere at the low point of the orbit (periapsis). The reduction in speed produced by the resulting friction drag brings the orbit closer to the planet, and careful steering and timing over many orbits will finally produce a circular orbit at the desired height. The fuel used for steering is much less than would be required to circularise the orbit without atmospheric drag. In the case of *Mars Reconnaissance Orbiter*, the saving due to aerobraking was estimated at nearly half a ton, about twice the mass of the fuel the spacecraft actually carried for orbit insertion and subsequent manoeuvres.

Aerocapture offers even larger savings, but is correspondingly more risky, and has not yet been used at Mars. Here, no retro-thrust is used at all; instead, the spacecraft is targeted very precisely to a quite deep level in the atmosphere where a single pass will produce enough drag to achieve orbit. Because Mars' atmosphere is quite thin, the trajectory required could actually result in a collision with a high feature such as Olympus Mons if it should get in the way. Unfortunately, the risk associated with aerocapture cannot be eliminated entirely by accurate targeting, since the density of the atmosphere varies quite considerably with season and meteorological conditions such as dust storms, so these capricious factors have to be taken into account as well.

Mars Reconnaissance Orbiter fired its engines for twenty-five minutes as it approached Mars. This reduced its speed by about one kilometre per second, just enough to allow the heavy spacecraft to be captured by the planet's gravity field. Contact was lost for a tense half an hour as the spacecraft passed behind the planet, but it emerged exactly where and when the mission controllers expected, indicating perfect performance. The ghosts of *Observer*, *Climate Orbiter* and the other lost Mars missions were being laid to rest.

Reconnaissance Orbiter's initial orbit, prior to aerobraking, had a period of about thirty-five hours with an extremely elliptical shape, the farthest point being about a hundred and fifty times more distant from the planet than the closest. The latter, the periapsis, was just four hundred and twenty-six kilometres above the surface of Mars, but still well above the effective top of the atmosphere. In order to initiate aerobraking, the rocket motor on the spacecraft made a short burn near periapsis on several successive orbits, reducing the closest point until the spacecraft was passing over the south polar region at a height of only about one hundred kilometres. Now, carefully calculated dips into the upper atmosphere could be used to generate friction that would slow the spacecraft down

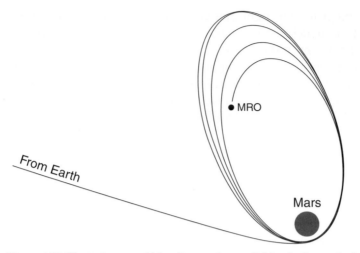

Figure 4.20 The trajectory of *Mars Reconnaissance Orbiter* during arrival and the first part of the aerobraking phase. The initial insertion into orbit was achieved in this case by a twenty-five-minute burn of an onboard rocket engine; in principle it could have achieved orbital capture by skimming the atmosphere on arrival and using the drag to achieve the required speed reduction of around one kilometre per second. This was considered too risky for *Reconnaissance Orbiter*, but the fuel savings are so great that the technique ('aerocapture') may well be employed by future missions.

and gradually bring it into a circular orbit over a six-month period, as shown in Figure 4.20. *Reconnaissance Orbiter* ended up circling Mars at a height that ranges from about two hundred and fifty-five kilometres over the south pole, to three hundred and twenty kilometres over the north pole.

Despite the accumulated experience with *Global Surveyor* and *Odyssey*, a conservative approach to aerobraking was still considered essential in order to avoid overheating or mechanical damage to the spacecraft. The atmospheric density can vary by a large factor at the drag altitude and an excessive reduction in the periapsis height might make the orbit decay rapidly, causing a crash landing on the planet. The dips into the atmosphere get more frequent as the orbital period gets shorter, occurring several times a day towards the end of the six-month operation, which terminates with a 'walk-out' manoeuvre to raise the periapsis above the atmosphere once more.

In addition to carrying out its own investigations, summarised in Figure 4.21, *Mars Reconnaissance Orbiter* was designed to serve as a communications relay satellite between spacecraft on the surface of Mars and the mission controllers on Earth. The first mission to use this capability was *Phoenix*, the stationary lander that studied Mars' north polar region from 25 May until 2 November 2008. The next is *Mars Science Laboratory*, the large rover now under development and intended to begin surface operations in 2012.

As we have seen, *Reconnaissance Orbiter* was conceived to make anticipated future missions to Mars safer and more productive. Since it does this principally

Figure 4.21 ***Mars Reconnaissance Orbiter* goals**

NASA's stated goals for the *Reconnaissance Orbiter* mission are:

Characterise the present climate of Mars and how the climate changes from season-to-season and year-to-year.

Characterise Mars' global atmosphere and monitor its weather.

Investigate complex terrain on Mars and identify water-related landforms.

Search for sites showing stratigraphic or compositional evidence of water or hydrothermal activity.

Probe beneath the surface for evidence of subsurface layering, water and ice, and profile the internal structure of the polar ice caps.

Identify and characterise sites with the highest potential for future missions that will land on Mars' surface, including possible missions to collect samples for returning to Earth.

Relay scientific information to Earth from Mars surface missions.

by examining the surface, subsurface and atmosphere in more detail than before, it aims also to produce at least an incremental advance in our understanding of Martian geology and meteorology. Of the six science instruments that examine the planet in various parts of the electromagnetic spectrum, from ultraviolet to radio waves, the camera is by far the largest. Although it is physically big, with an aperture of half a metre (not much smaller than that used by Lowell when observing canals on Mars from Arizona a century earlier), the camera uses lightweight optics and a carbon composite structure, resulting in a mass of only sixty-five kilograms.

Known officially as the *High Resolution Imaging Science Experiment* (*HiRISE*), the camera's resolution on objects on the surface of Mars is also of the order of half a metre. The downside of such high performance is that it takes a long time to map the whole planet, since ultra-high-resolution images can cover only a small area at a time. *HiRISE* is expected to image just one per cent of the surface of Mars during the mission lifetime. It will be quite easy to miss things, where one is not sure where to look; for example, *HiRISE* could take a picture of the wreck of *Beagle 2* on the surface, but only if it happens to hit the right spot. A smaller, lower-resolution 'context camera' is also carried on *Reconnaissance Orbiter* to take wider-swath pictures which can cover at least fifteen per cent of the Martian surface during

Figure 4.22 The Mountains of Mitchel reveal their secrets to the ultra-high-resolution camera on the *Mars Reconnaissance Orbiter*. From a height of just two hundred kilometres, a segment of the 'mountains' is seen at a resolution of twenty-five centimetres to contain just low, rolling hills and small craters.

the nominal mission. This lacks the resolution to detect *Beagle 2* (or the lost *Polar Lander*, or indeed the *Climate Orbiter*, which crashed on the surface in 1999), or unexpected small-scale natural features, so serendipity is still a major player.

Ormsby Mitchel would have been interested to see a close-up image of the mountains he discovered on Mars in 1846. *HiRISE* obtained several in June 2007, like that in Figure 4.22. Skimming just two hundred kilometres above the surface,

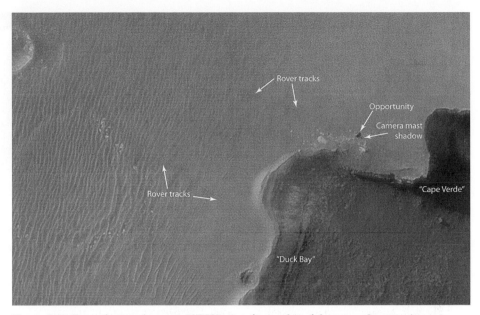

Figure 4.23 From the top down: A *HiRISE* view from orbit of the rover *Opportunity* sitting on the edge of Victoria Crater, into which it would later descend.

it surveyed a rather shallow and topographically uninteresting range of hills that give no real indication of why they should retain a snowy cover well after the rest of the polar cap has retreated all around. Certainly, it is not their exceptional height. Spectroscopy from *Global Surveyor* suggested that the carbon dioxide snow in this region is small-grained and bright, giving a high reflectivity that protects the ice from subliming for longer than the darker snows nearby. But it does not reveal what produces these small grains, uniquely in this region; some future astronauts may have to spend a season in the Mountains of Mitchel before anyone knows the answer to that.

Figure 4.23 shows another spectacular example of the capability of the *HiRISE* camera. On 3 October 2006 it snapped the tiny rover, *Opportunity*, on the edge of Victoria Crater, near a rocky outcrop called Cape Verde. The name of the crater comes from the ship in which Ferdinand Magellan and his crew first circumnavigated the globe, and the various internal features inside the crater have received names based on the places that were visited on that other historic voyage, such as Cape St Vincent, another rocky outcrop on the edge of Victoria where *Opportunity* began its descent. *Opportunity* had driven nine kilometres to get to this point, and shortly afterwards it drove down into the eight-hundred-metre wide crater and photographed Cape Verde itself from close by, obtaining the view shown in Figure 4.24. NASA's plan of coordinated surface and orbital surveillance of interesting features, like the dramatically layered rock face of Cape Verde, was working well.

However, the search for water-related minerals on Mars is bound to be limited in its ability to cover the possible sites of interest if only landers and rovers can be

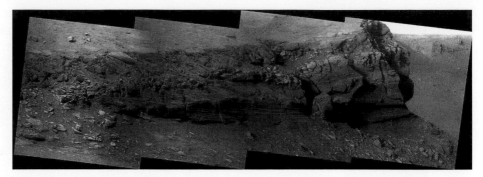

Figure 4.24 From the bottom up: the view from *Opportunity* of Cape Verde in Victoria Crater.

used to make sure a site is likely to be fruitful before setting the robot geologists down. A good site might be chosen while a great one lies ignored nearby. For widespread surveillance from orbit, imaging alone is not sufficient and spectral information is the key. The *CRISM*[21] spectrometer on *Reconnaissance Orbiter* can cover the whole planet at a resolution of two hundred metres, and selected regions at a resolution of a few metres. *CRISM* slowly builds up a map of the surface minerals, revealing large swathes of volcanic basalts in the mountainous regions and water-deposited clays in the valleys.

The main surprise, on a planet with a watery history and a carbon dioxide atmosphere, is the lack of evidence for more than a trace of carbonate minerals. The Earth is loaded with them. *CRISM* may have the answer, however, and that is a surprisingly large abundance of sulphates. These are produced by the erosion of rocks by sulphuric and sulphurous acid – like the familiar acid rain on Earth, but in Mars' case the sulphur must have come from volcanoes, rather than fossil fuel pollution. Chalky, alkaline carbonate rocks are particularly vulnerable to acid rain erosion and carbonate is unlikely to survive in significant amounts where a lot of sulphate is found. Mars volcanoes were still active when it was wet, and the air was full of hydrogen sulphide and sulphur dioxide and other gases that are washed out by rain to react with the rocky material on the surface.

In at least one place on the planet, a large deposit of carbonates appears to have escaped this fate. In December 2008 the *CRISM* team reported exposed slabs of carbonate-rich rocks ten kilometres or less across in the Nili Fossae region, where valleys in the ancient crust point to a watery past. The implication is that there must have been different rock-forming epochs, some of which were wet but not acidic. The new and additional complications this raises for the climate history of the planet are obvious and daunting.

A key detail is where the water ended up: much of it, along with a host of other interesting features of the past climate, is probably somewhere below the surface. Enter the *Shallow Subsurface Radar* (*SHARAD*) on *Reconnaissance Orbiter*, which

---

[21] *Compact Reconnaissance Imaging Spectrometer for Mars.*

uses reflected radio waves at a frequency that penetrates to a depth of up to a kilometre with about seven metres vertical resolution. This is being used to find underground layers of ice and rock and, perhaps, although not yet, aquifers containing liquid water. The *SHARAD* instrument was designed by the Italian Space Agency, who also provided the ground-penetrating radar on *Mars Express*. The latter works at a frequency ten times lower than the radar on the American spacecraft, and so achieves a correspondingly greater depth of penetration, up to five kilometres, although with lower vertical resolution.

The loss of *Mars Observer* in 1992 and *Mars Climate Orbiter* in 1999 left the atmosphere of Mars much less well explored than NASA or the scientific community had intended. Not only is there keen interest in comparing Martian weather systems to those on Earth, and learning how the climate can have changed so much, but atmospheric behaviour is as much of a hazard to landing as rough features on the surface. As the Russians found with *Mars 2* and *3*, parachuting into a dust storm is difficult and can be fatal, and large payloads descending slowly on retro-rockets in order to land upright on legs are always vulnerable to the sideways motions induced by strong winds, especially as they touch down.

One group of scientists who had reason to celebrate when *Reconnaissance Orbiter* arrived safely was the team operating the *Mars Climate Sounder*, an updated version of the instrument built, flown and lost twice before, once with *Mars Observer* in 1992 and then again with *Mars Climate Orbiter* in 1999. The team, of which the author was a member, had been working for over twenty years before it finally succeeded in its goal. It was trying to apply the remote sensing techniques used routinely by weather satellites to monitor Earth's weather, ozone and pollution, to the thin but dynamic atmosphere of Mars, something that would have been timely in the 1980s but which fate had decreed would not happen until twenty-five years later. The instrument, using designs successfully developed for Earth-orbiting instruments, combined with some innovations to produce a very compact device,[22] points both downwards and horizontally through the atmosphere, in order to quantify the global atmosphere's vertical variations of water vapour, dust and temperature. It operates alongside a small, wide-field camera, the *Mars Color Imager*, which produces global images to track visual changes in Martian weather produced by clouds and dust, as well as (through the use of ultraviolet filters) to examine variations in ozone. The results appear in more detail in Chapter 6, which deals with the Martian weather and climate.

In addition to the results from these instruments, the low orbit of *Reconnaissance Orbiter* meant it allowed better maps of Mars' gravity field to be produced by tracking small perturbations in the spacecraft's orbit. The seasonal change in the mass of Mars' polar ice caps is measurable by this effect, for instance. The payload for the mission also includes yet another camera, which compares observations

---

[22] Earth-orbiting instruments now typically weight hundreds of kilograms; size and weight-saving techniques usually have to be brought to bear to make a version that can travel to Mars, even on a two-ton spacecraft.

with the predicted positions of Mars' moons, Phobos and Deimos, as an aid to navigation. Finally, its ultra-high-frequency (UHF) radio is designed for relaying commands to landers on the Martian surface, and for returning science and engineering data back to Earth using the orbiter's telecommunications system. It is hoped that this relay will still be operational when the *Mars Science Laboratory* arrives on the surface in 2012.

4.12.1 The Mars *Scout* programme

In March 2001 NASA issued a call for proposals for the first phase of its *Scout* programme of small missions. The unique feature of this was the opportunity to propose, not just an experiment to go on a NASA spacecraft, but the whole mission, all under the direction of a lead scientist or principal investigator. The community's enthusiasm for *Scout* was somewhat tempered by the cost limit, which was set at a relatively modest three hundred million dollars, reminiscent of the unfortunate Faster, Better, Cheaper attitude. Nevertheless, there were plenty of proposals, and in June 2001 ten ideas were chosen for further study from forty-three submitted. Each team on the shortlist received a small budget to improve their proposal and answer queries, before the final selection two years later. The 'winners' in 2001 were:

1. *Sample Collection for Investigation of Mars*: To collect atmospheric dust and gas without landing by grabbing samples on a low pass through the atmosphere and using a 'free-return trajectory' to bring the samples back to Earth.
2. *KittyHawk*: Three gliders would swoop around the walls of Valles Marineris to explore the composition and stratigraphy in ways not possible for orbiters and landers (Plate 41).
3. *Urey*: A surface rover would allow the absolute ages of geological materials to be remotely determined for the first time on any planet.
4. *Mars Atmospheric Constellation Observatory*: A network of microsatellites around Mars would characterize the 3-D structure of the atmosphere and Martian climatology.
5. *Artemis*: Three small landers and micro-rovers on the Martian surface, with two directed to the polar regions, would explore the surface and shallow subsurface for water, organic materials and climate.
6. *Mars Environmental Observer*: To explore the role of water, dust, ice and other materials within the Martian atmosphere to understand parts of the hydrological cycle.
7. *Pascal*: A network of twenty-four weather stations on the Martian surface would provide more than two years of continuous monitoring of humidity, pressure and temperature.
8. *Mars Scout Radar*: An orbiter mission would use synthetic aperture radar imaging to map the surface geomorphology and very shallow subsurface (three to five metres deep), to detect buried water channels and other features.

9. *The Naiades*: Four landers would explore for subsurface liquid water using a novel low-frequency sounding method.
10. *CryoScout*: This mission would use heated water jets to burrow down through the Martian polar ice caps to depths of tens or even hundreds of metres, while measuring composition and searching for organic compounds.

In the event, none of these was successful in the final selection. Instead, NASA decided in 2003 to fly the *Mars Surveyor* 2001 lander, which had been in storage since its mission was cancelled in the wake of the failure of the 1999 *Climate Orbiter* and *Polar Lander* missions. Lockheed Martin, who built the spacecraft, had kept it in clean storage from which it was now retrieved and refitted for the new mission, appropriately named *Phoenix*.

4.12.2 *Phoenix*

Phoenix was selected not just because it had existing hardware that was burning a hole in NASA's pocket, metaphorically speaking, but because the gamma-ray spectrometer on the *Odyssey* orbiter had produced maps like that in Plate 7 showing large amounts of hydrogen under the surface of Mars, concentrated at high latitudes. This had to be H_2O, in the form of ice, and investigating this and any interesting impurities that it might contain fitted well with NASA's ongoing strategy to 'follow the water'. The refurbished spacecraft was launched from Cape Canaveral on 4 August 2007, and arrived safely on Mars on 25 May 2008. The name, of course, reflects that fact that this is much the same mission as the one that crashed in December 1999.

Phoenix, shown in Figure 4.25, consisted of a fixed lander with a robotic arm to dig for samples and search for signatures of life. While it resembles *Viking* in those

Figure 4.25 *Phoenix*, the first mission in the Mars *Scout* programme, and the first successful polar lander.

respects, a major difference is that *Phoenix* landed in an environment known to be water-rich, in the arctic plains at around 70° N, just 1200 kilometres from the north pole. It had a longer and more powerful arm than *Viking* did, one that can bring soil and ice samples from a depth of half a metre to instruments on the lander platform. There they were inspected by a microscope and heated for evolved gas analysis to understand whether the moist conditions below the surface make a habitable zone where microbes could grow and reproduce. Some samples are heated to as high as a thousand degrees centigrade, to address concerns expressed by some that *Viking*, by heating to only five hundred degrees, could have missed some promising organic materials that only break down at higher temperatures.

The date of the landing by *Phoenix* was chosen to coincide with Martian summer, when the polar cap was retreating and cold soil was exposed to sunlight after a long winter. *Phoenix* was equipped with three cameras – a descent imager to image the surrounding terrain on the way down and during the landing, a panoramic camera to look around after landing, and a magnifying camera on the arm to examine the specimens in detail. The pictures taken on the way down not only reveal interesting terrain, not previously seen close up, but also would have provided evidence, sadly lacking in the case of the 1999 *Polar Lander*, in the event of mission failure during the final and most difficult phase. The landing site, on the floor of the vast ocean that once filled the northern plains, contains sediments laid down in the form of mud and silt as well as the coarser grains of rounded and well-sorted sand that flowing water produces. The layering of different materials and their ice were examined to deduce the history of water in the region, while the meteorology package on board provides data for characterising Mars' present climate and weather processes for the first time at a landing site near one of the poles.

The planned length of the mission was ninety days; it managed a hundred and thirty, a considerable achievement because with the onset of the Martian autumn the temperature falls and the sunlight available for power drops off rapidly. The beginning of the end came in later October 2008, when the lander shut its non-essential operations down, entering so-called 'safe' mode, due to lack of power. The project management's report at this time is worth quoting:

Weather conditions at the landing site in the north polar region of Mars have deteriorated in recent days, with overnight temperatures falling to −141 °F, and daytime temperatures only as high as −50 °F, the lowest temperatures experienced so far in the mission.[23] A mild dust storm blowing through the area, along with water-ice clouds, further complicated the situation by reducing the amount of sunlight reaching the lander's solar arrays, thereby reducing the amount of power it could generate. Low temperatures caused the lander's battery heaters to turn on Tuesday for the first time, creating another drain on precious power supplies.

Reduced power operations were possible for a while, before the inevitable final shut-down in the cold and dark of the Martian winter on 2 November 2008.

---

[23] This range corresponds to from ninety-six to forty-five degrees centigrade below zero.

Further reading

Beagle: From Darwin's Epic Voyage to the British Mission to Mars, by Colin Pillinger, Faber and Faber, 2003.

Managing Martians, by Donna Shirley with Danelle Morton, Broadway Books, 1998.

Roving Mars: Spirit, Opportunity, and the Exploration of the Red Planet, by Steve Squyres, Hyperion, 2006.

Part II
The Big Science: motivation to continue the quest

As Mars exploration has progressed, the reasons for continuing have gradually evolved. In the last fifty years, the focus of attention has moved on from a broad desire to know what the planet is like, compared to Earth, and to search for obvious signs of life, to a set of more specific goals. These will be developed in detail in the next three chapters. They can be summarised as follows.

Firstly, understanding the origin and evolution of Mars, along with that of the family of planets that includes the Earth. This ancient quest is as old as astronomy itself, and astronomers are no more than halfway along the road to enlightenment.

Secondly, understanding the climate history of the planet, which appears so dramatically etched in its surface. Climate is a huge topic on Earth just now, as everyone knows. Mars must, and does, have a lot to teach us about global change on terrestrial planets with carbon dioxide, water and cloud-regulated 'greenhouse' warming.

Thirdly, finding out whether present-day Mars has habitable zones for lifeforms of any kind, and whether there are any traces of past or present life. The almost invariable selection and emphasis of this goal by the popular media when reporting any new investigation of Mars is eloquent testimony to the fascination it holds.

To these three basically scientific motivations for further exploration, we might add a fourth:

Solving the practical and political problems of manned interplanetary space flight and developing a capability for the establishment of manned bases and perhaps even colonies on Mars.

This last motivation is possibly the strongest, and most enduring, of all. This could be precisely because it is not really part of a scientific case; anyone from almost any background knows why exploring Mars is romantic and exciting in human terms. Natural curiosity alone can lead humankind to go to extraordinary efforts to explore, and looking forward to the day when someone from Earth sets foot on Mars is one of the most persistent fascinations of the human race.

Chapter 5
The origin and evolution of planet Mars

Current theories of the origin and evolution of the Sun and its family of planets all say that they came into existence at the same time some four and a half billion years ago, forming from a large cloud of dust and gas known as the protosolar nebula. It does not follow that they all should have the same composition, since there were temperature and pressure gradients in the cloud that caused some materials to condense, and others to escape, at different distances from the young Sun. What mix of elements went into each planet? If it was known how the bulk compositions of Mars, Venus and Earth differ, if indeed they do, it would be possible to reconstruct the distribution before they formed, and understand what happened during the condensation of the cloud to form the planets and their moons.

There are plenty of other open questions that bear on the history of the planets. For instance, why did Earth and Mars form at their respective distances from the Sun? Part of the answer lies in the dynamic stability of the system; the planets perturb each other through gravity, and the material that formed them accumulated in those orbits where it could remain for a long time without disruption. Then again, why is Mars so much smaller than Earth? Mars may have been robbed of material by the gravitational attraction of its giant neighbour, Jupiter, to a greater extent than Earth or Venus. But at present the details are missing. Is it mere coincidence that Mars and Earth have nearly the same rotational period and axial tilt at the present time? Probably, but even so the similarity is remarkable.

Mars is a small planet, with a mass that is only about one tenth that of the Earth.[1] When pictures of the two planets are compared at the same scale in Figure 5.1, Mars looks almost insignificant. Its diameter is only slightly more than half that of Earth, but since Mars has no oceans its land surface area is about the same. The terrain has major differences, however. Why are the Martian

---

[1] The latest values for the basic properties of Mars and its orbit are summarised in the tables in Appendix A.

Figure 5.1 Siblings: Earth and Mars formed at the same time, and from the same cloud of gas and dust surrounding the young Sun, about four and a half billion years ago.

volcanoes so massive, and the rift valleys so deep and steep, compared to their terrestrial counterparts? How can geologists account for the apparently very different ages of distinct regions of the surface? What role has been played by plate tectonics, the phenomenon that produces continental drift on Earth? What sequence of events produced the differences between the minerals that make up different regions, and how are they related to the apparent ages of those regions?

Thinking now about the interior of the planet, is Mars layered inside, like the Earth? What is the size and composition of the metallic core, and is it solid, liquid or both? What is the history of the Martian magnetic field, and of internal convection and plate tectonics, and how has this affected the evolution of the planet? What is the history of volcanism – the escape of gases and molten rock from the interior – on Mars? How does the density and composition of Mars' thin atmosphere, containing basically the same gases as Earth, but in quite different proportions, relate to the volcanism that has taken place, and to the properties of the interior, such as temperature versus depth?

Linked with the history of the planet is the story of life on Mars. It is possible that it could have come into existence long ago, when Mars seems to have been more hospitable, and then expired when the climate on the planet became too harsh. Conversely, the temporary presence of life may have affected the evolution of the planet as a whole. Earth's biosphere has had, and continues to have, a large effect on the composition of our atmosphere, affecting the climate, and in turn affecting the habitability of the planet. Venus seems to have evolved without life and is in a very different state from the Earth, partly for that very reason. What about Mars?

So many questions. These, and many other puzzles, are endlessly debated by the science groups and mission planners as they plot the course of the next fifty years of Mars exploration. A constructive approach is to review what is known with reasonable certainty already, what is being targeted by the latest generation of experiments on spacecraft currently operating at Mars, and then what problems are likely to remain to be addressed in the longer term.

5.1   Interior and magnetic field

Early in the exploration of Mars by spacecraft, it was shown that the planet had, at best, a very weak global magnetic field, and the very sensitive measurements by *Mars Global Surveyor* confirmed that there is no internal, dynamo-generated planet-wide field. However, *Global Surveyor* did find patterns of localised fields frozen into parts of the crust, the southern highlands in particular, which were strong enough to be detected by its magnetometers in orbit above. These 'magnetic anomalies' are more than ten times as intense as any similar features observed on the Earth, and are clear evidence that Mars had a strong field in its earlier history. This would have permanently magnetised iron-bearing rocks in that past era, leaving a record of the ancient magnetic field that is better preserved than the equivalent on the Earth, where ongoing dynamo activity in the core and active plate tectonics have scrambled the record.

In fact, a global map of the residual magnetism in the Martian crust shows strong evidence that Mars, too, once had plate tectonic activity, in which convection in the interior was vigorous enough to break the surface layer into plates and move them around. Secondary evidence comes from the straight-line alignment of the large Tharsis volcanoes, presumably along the fault line between plates where lava could escape from the interior more easily. Volcanoes on Earth often behave in that way. The alignment of the Valles Marineris is consistent with a tectonic origin for that feature also. The main gorge is aligned along the boundary between two of the plates revealed in the magnetic map, seeming to confirm that it is a huge rift valley produced when the plates moved apart.

The most strongly magnetised region, in the oldest terrain in the south, shows large bands where the residual field alternates in direction. It seems unlikely that such large plates could turn over, and more probable that the bands are evidence for the reversal of the global field on ancient Mars, a phenomenon that is well established as happening to the Earth's dynamo. If that is so, the plates with opposing frozen fields mark parts of the surface with different ages, when the field was in opposite directions. Except that plates with the strongest magnetism in both directions both appear to be very old, it is difficult to estimate the relative ages of the plates. That will have to wait for *in situ* geological investigations by robots or humans.

The absolute timeline that describes the origin and history of the Martian magnetic field is interesting, not only because of what it reveals about the evolution of the interior and the surface of the planet, but also because the field may have been strong enough to protect the surface environment from harmful radiation from space at the time when the climate was more conducive to life. It could also have protected the atmosphere from loss into space as a result of erosion by the solar wind. A key goal of future exploration must therefore be to discover the ages of the rocks that show different degrees of remnant magnetism.

Early in Mars' history the core would have been liquid and in convective motion, capable of forming a dynamo like that which creates the Earth's field. The early cessation of this activity, compared to the Earth, is probably due to the fact that the Martian interior cooled much more quickly because of the smaller size of the planet and the consequent higher ratio of surface area to volume. Plate tectonic activity may also have been important for cooling, since it is a much more efficient way of moving heat from the interior to the surface of the planet than any other. The convective activity that moves the plates is presumably related to that which generates the field, so it seems likely that the two stopped more or less together. According to modern theories of planetary field formation, however, the core does not have to cool to the point where it is completely solid for dynamo action to cease, since the rapid convection necessary to produce a measurable field occurs only during the earliest phase of rapid cooling. On Venus, the absence of a magnetic field, and the apparent high level of current volcanism, may both be attributable to the lack of plate tectonics as a mechanism for rapid interior cooling on that planet.

From measurements of the gravity field, made by observing the trajectory of a spacecraft passing near the planet, it is possible to make inferences about the internal distribution of mass. From these, drawing also on analogies with the interior of the Earth, basic models of the Martian interior like that in Figure 5.2 are proposed. The tidal bulge produced by the gravitational pull of the Sun on the solid body of Mars is only about one centimetre high, but its signature in Mars' gravity field can be followed as it moves by tracking the positions of spacecraft on the surface and in orbit simultaneously. The size of the tide measured in this way using *Pathfinder* and *Global Surveyor* is larger than expected, leading to the suggestion that the core of Mars is still at least partially liquid to this day. This would be consistent with the possibility that there is still some geothermal activity, if not full-blown volcanism, near the surface. This in turn could explain the evidence for recent seepage of liquid from exposed slopes, and for traces of methane in the atmosphere. Such a scenario is encouraging to scientists seeking life in subsurface warm-water habitats on Mars.

As might be expected, the geologically younger-looking parts of the surface tend to show much smaller magnetic fields, as do some of the large impact basins such as Hellas and Argyre, but not the oldest, like Daedalia and Ares. A possible interpretation is that, since the heat and shock from large impacts tends to demagnetse the rock in and around the collision site, some basins must have

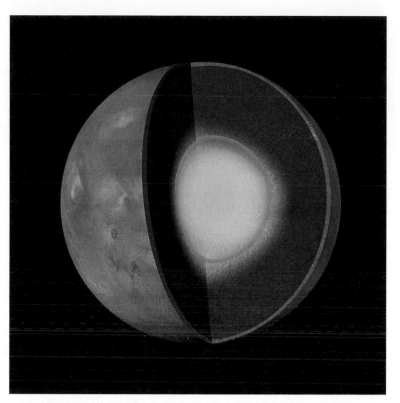

Figure 5.2 Tracking data from *Mars Global Surveyor* suggest that Mars has an iron core about half the size of the planet, and that it is at least partly liquid.

been formed when the field was still extant so it could be remagnetised, whereas the others formed too late for that. Since there are reasonable estimates of the age at which each of the large basins formed, it has been possible to estimate that the dynamo-produced field was gone after only a few million years. This matches the predictions of the latest theoretical models of the formation of Mars and the evolution of its internal dynamo.

5.2 The core of Mars

Mars must have an iron-rich core like the Earth, because a planet the size of Mars that was made entirely of rock would have a much smaller mass. Iron is the most common of the heavy metals, in the entire Universe so far as is known, and the best models of the formation of the Solar System tell us that Mars, like Earth, must contain huge amounts of iron. Next comes nickel, and then smaller proportions of the other metals. If it were possible to determine what those proportions are, and how much of the other heavy elements are present, this would provide not only a test of formation theories, but also a powerful constraint on Mars' evolution. Radioactive metals, like uranium, release heat and tend to keep the core molten, so the history of Mars' magnetic field, generated by movement in the conducting fluid interior, is related to the composition of its core. Direct sampling of the core, of course, is almost unimaginably far off in the future; the composition of the Earth's core is not known with any real certainty either.

The absence of a present-day magnetic field, together with evidence for residual magnetism in the surface rocks to testify that there was a field once, makes it tempting to assume that Mars used to have a liquid core that has since solidified. The necessary cooling would happen more quickly on a smaller planet like Mars than on the Earth, which does still have a partly liquid core. However, as we have seen there is some recent evidence from delicate measurements of the wobbling of Mars in its orbit that it, too, has a partly liquid core. If so, why is there not a larger magnetic field? If the core is partly molten but not internally active enough to produce a field, what does that imply for volcanic activity? The volcanoes seen on Mars today seem to be dormant, but they may still release enough heat to melt subsurface ice and even produce thermal springs or effusions of water from crater or canyon walls, as glimpsed in recent photographs from Mars orbit. If so, then these places are front-runners in the search for extant microbial life and linked to the history and current state of the interior.

Perhaps we can trace the origin and fate of the Martian dynamo from the history of the surface field captured in the remnant magnetism of the iron-bearing material in the crust. This is a fruitful research field on Earth, and the record on Mars is probably better preserved, thanks to the fact that the crust is no longer churned by plate tectonics, as it is here on Earth. The problem is that, while orbiting instruments can map the broad patterns in the magnetic field frozen into the surface rocks on Mars when Mars' outer shell solidified as part of the long-term cooling process, they do not do a very precise job of dating the giant plates which the field patterns mark out (we will discuss later how dating is done, and its limitations). Accurate dating lies in the future, when geologists on the surface of Mars (or in the laboratory on Earth, using carefully selected samples returned by robots) can use sophisticated techniques to determine the ages of magnetised rocks taken from geological strata laid down in different periods of time, going back to the very early history of the planet. They could then retrieve the timeline describing the behaviour of the metallic core over the life of the planet, covering the past four and a half billion years or so.

The measurements will need to cover the globe, obviously a formidable task for scientist-astronauts on Mars even if they are numerous and have good global transportation systems. The highest priority for magnetic mapping is the ancient southern highland region, where the oldest rocks on the surface are likely to be found. These contain iron- and nickel-bearing minerals that were magnetised by the field that existed when they solidified, becoming the permanent magnets that give rise to the very weak fields picked up by the magnetometer on *Global Surveyor*. If it is anything like Earth's, the internal field on Mars will not have simply declined steadily, but fluctuated and even reversed during the period that it was active. Thus, it will take a large number of samples in each area to try to separate the history of the field in various epochs, going back to the earliest times. Not only will the field depend on when the material solidified, but also on any subsequent movement of continental plates when the surface was still sluggishly mobile.

There are various other indirect ways to study the core, and to investigate its internal activity and large-scale structure. The likely presence of a liquid shell between the rocky mantle and the solid core that was inferred from the details of the planet's gravitational field can also be established by estimating how density varies with depth from the movement of the rotational axis with time. Seismometers, instruments sensitive to vibration similar to those that are routinely used for the detection of earthquakes on Earth, deployed at multiple sites on the surface of Mars could listen for the vibrations from natural marsquakes. The waves follow multiple paths, some of which involve reflection from the core and the various rock strata in the mantle, and so timing how they propagate through the interior of the planet to reach the various listening stations gives information about the interior structure of the planet. This is the source of most of what we know about what lies inside the Earth beneath our feet.

A network of small stations placed on Mars to listen for quakes can make other measurements; for example, they can place temperature sensors in boreholes to measure the heat flow from Mars' interior. This will probably vary from location to location, but with sufficient measurements around the globe to form a meaningful average it will be possible to estimate of how hot the core is at present. Multiple small stations also offer the chance to inspect a wide selection of terrains, and analyse their geochemistry and apparent climatic history, all over what is clearly now seen to be a very varied planet. This is not just good science; it also means that the expensive rover and sample return missions that follow will have a much better chance of finding the best sites at which to pursue their goals. The negligence of the space agencies in their failure to actually carry out a network mission long before now, despite countless studies and proposals, is one of the curious vagaries of the history of Mars exploration.

5.3   The composition of the surface

The little that is known about the chemical and mineral composition of the rocks and soil that make up the surface of Mars is derived from *in situ* analysis at the sites explored by a handful of landers and rovers, from inferences drawn from the analysis of meteorites, and from global infrared and gamma-ray spectroscopic mapping from orbit. This patchwork of measurements certainly reveals a planet that is extremely diverse and complicated, but the interpretations are equally patchy and there is no comprehensive understanding, even of which aspects are Earth-like and which unique to Mars.

The red colour of Mars is caused by iron oxide – rust – in the soil. Geophysicists estimate that Mars has about twice as much iron in its crust as the Earth does, probably because Earth was bigger and hotter in the days when much of the iron separated from the rock and formed the metallic core. This separation must have happened on both planets, but could have been less efficient on Mars. The highly oxidised nature of the iron in the surface material is to be expected given the solar-ultraviolet driven photochemistry in the lower layers of the atmosphere, part of the same family of reactions that turn carbon monoxide back into the dioxide, and produce the peroxides that sterilised the samples tested by *Viking*.

The high iron content makes the soil, dust and at least some of the rocks highly magnetic, and capable of trapping the record of the ancient field once generated in the core. The composition of the fine material has been estimated to be about 80% iron-rich clays, 10% magnesium sulphate, and 5% iron oxide (Fe_2O_3), with small amounts of other compounds, including silica (common sand), calcium carbonate (limestone) and calcium sulphate (gypsum or plaster of Paris). Some types of rocks and minerals form only in the presence of water; but are best studied in solid strata, since the soil is often the same material as the dust that is mixed all over the planet by the winds and storms, and so does not tell us much about what different types of rocks exist on the Martian surface in geologically different regions. Strata also allow the sedimentary (water-deposited) material to be dated relative to the layers above and below. This is the sort of study that has been dramatically advanced by the *Mars Exploration Rovers*, promising even more when advanced explorers can analyse layered material at the polar caps, or in the walls of Valles Marineris.

The complexity of the Martian surface is vividly illustrated by maps like that in Figure 5.3, made by the US Geological Survey to show the inferred surface

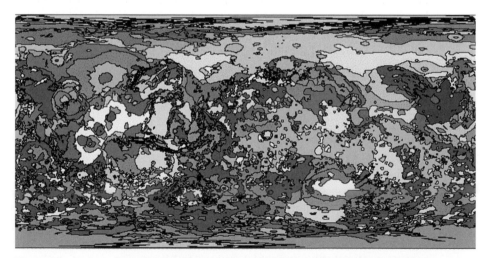

Figure 5.3 A mineralogical map of Mars by the US Geological Survey. Each shade represents a different type or composition of surface material. The considerable complexity of the map, despite the incomplete information that went into making it, illustrates the challenge facing those who try to understand the history of the evolution of Mars over the last four billion years.

mineralogy region by region. As would be expected from the amounts of air-borne dust seen on the planet, especially during the major storms, deposits of fine material cover large areas, particularly the lowlands. In general, except for the icy polar caps, the dusty plains are the regions that appear brightest when Mars is viewed through a telescope. The darker regions tend to be high ground, where the rocky surface is exposed. It is perhaps not surprising that the hilly regions seem to consist mainly of volcanic rock; many of the hills themselves are ancient volcanoes. The other main category, the sedimentary materials, would naturally occur in the low plains where water could collect, and over time be covered by dust deposits. The spectral characteristics of the solidified lava show evidence for two principal kinds, distributed geographically roughly by hemisphere. The southern highlands have a high concentration of ordinary basalt, the commonest kind of volcanic rock on the Earth and the Moon, while rock in the northern hemisphere appears much richer in silicon, causing it to be classified as andesite. This is a glassy material, produced at very high temperatures, and it is hard to see how there could be so much of it, so widely distributed, without plate tectonics to cycle it through the hot interior. It is possible, as we have seen above, that Mars did have a tectonic phase in its early history, when the interior of the planet was hot and convective. But why would the signature of this ancient activity be confined to the relatively young, relatively low, northern hemisphere?

Possibly the spectra have been misinterpreted – Figure 5.4 gives some feeling for how difficult it can be to identify minerals correctly from their spectra, even when they are not mixed with other types, which of course is usually the case on real planetary surfaces. Perhaps the material in the north is not andesite, but just basalt that has been modified in some way or contaminated with something that changes its spectrum. Since it occurs mainly in the basin once possibly occupied by the Northern Polar Ocean, it might be the output of volcanic sources that were actually under water. Basalt could have dissolved and then been redeposited as clay-like material, some of which can look a lot like andesite in low-resolution spectra.

Some limited regions in both hemispheres have spectra that are not like dust, basalt or andesite. Some of these seem clearly to feature olivine, a volcanic mineral that decomposes rapidly on exposure to standing water to form clays and iron oxides. The implication is that regions rich in olivine have not seen much exposure to water. Regions that did have long-standing liquid water should, on the other hand, be rich in carbonate minerals. These are formed when atmospheric carbon dioxide dissolved in liquid water reacts with other minerals on the surface, a very common process on the Earth and one that must have occurred on early Mars if it was warm and wet. Frustratingly for that hypothesis, there is only limited spectral evidence for carbonates, although small patches a few kilometres across have been found recently. The dusty regions also show signs of carbonates present at the small percentage level, by some estimates enough to account for the loss by weathering of a thick ancient atmosphere, supporting arguments in favour of a wet early Mars. Others claim

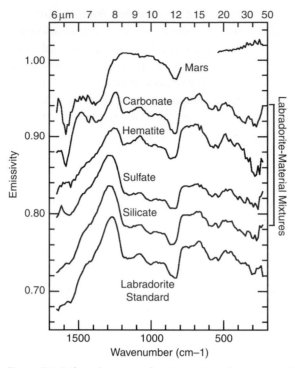

Figure 5.4 Infrared spectra of various minerals compared to the spectrum of a dusty region on Mars (top), observed by the *Thermal Emission Spectrometer* on the *Mars Global Surveyor* orbiter. While this is from a paper (Bandfield *et al.*, 2003) in which the analysis of the Mars spectrum is shown to be evidence for the presence of carbonates, it can be seen how difficult this kind of diagnosis is in practice.

that the amount observed could just as easily have been produced while the dust was airborne, through reaction with water vapour and carbon dioxide in the atmosphere. Whether or not this is the case, if Mars was actively volcanic when the surface water was present, it may be that carbonate formation was prevented, or that any carbonate that formed was subsequently dissolved, by reactions with sulphurous gases of volcanic origin that were plentiful in the early atmosphere. The high sulphate content recently identified in some regions by the *Exploration Rovers* is consistent with this. Again, it also remains possible that large deposits of carbonates exist in the low-lying plains that once were seas, and are now covered over by dust deposits so they cannot be observed from orbit. The dust on the surface, which moves with the wind, is basically powdered basalt, but the soil underneath is more consolidated. The component that provides the extra cohesion is again probably the magnesium and iron sulphate that formed out of the sulphur released during the volcanic phase, when 'acid rain' was a feature of the Martian climate.

As we saw earlier, when *Pathfinder* landed in the northern hemisphere it found rock with a large proportion of andesite, fitting the broad-brush picture obtained from the orbiting spectrometers. However, the *Vikings*, also in the northern hemisphere although thousands of kilometres away, found mainly basalt. The Mars meteorites, which have been shown by laboratory analysis all to be igneous rock, tend to have a lower silica content than any of the rocks analysed on the surface of Mars, which makes them basalts also. In their case it could imply that the SNCs originated in the southern highlands, where basaltic rock is more common. However, to complicate the picture, some of the igneous minerals making up the Martian meteorites belong to categories whose spectral signature is found only in a few special regions dotted around the planet. The famous Allen Hills meteorite AH84001, for example, has a spectrum that does not correspond to most of Mars but does match an area in Eos Chasma, part of the Valles Marineris, where it may have originated.

Other special regions show the spectral signature of haematite, a compound of iron that is usually produced in standing water on the Earth. The small spherical 'blueberries' pictured in Figure 5.5 that *Opportunity* found in sedimentary deposits after landing in such a region consist of haematite. Some of the rocks photographed at the *Pathfinder* site also have the appearance characteristic of rocks that formed by deposition of material in water. These include large fragments with layered cross-sections, rocks with embedded pebbles that look like conglomerates, and rounded stones lying on the ground. The smooth and polished appearance of some of the rocks may be further evidence that liquid water flowed for a long time, and that the climate was therefore stable in a state much warmer and wetter than at present. The cratering record in the Ares Vallis area where *Pathfinder* landed suggests that the flooding there occurred more than two billion years ago. Recently, the team operating the *CRISM* spectrometer on *Mars Reconnaissance Orbiter* reported the detection of large amounts of hydrated silica, chemically the same as the gemstone known as opal. On Mars it occurs in layers and outcrops that are geologically quite young, laid down not more than two billion years ago. This is considerably more recent than the regions where hydrated minerals had been found before – the clays and sulphates are thought to be three to four billion years old. The implication is that Mars may have been warm and wet for *billions* of years, if this first interpretation stands up to detailed scrutiny. The methods used for dating geological features of all kinds are obviously crucial for understanding the evolution of the planet as a system.

5.4   Determining the ages of regions and features

Establishing a chronology or time scale for the planet-wide changes that have evidently occurred on Mars is a top priority. This incorporates a host of key questions, for instance: when did the 'river valleys' form? When and for how long was there liquid water on the surface? When did active volcanism cease? And so on.

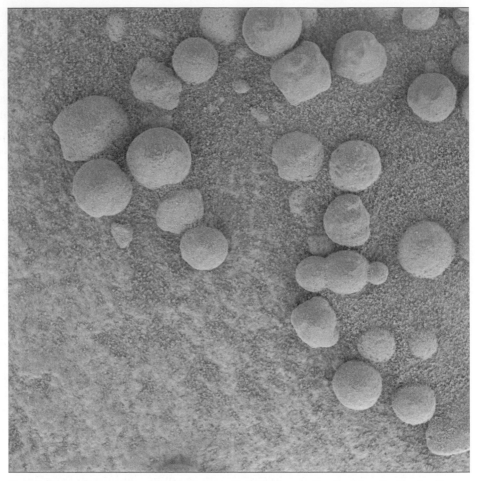

Figure 5.5 Small grey spherules like these (called 'blueberries' by the *Mars Exploration Rover* science teams) were found at many locations in the Meridiani region by the rover *Opportunity*. A few millimetres across, they consist mainly of the iron-bearing mineral haematite and must have been formed by large, persistent amounts of liquid water.

Dates that are quite approximate are often very valuable if they allow the sequence of events in the planet's history to be determined; for example, how wind, water, volcanism, tectonics, cratering and other processes acted successively or together to form and modify the Martian surface seen today. Apart from simple sequencing – noting that features that fall on top of other features must have been formed later – the most frequently applied method for dating is the 'crater-counting' approach. Surfaces throughout the Solar System have been dated on the basis of how heavily cratered the different regions are, using assumptions about impact rates as a function of time and distance from the Sun, and about any overturning, flooding or weathering that may have taken place subsequently. This method is best applied using photographic surveys made from orbit, where high-resolution pictures can be obtained that show the whole range of crater sizes, down to those less than a kilometre in diameter. Even then there is scope for a lot of error and uncertainty in this process, and while the resulting dates are sometimes treated as if they were precise, they cannot be considered reliable until they have been calibrated from samples made available for dating in the laboratory. At present, the historical geological record is at best uncertain and at worst confusing.

The dating of events can be separated into methods that give *relative* ages only, and those that give *absolute* dates. They can be further subdivided into methods that can be applied to images and spectra obtained from orbiting satellites – the easiest way to gather observations, and the only one that gives convenient global coverage – and those that require the deployment of instruments and eventually human geologists on the surface. Here is a summary, roughly in order of increasing precision:

Dating from orbit: relative ages

- Sequencing: using images that show features overlapping each other. For instance, a flow channel that crosses an ancient crater must be more recent than the event that formed the crater.
- Geochemistry: some events that affect the composition of the surface occurred in a known sequence, for instance those involving liquid water, those involving volcanic emissions, then those involving neither. Compositional maps can thus be used to set crude relative ages.
- Appearance: geological objects are subject to erosion, for instance by windblown dust. A crater or other topological feature that appears pristine is likely to be younger than one that is heavily eroded. Even the untrained eye can easily deduce which of the two regions in Figure 5.6 is the older.
- Crater counting: the surface of Mars retains a record of impact events dating back to the earliest times. The density of craters on Mars does vary from region to region and, since the rate of bombardment has slowed with time, it is reasonable to assume that the crater count depends principally on how old the terrain is. The regions with the highest crater densities are the oldest. Areas that have been covered by relatively recent lava flows, or by sedimentary deposits from ancient floods, have relatively few craters, and the average crater size is smaller.

Dating from orbit: absolute ages

- Crater counting: the crater densities can be calibrated against the Moon, where absolute ages have been obtained from *Apollo* samples.

In situ dating: relative ages

- Appearance: When viewed close up, the degree of erosion and modification (for instance by landslides or sapping) of crater walls and other features can be appreciated in ways not possible from orbit.
- Stratification: layered structures appear all over Mars, and it is generally safe to assume that the newer layers are on top of the older ones.

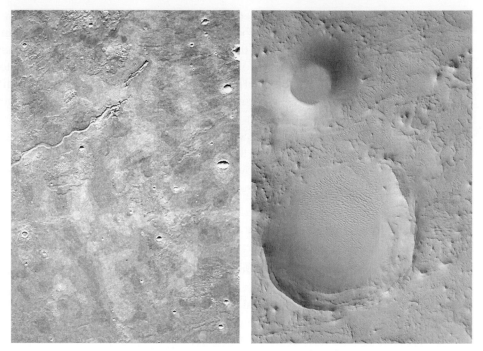

Figure 5.6 Old and young terrain: the northern flank of the volcano Elysium Mons on the left, and part of the region called Arabia Terra near the crater Schiaparelli on the right. Detailed crater counts, their diameters, and apparent states of degradation allow the ages of these regions to be estimated as forty to two hundred million years and three to four billion years respectively.

In situ dating: absolute ages

- Radioisotope dating: The ages of rocks can be determined with high precision in the laboratory (in principle the laboratory could be on Mars, although the technique will initially involve sample return to Earth) from the ratio of radioactive elemental isotopes that decay slowly from one form to the other (from potassium to argon, for example).

Application of the crater-counting technique for dating the surface of Mars led to the identification of three epochs in the planet's history, each corresponding to a characteristic part of the surface, shown simplistically in Figure 5.7. The youngest terrain is mostly in the northern hemisphere, above a boundary inclined at about thirty-five degrees to the equator. From the sparsely cratered nature of the volcanic plains found here the surface is considered to be relatively young, forming in the period comprising the last billion and a half years before the present. This falls within the period called the *Amazonian*, named after a typical and representative region, Amazonia Planitia.

South of the division the surface is heavily cratered, resembling the lunar uplands, and contains the huge multi-ringed basins Hellas and Argyre that were

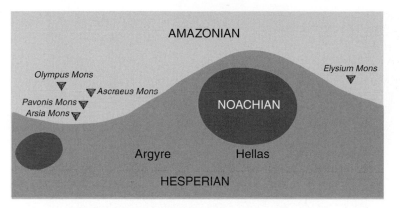

Figure 5.7 This simplified map of Mars shows the north-south dichotomy, between relatively young terrain (Amazonian) in the north, and older (Hesperian) in the south. The two large dark blobs mark the most ancient terrain, the Noachian. Their estimated ages are as follows:

Amazonian: Up to 2.9 billion years
Hesperian: 2.9 to 3.7 billion years
Noachian: More than 3.7 billion years

The same names are also given to the three principal geological epochs on Mars. The locations of the five largest volcanoes, and the two largest impact basins, are also shown for orientation. (Modified from Malin Space Science Systems website)

clearly produced by very large impacts. This is the terrain seen by the first three spacecraft to fly past Mars, when it suggested to the scientists analysing the data from those early missions that the planet had an ancient surface with negligible evidence for any recent evolution. Today the age of this part of Mars is conventionally divided into two distinct periods, the *Noachian*, named after a representative region of ancient cratered highland terrain formed during heavy asteroidal bombardment 3.7 to 4.6 billion years ago, and the *Hesperian*, a period of less heavy bombardment 2.9 to 3.7 billion years ago. Hesperia Planitia is an example of the cratered plains that date back to that time, shown in the image in Figure 5.8.

It is sometimes found convenient to further subdivide the oldest epoch, the Noachian, into three periods. The *early* Noachian extends from accretion of the planet to the time when its main features formed and most of the primitive atmosphere had escaped, estimated to range from 4.5 to 4.3 billion years ago. In the *middle* Noachian, (4.3 to 3.8 billion years ago), a secondary atmosphere was formed from gases exhaled from the interior, the internal magnetic field shut down, plate tectonics ceased, and an ocean may have occupied the large basin covering much of the northern hemisphere. Finally, in the *late* Noachian, from 3.8 to 3.7 billion years ago, the formation of the great valley networks was complete and climate change, leading to the loss of free water on the surface, occurred.

The team operating the *OMEGA* visible and infrared mapping spectrometer on board the *Mars Express* orbiter has proposed a different classification, based

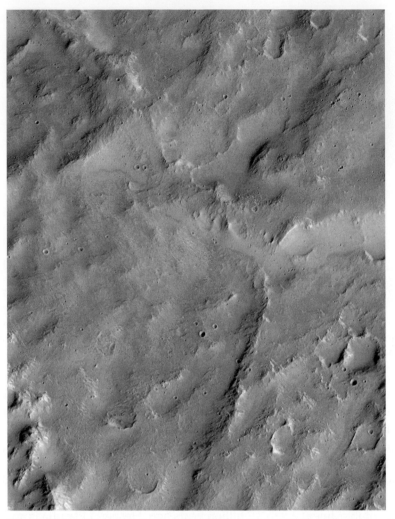

Figure 5.8 The Hesperian dates notionally from 2.9 to 3.7 billion years ago, although some estimates say it ended as recently as 1.8 billion years ago. Studies of Hesperia itself, like this one by the US Geological Survey, suggest that some areas may be even younger, possibly dating from only a few million years ago. Clearly, the dating of large swathes of the surface into epochs remains an approximate and uncertain business.

upon the predominant mineralogical composition of different regions, rather than their cratering record. This also divides the history of the planet into three epochs, which they call the Phyllosian, Theiikian and Siderikian. Figure 5.9 shows the periods of time covered by each, compared to the more usual description, from which it can be seen that the two different approaches lead to separation into roughly the same three eras. This is not a coincidence, of course, since it might be expected that mineral formation is related to the age of the surface through the different processes that occurred in different epochs.

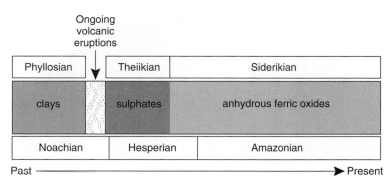

Figure 5.9 The alternative scheme for dating surface features on Mars proposed by the *Mars Express OMEGA* team in 2006, which is based on the formation of clays on a warm, wet Mars, followed by a period when volcanoes were active and sulphates formed, then the recent, drier era. The standard scheme, based on the age of the surface as inferred from its appearance and degree of cratering, is shown below.

The *Phyllosian* era is named after the clay-rich minerals called phyllosilicates that formed when the surface was warm and wet, presumed to be during the first five hundred million years of Martian history. The *Theiikian* is named after the Greek name for the sulphate minerals that were formed during the period of maximum volcanic activity, lasting through the next five hundred million years until about three and a half billion years ago. The *Siderikian*, from then until the present, saw the formation of the eponymous red iron oxides that give the planet its familiar colour by interactions between the atmosphere and the surface in the absence of liquid water or large quantities of volcanic effluent. This puts the main epochs in sequence and challenges experimenters to put more precise dates on when each occurred.

Obviously, the crater-counting method has a number of fundamental limitations. Some events may not be amenable to statistical analysis, for instance the very largest impacts like that which produced the vast Hellas basin, seen in Figure 5.10, or the one that appears to have removed much of the crust over the north of the planet, leaving it thinner than that in the south, as depicted in Figure 5.11. Even more seriously, several time-dependent phenomena, not just the impact rate, are often interleaved and separating them involves making assumptions about the very time scales under investigation. The surface elevations are lower in the north, with the north pole itself about six kilometres closer to the centre of the planet than the south pole. The lava from active volcanoes would tend to flow into the northern plains, reducing the height anomaly and at the same time covering up the early cratering. Liquid water may also have filled the entire region, further eroding the features on the surface and covering them with sediment. After the water disappeared and volcanism subsided the surface acquired a fresh population of craters, fewer in number and younger-looking than those that survive on the more ancient terrain in the south.

Figure 5.10 The Hellas basin has a diameter of two thousand, three hundred kilometres, is up to eight kilometres deep and is surrounded by an enormous ring of debris, piled two kilometres high and extending four thousand kilometres from the centre, a remnant of the massive collision that produced the crater.

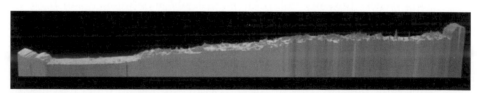

Figure 5.11 A pole-to-pole section of Martian topography from the north (left) to south (right) along the zero degree longitude line. The vertical scale is exaggerated over the horizontal scale by a factor of one thousand.

On the other hand, the processes that add confusion to the crater-counting method provide at the same time different opportunities to estimate relative ages. Since surface water vanished around the end of the Noachian period, and volcanism declined gradually right through the Amazonian, thermal and chemical erosion, cracking and landslides, and aeolian (wind-driven) processes have continually modified the surface everywhere. Large quantities of material are steadily moved to the poles, by the prevailing circulation of the atmosphere and by the enhanced deposition that occurs there, particularly in the very cold polar night when water and carbon dioxide freeze to form heavy layers of ice on airborne dust particles, causing them to precipitate. The layers that have accumulated in this way as thick sedimentary deposits under the icy polar caps must contain a detailed, and ultimately accessible, record of periodic and secular climate change during the history of Mars. Some of this laminated terrain can be glimpsed in high-resolution photography from orbit. Core samples will reveal much more, and the layers will contain bubbles of ancient atmosphere of identifiable age whose composition and isotopic ratios will provide a mine of information, just as similar samples have told us about the climate history of the Earth.

5.5 Towards an absolute chronology

Once geologists have samples of carefully determined provenance from the surface of Mars in their hands and in their laboratories, they can use radioisotope techniques to determine the age of many types of rock. Until then, they must rely principally on models of the rate of impact as a function of time, produced by extrapolating from the corresponding rates at the Moon. The absolute chronology that has been established for the Moon uses radiometric dating of rock samples collected by the *Apollo* astronauts. The rocks then have to be related to the impact crater densities observed in images of their collection sites.[2] The resulting curves that describe the cratering rate as a function of time can then be scaled to Mars, allowing for factors such as the difference in impact rates between the planetary bodies; the nature of crater-forming projectiles; and the effects of differing gravity, impact velocity and surface and atmospheric properties on impact crater formation. Figure 5.12 shows the general form of a plot of crater density versus crater size; each such curve is taken to be characteristic of a particular epoch when bombardment was taking place. Surfaces with few or no craters are 'young', but obviously cannot be dated any more precisely than that using this method alone.

The need for these corrections introduces further uncertainties. For instance, after the *Viking* orbiters had surveyed the planet it was believed that there were

---

[2] The coverage of the Moon is, of course, incomplete and inadequate and this is another source of error that has to be carried forward to the dating of Mars by the crater-counting method.

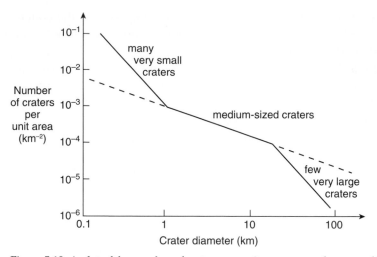

Figure 5.12 A plot of the number of craters per unit area versus the crater diameter can be used to estimate the age of a feature or region on cratered bodies like Mars. The basic data come mainly from the Moon: originally, when only telescope observations were available, it was thought that the crater density versus size followed the simple rule represented by the dashed line. As better data have been accumulated using spacecraft, it has become clear that large craters are less common and small craters are more common, on both Moon and Mars, than this would imply, and a three-branched curve is a better fit.

no craters on Mars smaller than fifty metres in diameter. This seemed rational, because calculations suggested that smaller bodies would burn up in the atmosphere before impacting. However, the higher resolution pictures from recent missions show plenty of small craters, some only a few metres in diameter. Lava flows, wind erosion and dust deposition can remove or conceal craters entirely, while researchers suspect that some surfaces were protected from cratering for an unknown amount of time and then uncovered, perhaps during major storms that lifted exceptionally large amounts of dust. Other potential pitfalls include counting 'secondary' craters produced by large solid fragments thrown out during the impact that formed larger craters nearby. Conversely, older, smaller craters may be covered up by the very extensive lobes and rays sometimes produced if the ejected material from a large impact is 'muddy' rather than rocky. Mountainous regions are more prone to the production of secondary craters, low-level plains with water-ice deposits below the surface, which melt on impact to generate mud.

For these and other more subtle reasons, the scale of Martian ages obtained by the crater counting method depends on the quality of the data available and the terrain being studied. Even under the most ideal conditions, the dates remain highly uncertain, and subject to regular revision and updating. For instance, the Hesperian was at one time proposed to be from 1.8 to 3.5 million years ago, rather than the more recent 2.9 to 3.7, and some recent work has suggested that it contains areas that have been resurfaced by volcanic lava much more recently

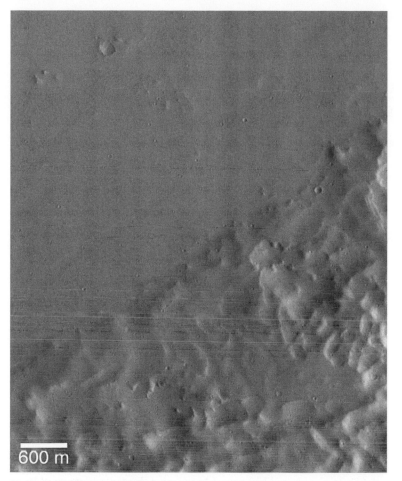

Figure 5.13 This *Mars Orbiter Camera* high-resolution image shows cratering on the Amazonis Planitia lowlands towards the top of the frame, and on the Lycus Sulci uplands below. The ridge that separates them is part of the putative shoreline that may once have contained the northern ocean.

than either of these ranges. As a more specific example, to show a typical range of uncertainties for an individual well-defined region, here is what an expert in the technique[3] wrote about features in Amazonis Planitia, part of which is shown in Figure 5.13: 'The 'older' lava flows appear to fit well the isochrons giving a model age of about one hundred million to one thousand million years, interpreted as the true age of the flows, and the younger flows give a model age of about nine million to ninety million years for the youngest flows.' The 'isochrons' are the curves of crater density versus crater diameter relating to a given age, shown in Figure 5.14.

---

[3] Dr William K. Hartmann of the Planetary Sciences Research Institute in Tucson, Arizona. See http://www.psi.edu/projects/mgs/chron04a.html.

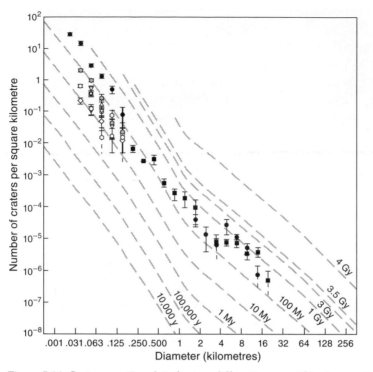

Figure 5.14 Crater-counting data for two different areas within Amazonis Planitia, plotted on 'isochrons' similar to those in Figure 5.12. The fits suggest older lavas (solid points) date back to a few hundred million years, and younger lavas (open points) date from a few tens of millions of years. (After W. K. Hartmann, Planetary Science Institute, Tucson)

The absolute geological history of Mars, and the time history of its evolution, will be known much better when crater counts and regional characteristics are tied to an absolute chronology derived from Mars itself and not the Moon. The only samples available at the present time, the SNC meteorites, are not very useful for this purpose because there is no information about which part of Mars they came from. They do show a wide range of radiometric ages: some, perhaps originating on the ancient cratered highlands, being as old as four and a half billion years, close to the date of formation of the planet. Others (recalling that all of the SNCs are clearly of volcanic origin from their mineral composition) provide evidence that there was some volcanic activity, at least, as recently as about 175 million years ago. This is consistent with the crater count for some lava flows that suggests they are less than ten million years old.

Potassium–argon dating is accurate for ages from one million years ago, back to the age of the planet. A related method, the analyses of changes induced by long exposure to cosmic-ray fluxes on the surface, can date rocks in the age range from a hundred thousand to ten million years ago. Radiometric dating can be applied to a wider range of decaying elements and their products, including the rubidium–strontium and uranium–thorium–lead sequences. Robots can be devised to make these determinations on Mars, but the experts believe that sample return will always offer more accurate and precise radiometric age determination than can ever be achieved *in situ* because the samples can undergo intensive laboratory examination and more refined methods can be used. *In situ* analyses will improve with time and experience, of course, but laboratory analysis will always be a jump ahead, at least until the time when manned laboratories on Mars have the same equipment as the latest facilities on Earth. This probably will have to be the case before there can be a full and detailed chronology for the geological evidence available across the whole planet and deep below the surface.

5.6   Volcanoes and volcanism

Because it is smaller, and possibly has less internal heat release, Mars probably cooled faster than Earth. The convective activity that still produces plate tectonics on the Earth apparently stopped early in the history of Mars. Volcanoes then took over as the main route for the remaining energy to escape from the interior, the accompanying lava flows building up the large volcanoes at the weakest points in the crust where there were cracks or vents.

Evidence for past Mars volcanism is found today over a wide range of latitudes, but is concentrated in three main regions. The most dramatic of these is Tharsis, with the three giant volcanoes Arsia Mons, Pavonis Mons and Ascraeus Mons aligned in a north-east–south-west direction across the equator on the highest part of the dome, from where they stretch up a further nineteen kilometres to their summits, each bearing a caldera several tens of kilometres across. Olympus Mons, on the north-western flank of Tharsis, is the largest known volcano in the Solar System, with a base diameter of about six hundred kilometres and a summit rising twenty-five kilometres above the surrounding plains.

The existence of the great Martian volcanoes, some of them ten to a hundred times larger than those on Earth, implies a greater thickness or rigidity of the crust on Mars so a stable base was available for their formation and support. They must also have formed from massive eruptions that were sustained over long periods of time. On Earth, the source regions for some kinds of volcanoes move laterally as lithospheric plates travel across the stationary mantle plumes beneath them. This plate tectonic action produces chains of volcanoes, with Hawaii the best-known example. Lacking this lateral movement, the Martian surface must have remained above the plume source, so that huge volumes of lava can accumulate over many millions of years of activity generating single shield volcanoes of enormous volume. Lava flows on the flanks of the larger volcanoes can be traced for hundreds of kilometres; to be able to flow for such distances the lava must have been extremely fluid. This is consistent with the finding that the Martian shields, like those on the Earth and in the lunar maria, are composed of basaltic rocks with a low silica content that flow easily.

The volcanic Elysium region is located on a broad dome about two thousand kilometres across located within the mid-latitude plains of the northern hemisphere. The two large shield volcanoes here, known as Elysium Mons and Hecates Tholus, are significantly smaller than those in Tharsis. The surrounding lava plains, which are generally elevated by several kilometres relative to the mean height of the Martian surface, cover an area about a quarter as great. To the south of the Elysium volcanic rise lies the Elysium basin, a depression three thousand kilometres wide resembling the site of a vast, deep lake, although whether of water or lava was unclear until recently. The floor of the basin looks young, with few fresh impact craters, suggesting that it became dry only relatively recently. It also appears to have drained from its east end, flowing north-east for hundreds of kilometres in a channel, Marte Vallis, which contains streamlined islands. Despite this apparent evidence for water flowing from the basin, the pictures from *Mars Odyssey* like that in Figure 5.15 show clearly that the channel floor is covered with giant plates that formed as the crust on a lake of molten lava.

The third principal region of volcanism lies within the ancient cratered terrain around the Hellas basin in the southern hemisphere. Here several volcanic structures are found that have central depressions and a radial pattern of grooves, each tens to hundreds of kilometres long, which clearly distinguishes them from impact craters and, at the same time, from the northern shield volcanoes. The concentration of impact craters and the greater degree of erosion show these volcanoes to be much older than those in Tharsis and Elysium. They also have much shallower profiles, perhaps because they date back to a time when the crust may have been thinner and possibly unable to support large mountains.

In addition to the three regions containing massive volcanoes, it is now apparent that the surface of Mars is studded with smaller vents, tubes and flows that were invisible in the earlier high-resolution images. So numerous are these that it seems likely that much of the resurfacing of the fluvial plains was due to lava from multiple small vents rather than from the relatively small number of giant volcanoes. It should be possible to understand these processes by measuring the composition and ages of solidified lava flows in and around the Martian volcanoes, the sort of task that could most efficiently be carried out by a human expedition. The results will also help to establish the extent to which volcanoes released water and atmospheric gases that may have supported past and present life. And of course, any sites they may find where there are still traces of volcanism will be prime places to look below the surface for signs of life.

5.7 Water on the surface

The presence of the two long-lived rovers, *Spirit* and *Opportunity*, on the surface of Mars provided a new perspective on its geology and history. Perhaps their

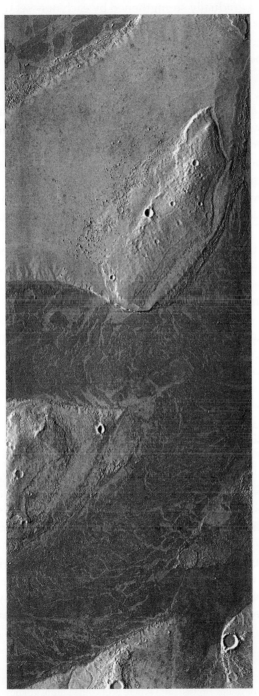

Figure 5.15 The lava flow in Marte Vallis, seen here in a visible-wavelength image about twenty kilometres across by the *THEMIS* (*Thermal Emission Imaging System*) instrument on *Mars Odyssey*, has frozen into large plates.

greatest achievement has been to confirm beyond doubt the impression gained decades ago from *Viking*, and gradually strengthened by each new set of images from orbit, that the surface of Mars has been extensively modified by the action of liquid water. The *Viking* scientists were sure, in the 1970s, that running water must have formed the channels, some very large, that they saw running from the southern highlands to drain into the northern lowlands. What was less clear is how long the wet era lasted and how the water cycle between surface, interior and atmosphere worked. On what scale were there large bodies of standing water, where on the planet were they, and were they 'flash floods', or did they endure for an extended a period of time? It is still uncertain whether the surface was exposed to a warm atmosphere at the time, or covered by an icy layer that insulated it from a cold climate similar to today's, and allowed water to flow underneath. There is some evidence for rainfall, which would support the warm scenario, but it is far from clear whether rain was a common occurrence in significant amounts.

The much greater abundance of sulphates compared to carbonates in the rocks examined by the rovers indicates that some, if not all, of the water in which they formed or evolved was acidic. This is a characteristic of groundwater driven from the regolith by internal heating, or of moisture that has been cycled through the atmosphere at a time of extensive volcanic activity. In addition to carbon dioxide and other gases, the volcanic emissions would have included water vapour and sulphur compounds, probably mostly sulphur dioxide, that then washed out of the atmosphere as acid rain, similar to the notorious terrestrial variety but more concentrated, and of natural rather than anthropogenic origin. On the surface, the acid solution would dissolve other minerals to form sulphates, which form solid deposits when the water evaporates leaving what one investigator called a 'bathtub ring' around the previously water-filled basins on Mars.

Perhaps the surface was only wet while the volcanoes were at their most active, contributing more water from the volatiles dissolved in the escaping magma, and releasing gases that kept the atmospheric pressure up. Certainly, the long, meandering outflow channels are easier to explain if the prevailing temperatures and pressures were much higher than they are now, even if the flowing water was driven from subsurface reservoirs by geothermal heating from below. Perhaps a warming trend induced by the slow, astronomical variations in Mars' orbit coincided with the period of high volcanic activity when trillions of tons of gases were emitted into the atmosphere,[4] maintaining a powerful greenhouse effect.

---

[4] The amount of gas that Martian volcanoes emitted, and when, is of course quite unknown, but rough estimates can be made. For instance, during the Permian extinction on Earth two hundred and fifty million years ago, the lava that covered part of modern Siberia is estimated to have been accompanied by the release of ten gigatonnes of carbon dioxide. A single sustained eruption of this

The *Spirit* landing site, Gusev Crater, is a bowl a hundred and fifty kilometres across that appears to have been the receptacle for water discharged from Ma'adim Valles, one of the largest of the dry 'river valleys' seen from orbit. Despite the apparent history of the crater, the rover found mainly olivine-bearing basalt of volcanic origin in Gusev, and no evidence for the clays that would mean flowing or standing water. When it climbed into the Columbia Hills, however, layered rocks were seen with spherules in the gaps, and deposits of iron and magnesium sulphate were found, both suggesting the action of water. This apparently contradictory finding is explained by crater counts, which show that the hills are a billion years or so older than the plains. The volcanic activity that produced the soil in the plains must have occurred after running and standing water had vanished from the scene, covering the evidence for water in the lowlands that remains exposed on the surface in the hills.

Opportunity landed on the other side of Mars in a small impact crater on the surface of Meridiani Planum, which is a plain about the size of England and Wales. In and around Eagle Crater the evidence for water was much more abundant than in Gusev, with spherules of grey haematite ('blueberries') on the surface, having apparently been deposited after the soft rocks in which they formed eroded away. Other spherules were seen inside layered deposits, in the cracks where they formed as water percolated through the existing rock. Later in *Opportunity*'s mission, as it climbed to higher ground, it found that the spherules were less common, suggesting that the higher terrain in this area formed later and is therefore younger. Meanwhile, deposits of blueberries were found by *Spirit* at the Gusev site as well. Evidence from the rovers suggests that the groundwater rose and subsided periodically, presumably related to the changing temperature during major climate cycles.

As the two *Exploration Rovers* soldiered on, far beyond their design lifetimes, the evidence for extensive, long-standing seas of surface water on ancient Mars accumulated. The minerals and concretions being found, including the huge amounts of sulphates, showed clearly that rocks formed by deposition in standing water, followed by evaporation, are common at some places on Mars. The debate goes on as to whether the water percolated upwards through the soil, perhaps through deep beds of volcanic ash deposits, or fell as rain and flowed as rivers into the basins where it evaporated to leave the sedimentary rocks, or of course both. In any case, the formation process evidently took many millions of years and might have provided a suitable habitat for life during this time.

Some of these habitats may be exposed on the surface, but it seems much more likely that they are buried at some depth. Ancient lake, river and sea beds are low-lying regions that get filled by wind-blown dust and eroded material from

---

size on early Mars would have released the equivalent of the present-day Martian atmosphere in about a thousand years, while a hundred of them would have raised the surface pressure on Mars to that of today's Earth in about the same period of time. Mars could well have been at least this volcanically active four billion years ago.

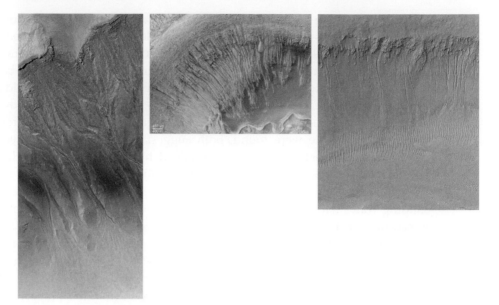

Figure 5.16 Evidence for near-surface water in recent erosion features found in three different locations: (left to right) Noachis, Newton Crater and Nirgal Vallis.

nearby higher ground. While there has been no rain, for billions of years at least, frost still occurs and helps to cement the deposits and assist slow chemical processes to transform dust and sand into soil and rock. The water that once flowed on the surface in large amounts is likely still to be present at depth, as permafrost and ice. The idea that some of it may be liquid water is supported by the discovery of erosion of canyon walls and gullies, apparently by liquid that relatively recently seeped down a cliff face. Such evidence has been spotted from orbit in several locations around Mars, three of which are shown in Figure 5.16, and it looks like gullies may be quite common. If so, there may be a lot of liquid water below the surface, some of it just a few metres deep, warmed by geo-thermal energy from residual volcanism. The *Mars Exploration Rovers* have not been near to any of these surface manifestations of hydrological activity, leaving them as likely targets for future robotic explorers, equipped with drills and other tools for subsurface exploration.

As discussed in more detail in the next chapter, massive seasonal cycles occur in the present-day Martian atmosphere, dominated by the behaviour of carbon dioxide, water vapour and dust. These include important interactions with the surface, resulting in the ongoing erosion, transport and deposition of massive amounts of solid and volatile material even in recent times. One spectacular result of this is the existence of the layered terrain in the polar regions and other ancient layered deposits of sedimentary origin, which indicates that climate cycles lasting longer than a Martian year are important. The amount of airborne dust is highly variable, with huge global dust storms that occur in some years but not in others. Triggered by unstable atmospheric conditions, feedback sustains them until a heavy pall of dust covers the entire planet, and then something suddenly switches them off. Why they stop as suddenly as they began, and what role they have had, over the ages, in determining the geography of the Martian surface, is still not clear.

Atmospheric water vapour is the single most important contributor to the greenhouse effect on present-day Earth, and it may have been on early Mars as well, if the atmosphere was warmer and more humid. However, the amounts of water required to explain Martian features like the large channels and the Northern Ocean are far greater than can be found today, whether as water vapour and ice in the atmosphere, seasonal water ice deposits at the surface, or permanent water ice deposits in the polar caps. The latter contain most of the observable water on Mars; according to thickness measurements by laser altimetry from the *Global Surveyor* orbiter, there is enough to make a global water layer only about thirty metres thick. This is about ten times less than the amount needed to fill the Northern Ocean basin up to its observed coastlines. However, this calculation does not take account of the unknown amounts of water that are hidden below the surface. Some is in the layered deposits surrounding the north and south polar caps, and probably much more is at lower latitudes in near-surface ground ice and permafrost similar to that found on Earth. The betting at present is that a final inventory will show that sufficient water is still present on the planet to explain the formation of rivers and oceans long ago. If the small gullies on crater walls that seem to show recent erosion by fluids mean there is an active water cycle on present-day Mars, features of intermediate age should be present as well and eventually will be found so the entire history can be traced.

Because it is so cold, the modern Martian atmosphere contains very little water even when the relative humidity is close to saturation. Thus, it can no longer play a huge role in transporting water around the planet, at least not on annual time scales. The amount of seasonal movement of water vapour, for example between the relatively warm equatorial zone and the polar 'cold traps', is still unclear. Some transfer between the summer and winter hemisphere apparently occurs, and studies with circulation models suggest that not all of it is reversible with season, and if so then the north polar cap is gradually getting more massive at the expense of that in the south. Along with trying to find out how much water is present today in hidden reservoirs, it is important to understand how well such reservoirs are connected with the surface and atmosphere. Measurements of the composition of the atmosphere, including humidity and isotopic abundances, plus maps of the global distribution of ice and liquid water in the regolith using gamma-ray spectroscopy and sounding radar, will provide answers when backed up by mobile platforms and drills at a selection of the most promising sites on the surface to search for aqueous minerals. These minerals include not just the more obvious ones such as clays, carbonates, sulphates, phosphates, silica and haematite, but also various sulphides, borates and halides. Water-formed geomorphic features such as shorelines and river deltas require detailed examination to establish when they formed and dried up, to date the sequence of hydrological activity, and to help to calibrate the geological time scale for the whole of Mars.

5.8 The polar caps

In addition to the permanent (also called the 'residual' or 'perennial') polar caps, which are mostly water ice, Mars has seasonal caps that are composed of frozen carbon dioxide ice ('dry ice'), the solid form of the gas that is the principal constituent of the atmosphere. Near the poles in winter, when the Sun remains below the horizon for months on end, the surface temperature plunges. It soon reaches a hundred and twenty-three degrees below zero centigrade, the condensation temperature of the carbon dioxide atmosphere at the mean surface pressure of six millibars, and the atmosphere begins to precipitate out as a blizzard of carbon dioxide snow. Spring arrives after about a quarter of the whole atmosphere has frozen out in this way, and the process begins to reverse, soon to start again at the opposite pole. The polar caps do not entirely disappear in summer since the relatively small perennial caps remain, because the water ice does not get warm enough to melt or sublime away. The southern perennial cap, in particular, has a complicated local meteorology that results in some solid carbon dioxide being permanently retained in places, along with the water ice. Both kinds of ice trap dust from the air in various amounts and this forms sedimentary layers of solid material when the seasonal caps evaporate. Longer-term climate change produces thicker layering effects that are etched on a sufficiently grand scale as to be visible in pictures taken from orbit.

5.9 The moons of Mars

Mars' satellites Phobos and Deimos do not have the key role in understanding the history of the evolution of their parent planet that Earth's large moon does. Visible only as points of light from Earth, even through a large telescope, photographs like those in Figure 5.17 from closely approaching spacecraft show them to be small, dark, irregular bodies. They are probably not original members of the Mars system at all, but rather asteroids captured by the planet in the distant past. Phobos has a low density, less than two grams per cubic centimetre, which is only about half that of Martian surface rocks, but typical of some classes of asteroid. The Martian moons were probably just two of many such objects that encountered Mars as they drifted through the inner Solar System. They survive as they do because they approached the planet on a trajectory that allowed them to enter orbit, rather than miss or crash into the planet as most such interlopers did (and, to a much lesser extent, still do).

Figure 5.17 Asteroid 951 Gaspra (top) compared with the Martian moons Phobos (right) and Deimos (left) at the same scale. The illuminated part of Gaspra is about seventeen kilometres long.

While they are of only indirect interest in the exploration of Mars, Phobos in particular may prove valuable as a base or fuelling station during missions to and from the surface of its adopted parent. Phobos is just over thirteen kilometres in its longest direction and Deimos about seven and a half; they orbit Mars within a degree or so of its equatorial plane with periods of a third of a day and about a day and a quarter respectively. Tidal forces keep the rotational periods the same as the orbits, and so, as with Earth's moon, the same side continually faces the planet, a useful feature for any future observatory or way station. The same forces are gradually moving Phobos closer to Mars, and Deimos further away, since they are respectively below and above the 'geostationary' height at which a body would orbit with the same point on the surface always directly below.

They may have little to do with Mars itself, but that is not to say that the Martian moons lack intrinsic scientific interest. We have seen that the Russians have been intrigued enough by Phobos to devote a large and expensive mission to studying it close-up, and although that failed, the interest has remained and many of the Russian scientists remain keen to try again. They point to the fact that they are primitive bodies that have been little modified, except by impact cratering, which helpfully exposes the subsurface. Spectrally, the moons look a great deal like each other, suggesting a similar composition, which appears to be close to that of the well-known, but relatively rare, type of meteorite known as carbonaceous chondrites. About five per cent of all known meteorites are of this type; they have low densities, and often contain a lot of water and organic material. What differences might be present between meteorites and a much larger body like Phobos formed out of some of the same material is the subject of some interesting speculation that will only be resolved by returning samples, preferably deep cores. Sample return from Phobos is much easier than from the surface of Mars, and the Russians have said they may make the attempt quite soon.

Phobos is small but so close to Mars that it would be seen by an observer on the surface, as it races overhead several times a day, as a disc nearly one-third of the diameter of the Moon as seen from the Earth. The tidal stresses of occupying such an orbit mean that Phobos cannot survive indefinitely; it has been estimated that it will break up or crash onto Mars in less than fifty million years from now. A large crater ten kilometres in diameter dominates the surface appearance of the moon. This has been named after Angeline Stickney, the wife of American astronomer Asaph Hall, who discovered Phobos in 1877. The impact that produced Stickney must have come close to breaking Phobos apart, and large cracks radiating from the crater can be seen in high-resolution images obtained from spacecraft that passed close to this moon while orbiting Mars.

The long-standing theory that the Martian moons were captured from the asteroid belt invites comparison to those current members of the asteroid belt for which there are close-up observations. Gaspra, an S-type asteroid[5] orbiting near the inner edge of the belt, was photographed from a distance of sixteen thousand kilometres by the *Galileo* spacecraft on 28 August 1993, when the spacecraft was on its way to orbit Jupiter. All three bodies are rocky, irregular and pitted, but otherwise rather dissimilar in appearance, perhaps reflecting the fact that they evolved in different parts of the Solar System. Also, Gaspra apparently has a quite different composition from the Martian moons.

A detailed study of the composition of Phobos and Deimos may in time reveal how and when they came to orbit Mars, and provide one more piece in the jigsaw that is the history of the Solar System.

Further reading

The Surface of Mars, by Michael H. Carr, Cambridge Planetary Science Series, Cambridge University Press, 2007.

---

[5] Gaspra was discovered in 1916 by a Russian, G. Neujamin, who named it after a Black Sea resort between Sevastapol and Yalta in the Crimea. An S-type asteroid is one with a high silicate rock content, as opposed to the carbonaceous or C-type, which both Phobos and Deimos resemble.

Mars: An Introduction to its Surface, Interior and Atmosphere, by Nadine G. Barlow, Cambridge Planetary Science Series, Cambridge University Press, 2008.

The Martian Surface: Composition, Mineralogy and Physical Properties, edited by Jim Bell, Cambridge Planetary Science Series, Cambridge University Press, 2008.

The Cambridge Photographic Atlas of the Planets, by Fredric W. Taylor, Cambridge University Press, 2001.

Spectroscopic identification of carbonate minerals in the Martian dust, by J. L. Bandfield, T. D. Glotch, and P. R. Christensen. *Science*, 301, 5636, 1084–1087, 2003.

Orbital identification of carbonate-bearing rocks on Mars, by B. L. Ehlmann, J. F. Mustard, S. L. Murchie, F. Poulet, J. L. Bishop, A. J. Brown, W. M. Calvin, R. N. Clark, D. J. D. Marais, R. E. Milliken, *et al.*, *Science*, 322, 1828–1832, 2008.

Global mineralogical and aqueous Mars history derived from OMEGA/Mars Express data, by J.-P. Bibring, Y. Langevin, J. F. Mustard, F. Poulet, R. Arvidson, *et al.*, *Science*, 312, 5772, 400–404, 2006.

Chapter 6
The changing climate of Mars

6.1   An introduction to the Martian climate

We saw in the previous chapter that many evocative features, like the example in Figure 6.1, are found on Mars that speak to us of a more 'Earth-like' climate in the past. However, exactly what this means is far from clear, except that the atmosphere must have been denser and the surface warmer, so that liquid water could have been a prominent feature of the surface environment. What details can be divined about the conditions that prevailed then, and how long they lasted? In particular, how long did the transition to the present state take, and what processes were involved in producing the changes? Finally, given that there has been climate change in Mars' history, it may still be going on. How stable is the climate today?

A good place to begin to unpick this mystery is by defining more carefully what is meant by 'climate'. On Earth, the term refers principally to the surface conditions, especially to the temperature and other factors that depend on the atmosphere, such as cloudiness and precipitation. Usually, climate data are based on seasonal or longer-term means, with the minor fluctuations that we call 'weather' averaged out. Mars has seasonal cycles similar to those on the Earth, with the summers and winters in each year resembling the same season in recent previous and successive years. There are important inter-annual variations on Mars, just as there can be, for example, unusually mild or dry winters at any particular location on Earth, but in general these variations are much smaller than the difference between summer and winter in any given year.

As recent problems predicting global change on the Earth have demonstrated, the climate on any planet is the result of a range of complex, interacting effects on a wide range of time scales. The Martian climate system is simpler than Earth's in that there are no liquid oceans, at the present time at least, no vegetation, and no volcanic or anthropogenic emissions. However, it still includes the surface, the atmosphere, the polar caps, and accessible regions of the subsurface. The interactions between these give rise to the fields of temperature, pressure, wind, humidity and dust loading that can be measured and, eventually, understood.

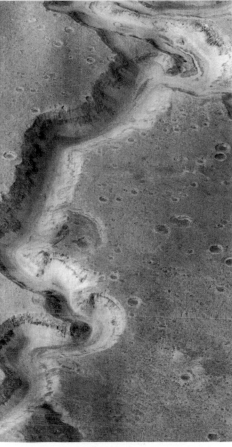

Figure 6.1 The canyon named Nanedi Vallis cuts through cratered plains in the Xanthe Terra region, and is over two kilometres wide. It seems inconceivable that this can be anything other than an ancient riverbed, cut by water that flowed in large quantities over an extended period of time. Such behaviour would not be possible on Mars today, leading us to ask when, and why, did the climate change?

One powerful approach is to compare observations to theory, using numerical simulations on computers to represent the latter. These simulations or 'models' of the climate on Mars can draw heavily on the vast amounts of work that have been done to try to understand the climate on the Earth, since the physical processes involved (including radiative transfer, molecular physics and thermo-dynamics) are the same. The actual climate that the planet experiences at any given time is the result of a complex balance between mechanisms that heat and cool the atmosphere, add or remove water or dust, create clouds and ice, and so on.

A question that arises whenever climate is considered, on the Earth, Mars or another planet (Venus is an interesting example), is how stable the climate is against gradual or sudden change (where 'sudden' may mean time scales of a few decades or centuries). It is believed that the terrestrial climate has stayed reasonably constant, within the familiar seasonal cycles, over the last few centuries at least. In contrast, the Earth experienced ice ages in the more distant past, until as recently as about ten thousand years ago, and it now appears that these sometimes involved almost incredibly sudden changes, with seasonally averaged temperatures changing by ten or twenty degrees in as many years. Recently it has gradually come to be accepted that human pollution of the air is driving a process of global warming that is expected to have serious effects on life on Earth before the present century ends. Mars is unlikely to have had the latter problem, but it clearly has had its own version of the former, and it would be very nice to know more of the details.

The evidence on Mars for climate variability can be considered on three time scales: the present-day climate, with its interannual variations and trends in polar cap ice cover; quasi-periodic climate change on scales of tens or hundreds of thousands of years, producing the layered terrain that can be seen even from orbit; and long-term change (gradual or sudden) over a billion years or more from the fascinating 'warm and wet' era of (probably) several billion years ago, to the cold, dry regime of more recent times.

Interannual (year-to-year) variability was seen on Mars long before the first spacecraft arrived. Telescopic records of what we now know were vast dust storms exist from before the time of Lowell and Schiaparelli. In modern times, multi-year surface pressure measurements acquired at the *Viking* landing sites, orbiter observations of the variations in the seasonal and residual polar caps, and large variations in atmospheric water vapour measured from the Earth, record the variations from one year to the next. Computer models of the atmosphere, primed with observational data, indicate that these are produced by interactions among the cycles of carbon dioxide, dust and water in Mars' present climate.

Climate change on the Earth is due at least in part to quasi-periodic variations in its orbital and axial elements produced by gravitational interactions with other Solar System bodies and having time scales of tens to hundreds of thousands of years. Mars has a similar behaviour, but the magnitudes of these variations for Mars are significantly greater, for various reasons including Mars' orbital eccentricity and relative closeness to the giant planet Jupiter. The consequent changes to the solar heating, especially at high latitudes, are bound to have caused significant changes in the seasonal cycles over the ages.

How this relates to the evidence exhibited by a wide range of surface features on Mars for warmer climatic conditions at an early date in the planet's history is not yet understood. While it is entirely possible that Mars has enough water hidden below the surface to produce a warm and wet climate, the problem is that at Mars' distance from the Sun the stable form of water is not as liquid on the surface or vapour in the atmosphere, but condensed into the solid form as ice. Conditions would have had to have been very different, and would probably have had to include much more carbon dioxide and other greenhouse gases, to raise temperatures sufficiently by initiating and sustaining a large greenhouse effect.

If the surface pressure on Mars was much higher before the changes that led to the present-day atmosphere, perhaps more like Earth's, the sequence that reduced it to less than one per cent of that at present was probably a combination of many processes. The most important – impact erosion by rocky meteorites and icy comets, where some of the atmosphere is literally blasted away – is likely to have occurred in the early period, when the Solar System contained more drifting debris than it does now. Collisions on all planets and moons have declined gradually over the history of the Solar System, as the drifting debris between their orbits either collided with something or found a stable orbit of its own, for example in the asteroid belt. Evidence of massive early impacts, such as the Argyre basin, can be studied on Mars today, along with evidence for a number of other, more subtle, factors, such as solar variability and magnetic field variations. Records of these are mostly lost on the Earth, but may be better preserved on Mars, precisely because of its thin, dry atmosphere and the absence of tectonic cycling into the crust over most of the intervening time.

The original Martian atmosphere was probably lost during the early Noachian era, more than four billion years ago. A secondary atmosphere would then have been produced later in the Noachian by volatiles released from the interior, continuing as long as there was volcanic activity. Once volcanism declined, this new atmosphere would also suffer decline due to losses during impacts, some of them massive. The end of the Noachian period, when Mars was about one-quarter of the way through its four-and-a-half-billion-year history, corresponded to a huge change in the climate and the distribution of water and ice. As the rates of movement and change in large-scale surface features lessened, valley network formation apparently practically ceased. The intrinsic magnetic field appears to have become very small at about that time, allowing increased erosion of the atmosphere in the flow of energetic charged particles from the Sun, the solar 'wind', in addition to the continued loss of atmospheric gases as a result of the impact of some of the remaining comets and asteroids whose orbits crossed that of Mars.

The mean atmospheric pressure on Mars will have varied over time due not only to loss from the top of the atmosphere, but also by the trapping of carbon dioxide and other atmospheric gases in the crust of the planet. This has undoubtedly taken place, particularly at times and places when Mars had free water on its surface, allowing carbon dioxide to dissolve and precipitate out to form carbonate rocks. We have seen that carbonates are rare in the exposed rocks on the surface, although small deposits have been found and the windblown dust seems to contain some, at the small percentage level. Much more may be buried under the present-day surface; finding carbonates and other minerals that react with the atmosphere and studying their distribution is a high priority, but understanding their history will probably require the analysis of core samples obtained by drilling.

As an alternative to the whole idea of a period of mild climate produced by a thick atmosphere, or in addition to it, the heating that produced running water on Mars may have been provided from the interior of the planet during the period when it was still volcanically active. There is evidence to support this possibility, including the appearance of some of the surface flows in which the water was apparently driven out from below. Of course, this does not preclude the presence of a denser, warmer atmosphere at the same time, and indeed this might be expected since water vapour was probably not the dominant gas that was released, but rather carbon dioxide. The high-resolution pictures from orbit that show discharges from cliffs and crater walls, which appear youthful because the features in the walls are sharp with little or no apparent erosion, may be an example of less extreme but more recent activity of this kind. The existence of these suggests to some investigators that there is still some geothermal activity on present-day Mars, sufficient to maintain liquid water reservoirs at an unknown depth below the surface. Others have proposed that the features are caused by melting of packed snow, or even by the escape of liquid carbon dioxide that formed under pressure below the surface.

Many of the features on Mars have a layered appearance because they are composed of sediments, deposited from standing water or in some cases by atmospheric transport. The sedimentary record can be read in exposed vertical faces at many locations, but is often difficult to interpret, especially with limited sampling, little chemical or mineralogical information and (except for the few visited by rovers) relatively poor resolution. The need for dating the layers by laboratory analysis of samples has already been emphasised. Three different mechanisms could be, and probably are, involved in producing the layers: global and regional redistribution of dust and sand by winds; volcanic flow and lava solidification; and the deposition of suspended mud particles and dissolved salts by water. The regular layers, several metres or more thick, that are seen at many sites record long-term, periodic climate cycles, sometimes punctuated by the more episodic water-based or volcanic processes. The challenge at any given location is to try to understand a history that includes secular and periodic change as well as sporadic and cataclysmic events, and to link these with other sites around the planet.

6.2 The orbit and the seasons

The axis about which Mars rotates has almost the same inclination to its orbital plane that the Earth's does, so the seasons, including the solstices and equinoxes, can be defined in a similar way. However, because Mars' orbit is much less circular than Earth's, as well as taking about twice as long to complete, it is more important in the case of Mars to consider where the planet is along its orbit when each season occurs. This is defined by an index called the *areocentric longitude* of the Sun, written L_S. The convention, depicted in Figure 6.2, is that $L_S = 0°$ at the vernal (spring) equinox, 90° at the summer solstice, 180° at the autumn equinox and 270° at the winter solstice, where summer and winter refer to the northern hemisphere. The lengths of the twelve months and four seasons on Mars in terms

Table 6.1

Mars month	Initial L_S (degrees)	Final L_S (degrees)	Length (sols)	Comment
1	0	30	61.2	N spring equinox at $L_S = 0°$
2	30	60	65.4	NORTHERN SPRING
3	60	90	66.7	Aphelion at $L_S = 71°$
4	90	120	64.5	N summer solstice at $L_S = 90°$
5	120	150	59.7	NORTHERN SUMMER
6	150	180	54.4	
7	180	210	49.7	N autumn equinox at $L_S = 180°$
				Dust storm season begins
8	210	240	46.9	NORTHERN AUTUMN
9	240	270	46.1	Perihelion at $L_S = 251°$
10	270	300	47.4	N winter solstice at $L_S = 270°$
11	300	330	50.9	NORTHERN WINTER
12	330	360	55.7	Dust storm season ends

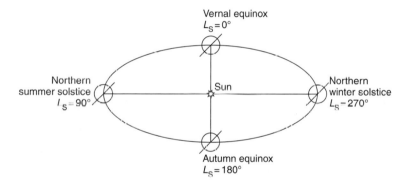

Figure 6.2 The sunfall anywhere on Mars is affected by the non-circular orbit, shown much exaggerated in this schematic. The actual distances from the Sun are 1.665 AU maximum, 1.381 AU minimum (1 AU = the mean Sun–Earth distance), and of course the irradiance is proportional to the square of this, so the eccentricity is large in climate terms. To take account of this, the position along the orbit, as well as inclination towards the Sun, has to be specified when defining the seasons. The solar longitude, L_S, is measured relative to the northern vernal equinox, defined as $L_S = 0$.

of Mars days (often called *sols*, each sol being twenty-four hours and thirty-nine minutes long) are then as shown in the Table 6.1

A month on Mars can have between forty-six and nearly sixty-seven days, making calendar-keeping a chore. This arises because perihelion, when Mars is closest to the Sun and travelling more rapidly (in accordance with Kepler's second law[1]) occurs near the beginning of the northern winter. Autumn and winter are then shorter than spring and summer in the northern hemisphere, with the opposite in the south. Something similar applies on the Earth, where perihelion falls in January, but the effect is much smaller here because Earth enjoys a near-circular orbit. The relatively large eccentricity of Mars' orbit is a major factor in its climate, leading to big differences in the seasonal behaviour of the hemispheres. The northern summer on Mars is about thirty degrees centigrade colder than southern summer, and the amplitude of the seasonal cycle is a hundred and ten in southern mid-latitudes but only fifty-five in the north. Even the smaller of these numbers is much larger than we are used to on the Earth. The reason the fluctuations of temperature between winter and summer are greater on Mars is not just because the seasons are longer. The thin atmosphere and absence of oceans to store and transport heat means that Mars responds faster to seasonal changes in the intensity of sunlight, and so reaches greater extremes. This dramatic seasonal cycle has a marked effect on the dynamics of the Martian atmosphere, and even on its composition.

In the summer, the polar regions are constantly illuminated and become warmer than the rest of the hemisphere, which leads to a stable situation without storms or large-scale wave motions. In the winter, the condition known as baroclinic instability,[2] the main cause of Earth's winter storms, is more likely, producing weather systems that, on Mars, tend to be fairly regular in space and time. The great dust storms also show a seasonal dependence, albeit a less regular and more complicated one.

In the winter the polar region enters a long period of permanent darkness, in which the Sun does not rise above the horizon for several months. Once the temperature falls below the freezing point of carbon dioxide, a hundred and twenty-three degrees below zero centigrade, the atmosphere begins to freeze out onto the permanent polar cap. Nearly a third of its total mass suffers this fate before spring arrives. During the solstices, one polar cap is vaporising while the other is condensing, so there is a large flow of air across the equator each year, first one way and then the other. Water vapour follows a similar pattern and, since this is not a symmetric process, some of it may actually be transferred systematically from one hemisphere to the other as the large eccentricity of Mars' orbit affects the seasons. In the very long term, any such transfer will tend to reverse, along with the precession of the equinoxes.

---

[1] Kepler's second law states that the line from a planet to the Sun sweeps over the same area in a given time everywhere along the orbit. To do this, the planet must move faster when nearer the Sun.

[2] Baroclinic instability occurs when the temperature contrast between equator and pole becomes so large that the initially smooth movement of air trying to reduce the gradient becomes unstable and forms turbulent eddies. In summer, both equator and pole are sunlit so the contrast is relatively small compared to winter conditions.

6.3   Climate models

Mathematical representations of the climate on Mars need to simulate the seasonal cycles of dust, water and carbon dioxide, the local disturbances in space and time (weather) superimposed on these, and the orbital variations which change slowly over long periods of time. The resulting climate models are one of the most useful tools for interpreting measurements and specifying the new and better observations needed for getting to grips with the climate and its mysteries. Data and models test and improve each other and both are essential to making progress.

Non-scientists sometimes have trouble with the concept of models, for good reason as they can often be large and complicated, although some are quite simple. The basic idea can be appreciated by considering where the name came from. In the days before computers were available, engineers would make models – literally, out of metal, wood and other materials – of complex structures like ships or bridges, and calculate the stress and drag factors they needed to evaluate the design by using known laws of physics to scale the model to the size of the real thing. When it became possible to make millions of calculations in a few seconds, the need for the physical model gradually lessened as it became implicitly represented by a mass of equations that represented its properties more precisely than even a full-scale mock-up.

Best of all, with modern computers, the model can be *time dependent*. The design for a new car, say, can be built up inside a computer and its behaviour on the road simulated by including the laws of motion and other dynamic factors and having the computer solve the relevant equations. The application of models to atmospheric motions, with the goal of making purely or partially physics-based weather forecasts, actually dates back to the work of Lewis Richardson in the 1920s, long before powerful computers were foreseen. His approach was to picture a room full of people solving the equations by hand, a groundbreaking but (we now know) hardly practical proposition.

Nowadays, comprehensive models of the atmosphere, run on large, fast computers, are at the heart of our everyday radio and TV weather forecasts, and are an essential tool for climate studies. The largest, usually known as *general circulation models*, are mathematical models that simulate the climate system on a large, fast computer.[3] They solve the time-dependent formulae that represent the myriad physical processes involved, and predict changes over time of quantities such as temperature, humidity and winds everywhere in the atmosphere in three dimensions (usually height, latitude and longitude).

---

[3] The 'general circulation' part of the name comes from the fact that the early models were designed to simulate only the overall, large-scale, slowly varying movement of the atmosphere, known as the general circulation to meteorologists. Modern models still cover the globe, but include a lot more detail, and are sometimes described as 'global circulation models'. The abbreviation GCM is commonly used for either.

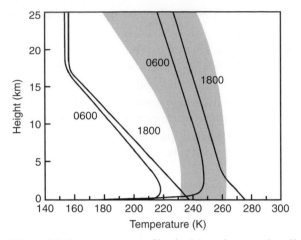

Figure 6.3 Temperature profiles for Mars showing the effect of solar heating due to dust absorption, from calculations made by Peter Gierasch and Richard Goody in 1972. The left-hand profiles are for a clear atmosphere, at the local times shown, and the right-hand pair for a dusty atmosphere. The shaded area shows the range covered by the measurements of *Mariner 6* and 7. The temperatures are on the absolute or Kelvin scale where 273 K is zero centigrade.

For Mars, models can be used to explain what is observed there, with the modellers adding or refining the representation of processes until they achieve a fit. A classic example of this is the sequence of events that first revealed the important effect of airborne dust on radiative transfer in Mars' atmosphere. Early modellers were not aware of the size of this effect, and omitted it, producing initially very puzzling results, which bore little resemblance to the real behaviour of Mars as observed by the early space probes. For instance, calculations of the atmospheric temperature profile were coming out nearly a hundred degrees too cold, as Figure 6.3 illustrates. They soon recognised and fixed the problem, and dust amounts and dust-lifting processes, along with some assumptions about the optical properties (such as refractive index) of the individual grains, are today a key part of any models that achieve a good representation of the climate on Mars, including those that are being used to investigate the ancient climate. For early Mars, the particles in condensation clouds and volcanic aerosols also have to be included in models and the signs are that they may have had an important role in maintaining a warmer climate.

The mathematical equations that comprise a general circulation model can be broken down into three separate but linked categories. Firstly, the dynamical forces in the atmosphere that describe the large-scale movement of air masses and transport of energy and momentum. Next, radiative transfer codes describing the processes involved in the transmission of energy through the atmosphere, its absorption and re-emission, and therefore heating and cooling cycles. Finally, the physics introduced by other factors such as topography, condensation, adsorption, and dust and soil properties. The equations are based on fluid dynamics, quantum mechanics and general laws such as the conservation of energy and mass. Where processes are too complicated to be represented by basic physics, the model must fall back on empirical relationships based on observations, such as those that govern the turbulent transfer of heat, water vapour and aerosol particles.

Another convenient way to picture general circulation models is by dividing them into two main parts, the *hydrodynamical code* that computes the large-scale atmospheric motions, and *parameterisation schemes* that are used to include other processes such as heating by the Sun, cloud formation, and so on. The hydrodynamical part uses a set of grid points in space and time on which to specify the atmospheric variables, and a finite-difference method to solve the dynamical equations connecting them. Alternatively, a spectral approach based on a decomposition of the horizontal fields into spherical harmonics[4] can be used. In at least one example (the 'EuroMars' model, see below) both approaches have been used to address the same problem, which is expensive in terms of effort but allows an assessment of model limitations and errors to be made.

The parameterisation schemes model the physical processes that force the general circulation and are used to compute the impact of those processes that are not resolved by the hydrodynamical code. For instance, the radiative heating and cooling of the atmosphere involves integrating the absorption and emission of infrared radiation by atmospheric gases and dust across the entire range of wavelengths, a vast computational task that is invariably replaced by simplified schemes. Similarly, the sub-grid scale atmospheric motions (turbulence in the boundary layer, convection, surface drag, deflection by topography, gravity waves, and so on) cannot be resolved by the model and the exact expressions have to be replaced by empirical approximations. 'Simplified schemes' and 'empirical approximations' are devised and tested in all sorts of ways until something is found that seems to work.

With all of the key processes in place and sufficient computer power, the general circulation models can calculate how temperature, wind, and tracers such as dust and water vapour should be varying on Mars in response to the known laws of atmospheric physics. Models that have been 'validated' by successfully representing the observed climate and its variations can then be (cautiously) extended to investigate the past and predict future changes. Essentially, this is the same approach that is the basis for modern weather and climate forecasting for the Earth. For Mars, general circulation models have an additional use, which is that they supplement (with theory and statistics) the very limited observations of the atmosphere that are available, as well as predicting atmospheric properties for which there is as yet no data. This of course introduces uncertainties that strain credulity even further when the models are

---

[4] Most mathematical functions, including those that describe the motions of a layer of gases on a sphere, can be expressed as the sum of a series of simpler functions. This makes them easier to manipulate on a computer.

used with key parameters (such as the axial tilt of the planet or the mass and composition of the atmosphere) changed to simulate the detailed climate of Mars in the past or in the future. The models undoubtedly help us to ask questions about changes in the global environment of Mars throughout geological time, but they have a long way to go before they provide unreservedly believable answers.

Building a Mars general circulation model is a considerable task and it can take a team several years of model-building, usually starting with a large amount of code inherited from terrestrial forecasting models, before they begin to get useful results. A number of different Mars models exist; the two most sophisticated are those maintained by teams at the NASA Ames Research Center in Mountain View, California,[5] and by a European consortium originally set up in Paris and Oxford. The latter has produced the Mars Climate Database, which describes the mean state of the Martian atmosphere as a function of season, dust amount and other parameters obtained from multi-annual runs of their general circulation models.[6]

Figure 6.4 shows examples of some representative output from the European model, depicting the temperature and the zonal wind (the velocity parallel to the equator) in the atmosphere as a function of height and latitude. These depend not only on the time of year, but also on other factors, in particular the amount of dust present. The calculations shown in the figure are for a period in the Martian winter ($L_S = 270$ to 300 degrees) with the dust set at the level observed during this season by the *Viking* mission in 1976. As would be expected, the surface and lower-air temperatures are higher in the south, where it is summer. Less obvious features are the higher temperatures over the *winter* pole in the upper atmosphere, where there is no solar heating at all. This is a feature of the circulation of the atmosphere, one that has been confirmed by observations, as we discuss below. The most prominent feature of the wind field predicted by the model is a stream of high winds, called a jet by meteorologists, also in the winter hemisphere. Instabilities in this jet produce stormy weather in the Martian winters, in contrast to its cool, calm summers.

6.4 The present climate

Another advantage of studying the climate with a model is that the factors affecting the three global cycles – water, carbon dioxide and dust – that act seasonally on present-day Mars can be varied separately as well as together. The fluxes of each quantity can be followed, in response to the changing seasons, and their regular, annual exchange between the atmosphere, surface and subsurface quantified. Consider the example of the water cycle. Figure 6.5 shows how the amount of water vapour in the atmosphere was observed by *Viking* to vary

[5] http://www-mgcm.arc.nasa.gov/MGCM.html   [6] http://www-mars.lmd.jussieu.fr/mars.html

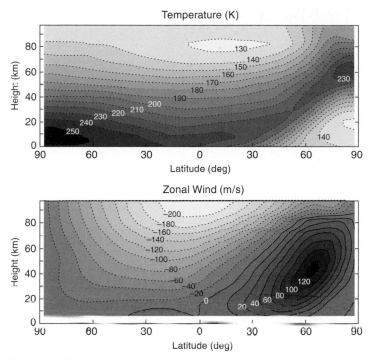

Figure 6.4 These temperature and wind fields from the European Mars model have been averaged around circles of constant latitude and over all times of day to show the zonal mean equator-to-pole temperature structure during a thirty-day period in the early part of the northern winter (L_S = 270–300 degrees).

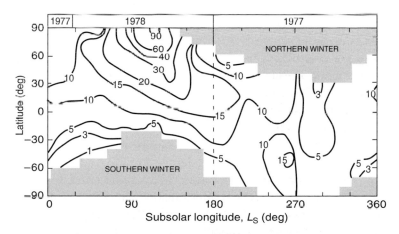

Figure 6.5 The water content of the atmosphere, over one Mars year, as observed by the *Mars Atmospheric Water Detector* on the *Viking* orbiters. Note the much higher concentrations in the northern summer (near L_S = 90 degrees), relative to the south. The unit is precipitable micrometres, the thickness of the liquid layer that would result if all of the water in the atmosphere condensed on the surface. Earth's atmosphere typically holds about one thousand precipitable micrometres.

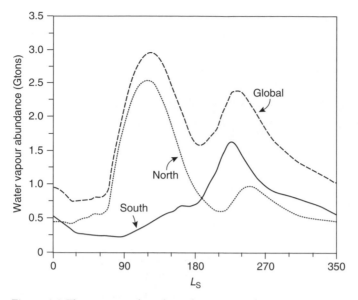

Figure 6.6 These curves show how the amount of water present in the atmosphere, as vapour or clouds, varies as a function of time, according to the Mars general circulation model developed at NASA Ames Research Center. The maximum, in the northern summer, for the whole planet is three billion tons of water, nearly all of it in the northern hemisphere. The southern maximum is smaller, but again the bulk of the water is in the summer hemisphere as would be expected. The asymmetry between the hemispheres is something that probably varies with the long-term climate cycles driven by cyclical changes in Mars' orbit.

during a Martian year; Figure 6.6 shows the total quantity of water in the atmosphere that corresponds to this, according to a model.

The most important source of water vapour in the entire Martian atmosphere is the north residual polar cap, from which large quantities of water vapour sublime during late spring and early summer, producing a maximum in the humidity of the air. The southern polar cap, by contrast, produces only about half as much, one of several interesting differences between the hemispheres. Although about three billion tons of water vapour is released in the northern summer, this is only enough to cover the planet to a depth of less than one hundredth of a centimetre if it were condensed to form liquid water. This is far less than on Earth, mainly because the lower temperatures and densities on Mars mean that the atmosphere can hold and transport much less water, even if it is in contact with a plentiful supply on or underneath the surface. In fact the air on Mars is often at, or close to, saturation.

The amount of water vapour in the atmosphere in the northern summer varies with latitude, with four times as much in a vertical column at 60° N as at 30° N, with smaller variations in the tropics. Comparisons between observations and models of the annual cycle of water suggest that there must be another significant reservoir of water on Mars besides the polar caps and the atmosphere. Water vapour increases in the spring near 30° N long before the permanent water ice cap receives enough heating to supply a significant amount of vapour at this latitude, which would seem to confirm that it comes from a source on or just below the surface. It has been estimated that the amount of water that is adsorbed in the top few centimetres of the Martian soil is ten times the total load in the atmosphere, in which case it must contribute to the mass of water that transfers permanently between the hemispheres during the course of the Martian year.

In addition to water adsorbed in the soil or regolith, there are likely to be subsurface deposits of more or less pure ice as well. Some of this ice may be seasonal, melting and contributing vapour to the atmosphere in the summer; but most of it is probably permanently frozen and so lost to the modern climate system. Calculations show that ground ice is stable a metre or so below the surface everywhere polewards of about 40° latitude. The models that predict this depend strongly on the nature of the soil, however, especially its thermal inertia, the quantity that measures how rapidly a material tends to heat and cool when exposed to or removed from heating by the Sun. If the Martian subsurface has a low thermal inertia, corresponding for example to a fairly loose aggregate soil, then ice is stable just below the surface at all except tropical latitudes in the models.

Before the *Phoenix* landed, the only experimental data that tested these theoretical expectations were the observations of splashy deposits around certain impact craters, like that in Figure 6.7, and of glacier-like flows in some shallow valleys. This constitutes good evidence for substantial subsurface ice deposits, in non-polar as well as polar regions, but does not provide enough information when it comes to estimating the total amount on the planet. *Phoenix* did not have to dig very far to uncover ice very near to the surface at its high-latitude landing site (68° N), although it could not address the important question of how far down it extends, nor give information about the local variations that must exist. Some day there will be many such measurements, all over the planet and to much greater depths. Then it will be known how much subsurface water exists across the planet as a whole, and whether it is enough to match the amounts that seem once to have flowed for a time as liquid on the surface.

Meanwhile, there are alternatives to digging and drilling that, although subject to uncertainties of interpretation, do span large areas of the planet with ease. There is evidence for large amounts of ice in the southern hemisphere of Mars from three different instruments on board the 2001 *Mars Odyssey* spacecraft. These experiments, measuring gamma radiation and energetic neutrons, are sensitive, not to water itself but to hydrogen in any form in the soil, by its effect in slowing down or absorbing the emission produced when cosmic rays impact the Martian surface. Some of the neutrons are absorbed and then emit a gamma ray of a specific and characteristic energy, which is detected by the spacecraft. Water, H_2O, must be by far the most common hydrogen-bearing molecule on

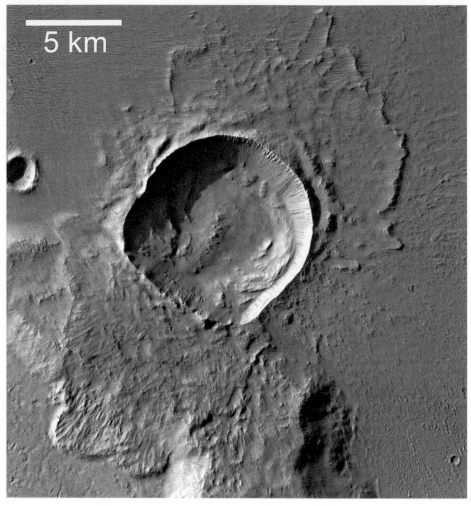

Figure 6.7 This small crater in the Medusae Fossae feature, near the Martian equator, shows unmistakable signs of subsurface ice that melted during the impact and produced a muddy splash several times the size of the crater.

Mars, so the lack of specificity in any detection of large amounts of hydrogen is not too much of a problem.

The gamma-ray results pertain principally to the top metre or two of the subsurface. The data imply that the Martian soil is relatively dry near the surface, with large quantities of hydrogen, presumed to be H_2O in the form of ice, present below a depth that varies from pole to equator. To match the data the water-rich layer contains a fifth to a half of ice by mass, and since ice is less dense than soil the material is mostly ice by volume. It is found close to the surface near the pole, with its upper boundary moving progressively deeper towards mid-latitudes. By latitude 40° S, the ice-rich layer is still present, but the overlying ice-free layer is more than a metre deep.

6.4.1 Water in the polar caps

The north residual polar cap, that is, the part that remains at the end of the summer after the seasonal cover of carbon dioxide frost has evaporated, is made primarily of water ice. This much is known mainly from mapping its temperature from orbit. The amount of water that it actually contains is also quite well known now, from measurements of its height profile by the laser altimeter on the *Global Surveyor* orbiter, shown in Figure 6.8 and Plate 16. It works out at a little over one million cubic kilometres, a formidable amount of ice, but only a small percentage of the estimated content of the Antarctic ice sheet on Earth, and only about one-tenth of the volume required to fill the ancient Martian ocean basins (if that is indeed what they are), even if the whole cap were melted. The south residual polar cap is colder than the north and most of it, on the surface at least, is carbon dioxide ice. Water ice is also trapped in the cap, probably in large amounts, but even if the cap were nearly all water its volume is only slightly greater than that of its northern counterpart so it cannot account for more than another ten per cent of the capacity of the putative ancient oceans. If the oceans were really there, and if all of the water is still on Mars, then something like eighty per cent of it is hidden below the surface at latitudes away from the polar caps.

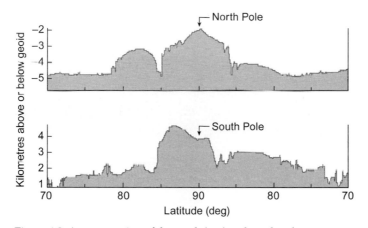

Figure 6.8 A cross-section of the north (top) and south polar caps, measured from Mars orbit by the laser altimeter on *Global Surveyor*. The sections are at longitude zero degrees and extend in latitude from the pole to seventy degrees north and south respectively. The vertical scale, exaggerated a hundred times relative to the horizontal scale, is in kilometres. From these data, it has been estimated that the two caps combined contain enough water to cover the globe of Mars to a mean depth of around thirty metres – much shallower than Earth's oceans if this were the only water there is on Mars, and not enough to fill Mars' putative Northern Ocean (see Figure 6.16). Note the chasms, which spiral around the pole and which are more than a kilometre deep, extending in places down to the level of the surrounding plains deep inside the polar cap – a fascinating objective for future explorers on Mars.

6.4.2 The dust cycle

Well before spacecraft exploration of Mars began in the 1960s, astronomers reported the occasional obscuration of features on Mars by extensive banks of yellow clouds. This is a global dust storm, part of the same phenomenon seen for centuries as the 'wave of darkening' observed in the spring, and once thought to be evidence for the seasonal growth and decay of vegetation. Most of the surface of Mars is covered by fine dust, and some of this is also present suspended in the air as a reddish haze even under quiescent conditions (see, for example, the view in Plate 5b). The atmosphere takes up much more dust during the major storms that break out periodically, whipping up larger particles from the dusty plains and producing an opaque blanket that sometimes covers the whole planet. Figure 6.9 shows the telescopic view of Mars before and during such an event. These global storms usually occur when Mars is closest to the Sun (at perihelion), presumably related to the fact that the solar heating is about forty per cent greater then than it is when Mars is furthest from the Sun (at aphelion).

The high winds and extreme temperature ranges that the surface of Mars experiences can lead to high rates of erosion of rock and soil, so it is not surprising that large amounts of dust are present on the surface and suspended in the atmosphere. The dust is extremely mobile because, despite the large amounts of frozen water present in the polar caps and buried in the crust, and the near-saturated state of the atmosphere, the surface environment is very dry. Some of

June 26, 2001 September 4, 2001

Figure 6.9 Two *Hubble Space Telescope* views of Mars, before and after the onset of a global dust storm in 2001. Dust transport on such a massive scale has played a major role in the evolution of the Martian surface.

Figure 6.10 A large dust devil, approximately one kilometre in height, observed by the *Spirit* rover. The dust it contains is opaque enough to cast a shadow, and its passage has left a trail across the surface.

the particles are very small and could stay airborne for a long time provided the temperature did not change too much. However, they have to contend with the plunging temperatures at night, when the settling rate is augmented by water and carbon dioxide freezing onto the airborne dust particles, adding to their mass and increasing their cohesiveness. The fact that the Martian atmosphere seems to maintain a substantial level of airborne dust at all times despite these losses in the diurnal cycle indicates that the background wind field and local disturbances such as the small tornadoes known as dust devils are energetic enough to raise dust into the air more or less continuously. One can be seen at work in Figure 6.10.

The fact that the Martian atmosphere is very dry, mostly cloud free, but rarely without significant amounts of airborne dust leads to an important difference in the nature of the greenhouse effect on Mars and Earth. On Earth, water vapour is the most powerful greenhouse gas, and its condensation into cloud, changing the opacity and albedo of the atmosphere and absorbing or releasing large amounts of latent heat, is the largest contributor to atmospheric variability. On Mars, dust is the largest variable opacity source in the atmosphere. Changes in the dust distribution can take place very rapidly, affecting the absorption of radiation and consequently the atmospheric temperatures, pressures and winds. These changes further affect the dust transport within the atmosphere, and the processes that lift particles from the surface. Hence the dust distribution is changed again, and if conditions are right, a small disturbance can grow into a major storm. This is a more powerful feedback process than any manifested on the Earth, and helps to explain why Martian storms can grow to envelop the whole planet in just a few days.

Dust affects atmospheric heating by absorbing and scattering solar radiation, preventing it from reaching the surface, which becomes cooler while the atmosphere gets warmer. The effect of this is to make the atmosphere more stable with respect to convection in the regions with large amounts of airborne dust, and also to reduce horizontal temperature gradients and winds inside the dust cloud. However, the winds, especially near the surface, are likely to be strong near the edge of a dust cloud due to the large temperature gradients there. The net effect is to increase the extent of the disturbance, again helping to explain why it sometimes spreads at great speed until it covers the entire planet. Nothing like this happens on Earth, despite surface wind stresses that are generally larger than those on Mars, mainly because the dust is more quickly removed from the atmosphere by coagulation in wet environments or by forming condensation nuclei for clouds and precipitation. Even if all of the land on Earth were dry and dusty, the oceans would make it difficult for Martian-style global dust storms to propagate.

Calculations of the effect of airborne dust in models of the Martian climate require some knowledge or assumptions about the properties of the particles. Estimates have been made using data from cameras, spectrometers and other instruments on the *Mariner*, *Viking* and *Phobos* spacecraft of the particle size distributions, concentrations and optical properties of the material in the atmosphere between and during dust storms of various intensities. The most important optical properties are the refractive index and the absorption coefficient, which describe the way in which dust scatters and absorbs the incoming sunlight and outgoing planetary infrared radiation, and which are a strong function of the composition of the material. From the way the dust was produced, it must consist of a variety of minerals eroded from different geological units on Mars over the ages. It must also have been thoroughly mixed by the dust storms, so that the same mixture is found everywhere on the planet, as with the sand, soil and other fine material on the surface.

The evidence so far is consistent with a heterogeneous mixture. For instance, magnesium carbonate and magnesium sulphate have been identified, although their formation processes tend not to be mutually compatible, suggesting the dust was produced by mixing material originally laid down in different regions or layers. Geological arguments, along with measurements of the optical and magnetic properties of the dust, have been interpreted as showing that it contains smectites, which are clays rich in iron oxide, and palagonite, a glassy mineral that could have been formed either in volcanoes or during meteoritic impacts. These sand-like silicate particles would not show the observed tendency of the Martian soil to form clumps without traces of an additional component such as magnesium sulphate to provide adhesion between particles.

The size distribution of the airborne particles is another key parameter when calculating their effect on the transfer of solar and infrared radiation in the atmosphere. This not easy to determine, and in any case it must vary a lot in space and time. Not surprisingly, sample particles that have been studied, including some recent images of captured dust gains under the microscope on board the *Phoenix* lander, are found to be very fine, typically about one or two micrometres[7] across, just a little larger than the wavelength of visible light. Larger particles than this would be expected to become airborne under windy conditions, with some around ten times larger (a thousand times more massive) contributing to the dust cloud for shorter periods during vigorous storms. A great deal remains poorly understood and learning more about what goes on in the great Martian dust storms, and about the atmospheric distribution, surface sources, sinks and fluxes of dust, is a high priority for new missions and studies in the near term exploration of Mars. Then it might be possible to say why they occur in some years and not others, and to be able forecast their occurrence, an important step for practical as well as scientific reasons since dust storms are not just important facets of the Martian weather and climate; they also present a major threat to manned and unmanned missions alike.

6.4.3 The carbon dioxide cycle

The existence of a carbon dioxide cycle on Mars is a consequence of the heat balance at the poles. During the winter, the region surrounding the pole is in permanent darkness for most of the winter. Without direct solar heating, and with only a thin atmosphere to transport heat into the region, the temperature rapidly falls below the condensation point for carbon dioxide. This is a phenomenon that, except for some very cold regions in the upper atmosphere where thin carbon dioxide 'cirrus' clouds can condense at any latitude or time of the year, only occurs in the polar night. The average temperature on Mars is about minus sixty degrees centigrade; at the *Pathfinder* landing site, about minus seventy-six was measured just before dawn. These and other typical non-polar temperatures

---

[7] A micrometre or micron, written μm, is one thousandth of a millimetre, close to the wavelength of visible light (the visible spectrum ranges approximately from 0.38 to 0.75 μm).

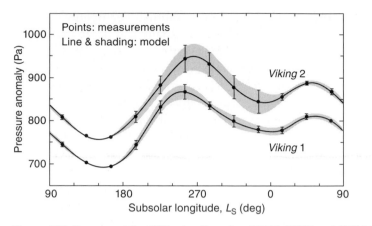

Figure 6.11 Pressure at the *Viking* landing sites (22° N, 48° W and 48° N, 226° W) as simulated by the European Mars model, compared to the measurements made by the landers themselves. The shaded bands are the model results; the width of these, and the bars on the measurements, show the variability due to weather patterns, mainly fronts like those we experience on the Earth, and which are more active in the northern winter. The vertical displacement of the two curves is simply due to the fact that the *Viking 2* site is lower than *Viking 1*, so the mean pressure is higher.

are well above the carbon dioxide freezing point of minus a hundred and twenty-three. The winter polar cap is a cold 'finger' where virtually the entire atmosphere tries to condense, resulting in the big seasonal change in surface pressure all over the planet. Figure 6.11 shows measurements of this, compared with theoretical model predictions and showing good agreement between the two.

On Earth, we experience only quite small variations in atmospheric pressure, of the order of a few percent, mostly associated with dynamical effects in migrating weather systems. On Mars, however, surface pressure measurements show that in addition to similar weather-induced fluctuations, there is a seasonal pressure swing of around a third of the mean value. Thus the same fraction of the mass of the atmosphere is condensing onto the winter polar cap. If the winter were longer, eventually all of the carbon dioxide would condense, leaving the planet with an even thinner atmosphere consisting mainly of argon and nitrogen. As it is, the sunlight returns to the region and begins to reverse the process before it reaches this stage. The global pressure increases throughout Martian spring in either hemisphere, and then starts to fall again as the opposite pole goes into its long winter night. When the pressure is highest, the north polar cap seems to have lost most or all of its frozen carbon dioxide, exposing a surface of water ice. At the south pole, however, the residual cap exhibits much lower surface temperatures, showing that frozen carbon dioxide is still present even at the end of the summer. There is also a lot of water ice there – a recent estimate based on surface-penetrating radar measurements placed the amount in the south polar cap alone at the equivalent of a global ocean of water eleven metres deep.

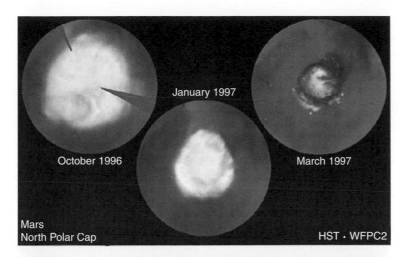

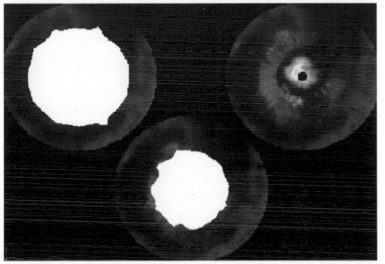

Figure 6.12 The top frame shows a polar projection of *Hubble Space Telescope* views of the Martian north polar cap. These can be compared with the carbon dioxide ice fields for the corresponding seasons as predicted by the European Mars general circulation model; areas shaded white indicate a carbon dioxide ice cover of greater than fifty kilograms per square metre.

Models also do a good job of predicting the expansion and contraction of the north polar cap, as illustrated in Figure 6.12. This shows that the expected rates of deposition of carbon dioxide ice, allowing for the release of latent heat, radiation of heat to space, and heat transport by dynamics, are quite well represented in the model. To explain the different behaviour of the south polar cap, the radiative energy balance must be different, and this has proved harder to reproduce in models, suggesting that additional processes are involved. Possibly, the answer lies in the amount of dust that is mixed into, or lies on top of, the two polar caps as a result of the different circulation patterns and dust storm activity levels. The amount of dust determines the reflectivity and the infrared emissivity of the ice, affecting the time it takes to sublime away in the springtime. Another possibility is that poorly understood dynamical factors cause more carbon dioxide ice to be deposited on the south pole in the first place, making a thicker covering that takes longer to vanish, so that in fact it never does.

Whatever the reason for the persistence of carbon dioxide ice at the south pole, it has to overcome the effect of the bias in heating due to Mars' orbital eccentricity. This might have provided a convenient explanation for the north–south irregularity, except that it works in the wrong direction because the planet is closer to the Sun in the southern spring. The asymmetry that can be seen in the annual pressure cycle – the pressure minimum during late northern winter is less deep than the corresponding minimum during late southern winter – is also related to the different behaviour of the two poles with regard to condensation and release of carbon dioxide. Here, however, the behaviour is at least in line with simple intuition, since the higher pressures occur near perihelion.

6.5   Winds, the general circulation of the atmosphere and the weather

Obviously, the climate on any planet, and specifically the three great cycles just described that dominate the climate of Mars, are all closely linked with the dynamical behaviour of the atmosphere. This is often separated into two areas of study: the time-averaged, seasonally varying motions, the so-called 'general circulation', and the 'weather', made up of short-term variations in the general circulation, waves, turbulence and eddies. The particular events that start dust storms, form clouds or initiate precipitation are part of an enormous subject that is not very well understood even on Earth, where there is a lot more data, but still major surprises like the great storm of 1987 in England, or Hurricane Katrina over New Orleans in 2005. Martian weather has plenty of common features with its terrestrial equivalent, including chaotic behaviour that will need careful monitoring.

We have seen that the Martian seasons also have common features with Earth but are relatively extreme because of the thin atmosphere, the absence of the oceans and the stabilising effect of their large heat capacity, and the non-circular Martian orbit. In the summer on Mars there is only a small equator-to-pole temperature contrast, and at the solstice the illuminated pole actually receives more solar energy per unit area than the equator. Winds are generated by the differences in heating due to the different elevations and albedos of the various regions across the globe, and the subsequent flow patterns are affected and modified by the mountains and other large obstacles. The surface topography on Mars is more extreme than on Earth, varying by about thirty kilometres and reaching heights where the atmospheric pressure is almost ten times less than at the surface. The missing Martian oceans, on the other hand, simplify the dynamics of the atmosphere, since on Earth they store and transport heat and provide unlimited amounts of condensable moisture with its important release and absorption of latent heat. The thin, dusty Martian atmosphere heats and cools rapidly, leading to a much closer coupling of surface and air temperatures than on Earth. The daily variation in solar heating leads to a large temperature and pressure cycle, and winds that rotate in direction during the day, and increase in strength with height.

In the winter there is a strong equator-to-pole temperature gradient and the flows associated with the huge transfer of mass to the pole as the atmospheric carbon dioxide condenses there. The result actually resembles an Earth-like winter, with prevailing westerly winds, a mid-latitude jet stream, and frequent localised storms. Heat is transferred polewards by tongues of warm air which can develop instabilities, producing vast cyclonic weather systems and fronts with associated cloud patterns. The latter have been photographed from orbit by satellites, and recognised by meteorologists, just as on Earth; Figure 6.13 shows a cyclonic system that, at first glance, could be on either planet. At night the lower atmosphere cools rapidly until fog forms, along with a light frost on the surface, both of which can linger for an hour or two after daybreak.

Very detailed and regular measurements are required planet-wide in order to get to grips with the meteorology on any planet and to understand the smaller-scale processes that govern the general circulation and the weather. Landed stations like *Viking* and *Pathfinder* have measured temperature and pressure cycles and also near-surface winds, but only at the isolated sites where they

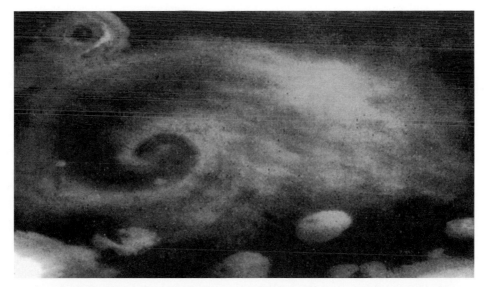

Figure 6.13 A cyclonic storm system, nearly four hundred kilometres across, seen on Mars by *Viking 1* orbiter. The rotating wind system is traced by dust lifted from the surface below and water ice clouds above.

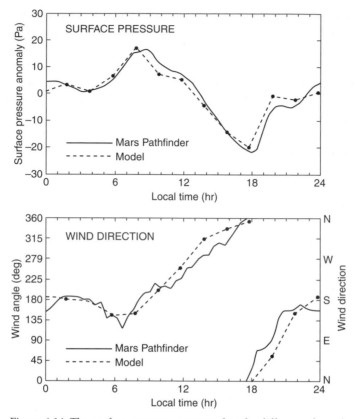

Figure 6.14 The surface pressure, expressed as the difference from the annual average (above), and the wind direction (below), at the *Mars Pathfinder* site as a function of local time of day for the first thirty days of the mission. For comparison, the dotted lines are the predictions of the European Mars model at the same location and time of year.

were deployed, and only near the surface. *Viking 1*, on the western side of the Chryse basin, recorded a light daytime wind from the south with anti-cyclonic motion around high pressure, and gentle night-time winds from the north. *Pathfinder* landed in the middle of the summer and measured temperatures that ranged daily from highs of about thirteen degrees centigrade below freezing to lows of minus eighty, as well as diurnal pressure and wind variations, all depicted in Figure 6.14. These data are valuable, of course, but there are never going to be enough of them acquired using surface stations alone.

Remote sensing measurements from satellites are routinely used to gather global meteorological data for the Earth, and the same techniques can be deployed at Mars. Surface temperature is a fairly simple quantity to monitor using infrared instruments that measure the heat radiated to space, using wavelengths where the atmospheric absorption is low. Pressure and wind are much more difficult to obtain remotely in a direct way, although pressure profiles as a function of height above the surface can be retrieved from temperature measurements at several different wavelengths, especially if the observations are made at the limb, i.e. viewing the atmosphere side-on rather than in the downwards direction.

The behaviour of the winds on Mars is hard to measure remotely, except in a limited way by tracking clouds. It is also in general considerably more difficult to understand, model and predict than temperature and pressure, and although the three variables are related to each other the relationships are difficult to express mathematically in a real and complex situation. Part of the problem is that the winds are mostly associated with travelling weather systems that need to be measured well away from, as well as at, the site where the wind is observed before a completely meaningful analysis is possible. The models are making progress, however, and surface winds during a dust storm episode, simulated by the European model, are clearly related to streak-like markings on the surface observed with the *Viking* cameras. The streaks are thought to have been produced during the decay of the global dust storms in 1977, so are a rough proxy for the mean surface wind direction. The correspondence between the two is encouraging but far from perfect, and will probably not improve very much until global surface winds are measured directly by a network of surface stations. Meanwhile, the high-resolution camera on *Reconnaissance Orbiter* is collecting data on streaks and other wind-formed features with much better detail than before. This will lead to a database on surface weather phenomena and winds that is too complex to compare with models, since there is a serious mismatch between the spatial resolution of the measurements (a metre or so) and the models (typically tens of kilometres). It is also hard to tell when wind streaks formed, even in relative terms; markings close to each other might be interpreted as part of the same major storm event when in fact they formed years or even centuries apart.

The general circulation cannot be measured directly at all. The approach has to be to measure temperature, and where possible pressure, fields, and use these to constrain the parameters in a general circulation model. This will calculate the wind fields that are consistent with the measured pressure and temperature data, and, if the model physics is sound, this will be the real wind field in the atmosphere. Of course, it will be a smoothed-out version, with spatial resolution that can be no better than that of the model grid or the spacing of the measurements (in time, as well as altitude and longitude). This procedure (called data assimilation) is highly developed for Earth's atmosphere, for weather forecasting and research, and attempts are being made to use it for Mars.

Before this is possible, obviously suitable data must be obtained from a Martian weather satellite. This was an early goal for NASA, with the first useful measurements coming from *Mariner 9* in 1971. The first attempt to get high-resolution data, including measurements at the atmospheric limb for good vertical coverage, was to have been made by an instrument on *Mars Observer* in 1992. We have already seen (in Chapter 4) how *Observer* was lost, and so was *Mars Climate Orbiter* in 1999, carrying a duplicate of the atmospheric sounding instrument. There was not to be a Mars weather satellite worthy of the name until 2006, when the *Mars Climate Sounder* arrived.

The *Climate Sounder* is an infrared radiometer optimised for remote sounding of the Martian atmosphere from orbit using principles that are well tried and tested on the Earth. Its observations are made from the *Mars Reconnaissance Orbiter*, which has been in a near-circular, sun-synchronous polar orbit around Mars at an average altitude of about two hundred and fifty kilometres since September 2006. The dedicated use of the limb-scanning technique, in which the atmosphere is viewed at a tangent to the surface, is used to obtain high vertical resolution, in eight narrow channels in the infrared spectrum chosen to be sensitive to temperature, water and dust, and one wide channel collecting reflected sunlight across its full wavelength range. Nadir (vertical) measurements of the reflected and emitted radiation from the surface improve the temperature profile retrievals near the ground, measure the seasonal energy budget of the polar regime, and map the inventory and behaviour of airborne water and carbon dioxide condensates in the polar night. The retrievals of atmospheric temperature use measurements of radiance emitted along the limb path by carbon dioxide, from the surface to approximately eighty kilometres altitude using a twenty-one element linear array of thermopile detectors projected on the limb in each of the nine channels. Radiance measurements in the thermal infrared are calibrated at two points by regular views of a reference blackbody of known temperature, and views of space above the limb, at regular, programmable intervals. A diffusion plate illuminated by the Sun is used to calibrate the measurements of the reflected solar radiation from the dayside of Mars.[8]

Several thousand profiles of atmospheric temperature as a function of pressure, distributed all over the globe, are obtained every day, offering much better vertical coverage and resolution than had previously been achieved. During the first few months of operation scientific interest focused on the south polar region, which was coming out of the permanent darkness of winter at the time. At the latitudes from 40°S to the pole, many of the profiles contain the signatures of variable amounts and layering of airborne dust and water and ice cloud particles, but a small number of profiles, less than one per cent of the total, is clear of significant opacity due to any kind of aerosol. These clear profiles are found to occur mainly near the boundary of polar night and are seen to follow its seasonal retreat. It is not known yet why clearer air is confined to this zone; it might be due to the scavenging effect of carbon dioxide ice formation in the boundary region between the 'snowfall' inside the polar night and the 'normal' dustiness of the atmosphere further away from the pole. Some as yet obscure dynamical behaviour may also be involved: the edge of the polar night is expected to be a region with strong winds.

The polar temperature profiles, like the examples in Figure 6.15, feature stratospheric temperature inversions with maxima near fifty kilometres altitude

---

[8] See the paper by McCleese *et al.*, 2007, listed in the further reading, for a full description of the *Mars Climate Sounder*.

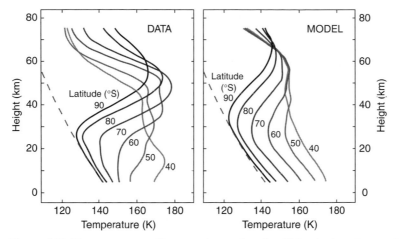

Figure 6.15 Temperature profiles over the south pole of Mars in the winter, when latitudes greater than 60° are in permanent darkness. The profiles on the left were measured by the *Mars Climate Sounder* on *Mars Reconnaissance Orbiter*, while those on the right are those predicted by the European Mars general circulation model for the same place and season. The warming aloft, produced by dynamical heating (compression) in the Hadley cell that dominates the atmospheric circulation, is much stronger in reality than the models predicted, indicating a stronger equator to pole overturning. Near the pole, the profiles follow the sublimation temperature of carbon dioxide ice (dashed line), indicating that atmospheric carbon dioxide is snowing out onto the surface. (After McCleese *et al.*, 2008)

that are as much as forty degrees centigrade above the surface value, and that evolve as southern spring on Mars progresses. Since the region is in darkness for the duration of the winter season, this can only be due to heating by dynamical compression, the result of mean motions that are consistently downwards over the pole. Such downwelling is to be expected: the most strongly heated surface is in the low-latitude zones, and the air there must become less dense and generally rise. It has to then come down somewhere, and the coldest, densest place starts out being over the pole. The compression there produces heating, like the familiar bicycle pump effect, working towards an overall dynamic balance in which the whole hemispheric atmosphere cycles from equator to pole and back in around a hundred Martian days. Something similar happens on the Earth, but in our denser atmosphere it takes more than a year. This behaviour was predicted as long ago as 1735 by George Hadley, and is known as a Hadley circulation or Hadley cell. Needless to say, the real behaviour is more complicated, but Hadley's idea remains a useful concept, on Mars as on Earth.

6.6 The ancient climate: was it really warm and wet?

If the fluvial features on Mars were produced by liquid water, kept from freezing by a thicker atmosphere of carbon dioxide, the surface pressure required would be more than a hundred times higher than it is now, somewhere between one and about ten times that on present-day Earth. This factor is so large partly because of the current belief by solar astrophysicists that the young Sun was around twenty-five per cent less bright than it is at present. It is not inconceivable that the pressure on early Mars was more than it is on present-day Earth; our own atmosphere probably had a higher surface pressure billions of years ago when there was much more volcanism and gases like carbon dioxide had not yet precipitated as carbonates in the ocean. Saturn's satellite Titan, smaller and colder than Mars, currently has a nitrogen atmosphere with an atmospheric pressure at its surface that is fifty per cent higher than Earth's.

If the atmosphere on Mars were once at a higher pressure, the tendency for carbon dioxide to condense and form clouds would have been greater. The clouds themselves provide part of the greenhouse effect; on Venus, they are an important part of the mechanism that produces baking temperatures on the surface of that planet. On Earth, clouds can either warm or cool the surface depending on their type and altitude, and it has been shown that the same is true in models of early Mars. With an atmospheric composition of carbon dioxide and water vapour, clouds containing large crystals of frozen carbon dioxide and water could have contributed to a surface temperature warm enough for liquid water to be stable. On the other hand, if smaller cloud particles were present they would tend to reflect sunlight and restrict the warming at the surface to something well below the freezing point of water, even with a much denser atmosphere.

It is possible that other gases that have since vanished might have augmented the basic warming effect due to carbon dioxide and water vapour. Methane, carbon monoxide and ammonia are all powerful greenhouse gases that are common in primitive atmospheres like those still to be found in the outer Solar System. The problem with invoking these primitive gases is that it is hard to see how they could have survived long enough to be effective on one of the terrestrial planets. Unlike the gas giant planets, the relatively thin atmospheres of Mars, Venus and Earth were exposed in their entirety to the ultraviolet rays of the Sun, which dissociate ammonia and methane very efficiently. If this process could be slowed down somehow, for instance if there was a steady supply of these gases for a time, which has since dried up, this provides a neat solution: gases that warm the surface of early Mars, but then vanish after a billion years or so and are not found now, except in trace quantities too small to affect the climate. Clouds might make an important contribution in this scenario too, since the photochemistry of methane and ammonia produces compounds that can condense and make a thick haze. This can be seen happening today on Titan, the large satellite of Saturn, which has an atmosphere about as thick as that hypothesised for early Mars. The ammonia is all gone on Titan, leaving nitrogen as the main gas in the atmosphere, but with lots of methane, and a deep haze of methane photochemical products in the form of condensed hydrocarbons like ethane and ethylene. It is something of a mystery why Titan still has any atmospheric methane; probably it is continuing to leak out from the interior through a kind of cold volcanic activity. As mentioned above in Chapter 4, and further discussed below, traces of this kind of behaviour may recently have been detected on Mars as well.

In summary, the situation is essentially that the geological and other evidence seems to be saying quite clearly that Mars was warm and wet. At the same time, the modellers and theoreticians are telling us that it is quite difficult to explain how it happened. They do leave us some loopholes, but before these can be seriously invoked they need to be backed up by data that show such processes really have been at work, and most of that evidence is still lacking. Let us look at the conundrum in more detail.

6.7 Evidence for climate change

6.7.1 Ancient oceans, lakes and rivers

Since the work of the *Mars Exploration Rovers* on the surface dispelled the remaining doubts, the scientific community is virtually unanimous in accepting that at some time water has been a significant force in shaping the Martian surface. However, there remains great controversy over how and when the various specific features formed, such as the giant outflow channels and valley networks, and the low-lying locations that may have been lakes and seas. Features resembling shorelines in the northern hemisphere have been interpreted as the boundary of a vast ancient ocean covering the north polar region, shown in Figure 6.16. In support of this, altimeter results from the *Mars Global Surveyor* mission showed that the northern plains are extremely smooth, consistent with a dried-out ocean sediment bed, and that most of the apparent shoreline is at a constant height that may define the erstwhile sea level. The area it contains is the lowest region on the planet, and most of the big outflow channels drain into it. The parts of the 'shoreline' that do not line up with the rest could have been altered by geological activity of some kind in the intervening few billions of years, although this is controversial. One suggestion is that the variations in height might be explained in terms of distortions of the surface due to the shifting of Mars' rotational poles. The loss of a massive ocean in the northern basin could itself have contributed to such 'polar wandering', by as much as fifty degrees of latitude.

The largest channels, some of them as wide as several tens of kilometres and hundreds to thousands of kilometres in length, have been dated to the Hesperian era. Very streamlined islands and terraced walls that strongly suggest scouring by massive amounts of water flowing in enormous floods can be seen in Figure 6.17, in contrast to the gradual erosion by rain and seepage more common on Earth. They originate either in canyons containing layered deposits, or in crater-like features that appear to have been produced by the collapse of part of the surface. It does not seem likely that any kind of Earth-like hydrological cycle, involving rain and runoff, or melting winter ice, could produce

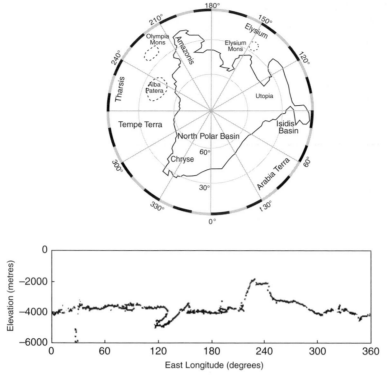

Figure 6.16 Measurements of the topography of the northern hemisphere of Mars, from *Mars Global Surveyor* laser altimeter data. In the upper part, the black line in the map indicates a possible ancient shoreline surrounding a vast, smooth-bottomed basin with a depth of nearly four kilometres below the global surface average. This has about the same area as the Mediterranean Sea on Earth, and at least six large outflow channels disgorge into it from the surrounding highlands. The near-absence of cratering supports the idea that it was once filled with water, protecting the surface and filling any impact features that did occur with sediment. The actual elevations of the trace in the upper figure are shown below, revealing it to be a likely, although not perfect, candidate as a coastline. (After Head *et al.*, 1999)

these. A more likely mechanism is that large volumes of water that were stored in the crust were released by giant impacts or volcanic events to flow quickly towards the lowest part of the planet.

In contrast, the smaller valley networks like those in Figure 6.18 do resemble the branching patterns of terrestrial river systems and could therefore be the result of surface runoff from rain occurring in a warm, wet climate. The evidence suggests that some of them formed in the Naochian era, i.e. that they predate the outflow channels and hence were not necessarily produced by the same kind of process. However, the resemblance to terrestrial systems may be only superficial, and some have argued that here also groundwater sapping is more likely than surface runoff as the major process in the formation of valley heads. The fact that the youngest group of networks, probably dating from the Amazonian, is found

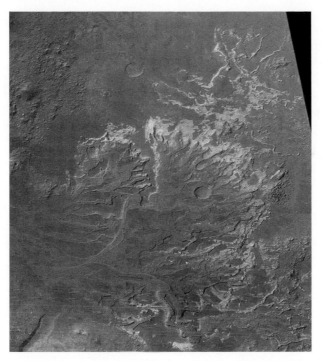

Figure 6.17 Patterns near Holden crater resemble alluvial fans on Earth, suggestive of the transport of soil and rocks by the rapid flow of large amounts of liquid water.

near the young volcanoes Alba Patera, Ceraunius Tholus and Hecates Tholus, lends support to the idea of a geothermal source for the water, as does the recent discovery of apparently very young hydrological gullies on steep slopes polewards of 30° latitude.

6.7.2 Aqueous mineral deposits

If there were long-lived bodies of water in the northern plains or in the craters and basins with valleys running into them, they will have left deposits of sediment that form a record of the climate in that vicinity and globally that can eventually be explored and interpreted. Several of the large canyons have layered material in their walls that probably formed in large standing bodies of water, although it is hard without closer study to rule out some kind of periodic wind-driven or ice-driven process (or two or all three of these working together). The best evidence for the involvement of liquid water is the presence of soluble minerals, which must have been precipitated when the water evaporated, such as the sulphates and the haematite 'blueberries' that have been found in abundance by the rovers.

Carbonates should also be present, especially if weathering was responsible for the removal of most of the atmosphere since the early, warmer epoch, but

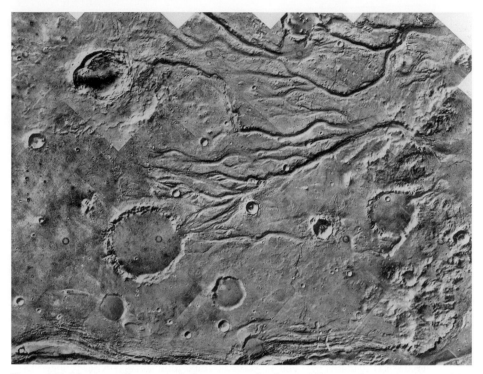

Figure 6.18 Martian valley networks have a superficial resemblance to drainage systems of this scale on the Earth, but on closer inspection they have fewer small-scale streams feeding into the larger valleys, suggesting they were produced more quickly by catastrophic amounts of water, possibly originating underground rather than as rain.

as we have seen these have remained fairly elusive. The preponderance of sulphates at the sites explored so far is consistent with the idea that the liquid water contained dissolved acids that eliminated, or prevented the formation of, carbonates exposed to it. This explains the absence of carbonates where sulphates are found, but makes it harder to understand where the carbon dioxide in the early atmosphere went. Carbonates, or some other sink for carbon dioxide, still have to be present somewhere, perhaps in many small, specific locations like the one recently identified in the Nili Fossae region, or still undiscovered in large quantities below the surface. The evidence that carbonates are a minor component of the wind-blown dust found everywhere on the planet supports the idea that there are extensive, exposed sources on the planet, perhaps in the form of chalk cliffs, the Martian equivalent of the white cliffs of Dover. The rest could easily be buried beneath accumulated dust in the former ocean basins. If not, it lends support to the idea that climate change occurred catastrophically when the early atmosphere was mostly blown away in a catastrophic collision between Mars and a large meteorite, an asteroid or a comet, since it is difficult to identify any other process that would remove large amounts of carbon dioxide.

6.7.3 Layered terrain

The 'layered terrain' that prominently surrounds both poles on Mars, and also appears in various forms all over the planet like the example in Figure 6.19, records changes in materials or deposition processes that have operated over long, geological time scales. The regions near the poles are virtually unmarked

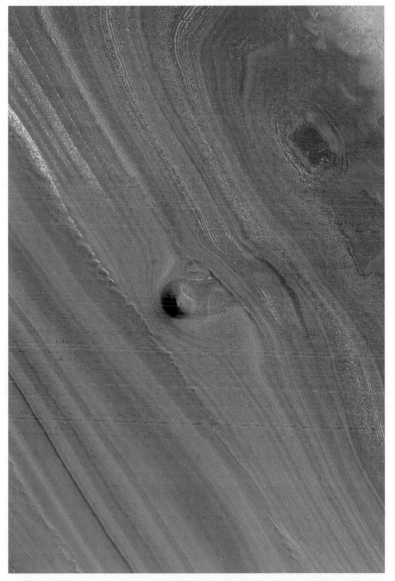

Figure 6.19 Layered terrain near the north pole (latitude 82.6 degrees) imaged by *Mars Reconnaissance Orbiter* in September 2008. The rather mysterious small conical hill about forty metres high near the centre of the image is a mound of polygonal blocks, probably made up of dust cemented by ice.

by craters, suggesting they are the youngest terrain on Mars, and they overlap cratered terrain in the south and the plains in the north. The main process at work is the transport of airborne dust, composed of material eroded from the surface at low and middle latitudes, towards the poles by the equator-to-pole overturning of the atmosphere in its 'Hadley cell' type circulation. At latitudes higher than about 75°, the dust is trapped by sedimentation as water and carbon dioxide freeze out at the poles, the deposition process showing cycles on time scales ranging from seasonal to millennial to millions of years. The main difference between lighter and darker layers may be the amount of dust they contain, but this is not likely to be known for certain without direct sampling, which no doubt will reveal other differences in the climate over time, conveniently dated by the layering.

Where the vertical cross-section is exposed along the sinuous outer margins of the polar caps, the layers revealed in the highest resolution pictures available so far range from a few metres to hundreds of metres in thickness, and show light and dark bands. The cliffs and terraces along these margins, and in the valleys that cut into the deposits, are sometimes scalloped or rugged and sometimes relatively smooth, suggesting that the layered polar cap material varies in hardness or integrity from place to place. Much finer layers may be present, of course; not only those thinner than the resolution of the camera, around half a metre at best from orbit, but also layers of similar colour will not be discriminated.

The layers both on and off the residual caps are different in the north relative to the south, and different from those in other parts of Mars like the deep canyons in the Valles Marineris. The climates at the two polar regions may have been quite different from each other over the entire history of the planet. The residual cap at the north pole – that left in the summer after the winter deposits have melted – is larger than that in the south, but the area of layered terrain is smaller. Sand dune fields, which surround the layered terrain at both poles, are more extensive in the north.

The puzzling differences between north and south are accentuated by the detailed pictures from *Global Surveyor*, which revealed the 'Swiss cheese' morphology of much of the south residual cap in Figure 6.20. This contrasts with the corresponding region in the north, which is flatter but heavily pitted on a much smaller scale. The Swiss cheese effect is due to the sublimation of parts of the carbon dioxide ice covering on the cap, revealing water ice below. It is a long-term and not a seasonal feature, and evidence of secular change is emerging, suggesting that moderate climate change (at least) is a feature of present-day Mars as well.

The dark bands that occur in each residual cap also differ between the poles, as can be seen by putting them in alongside each other Figure 6.21. In the north, the equator-ward slopes form a sequence of ridges and depressions, while southern layers tend to have a more stepped appearance. Both kinds sometimes have sand dunes deposited on them, suggesting that the layers form a relatively hard and glossy substrate across which sand can be transported and accumulated without much erosion of the ice below.

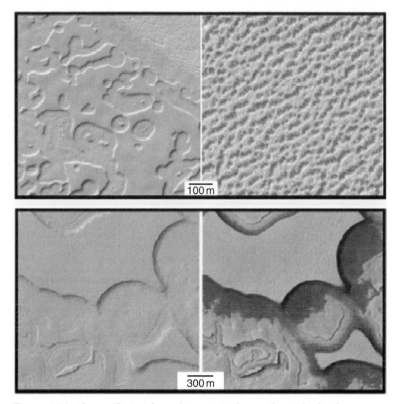

Figure 6.20 Above: 'Swiss cheese' terrain in the south residual polar cap, compared to pitted terrain typical of the north residual cap. Below: Λ segment of 'Swiss cheese' terrain in two different seasons, showing how the brightness varies with frost cover but the features remain permanent.

Figure 6.21 Layered terrain in the north (top) and south (bottom) polar regions, showing the different characteristics that tend to be typical of the layers near each pole.

Both polar caps are several kilometres higher than the surrounding regions, as shows clearly in Plate 16, and the layered terrains extend beyond the present residual polar caps. This supports the idea that they consist of deposited material, and also suggests movement of the poles and perhaps larger caps, corresponding to lower polar temperatures, in the past. The smaller extent of the northern layered terrain, and the extensive sand dune field around it, could be further evidence that this low-lying region was once an ocean basin. A detailed

203

Figure 6.22 This remarkable complex of rectilinear intersecting ridges near the south pole appears to have been formed by subsidence of icy plates typically a few kilometres on a side.

record of all of this behaviour, and of the Martian climate as a whole through time, seems likely to be recorded in the layered deposits, making them an attractive target for future exploration.

The morphology of the surface near the poles is highly complex in some places. Near the south pole, *Global Surveyor* discovered the remarkable set of rectilinear ridges shown in Figure 6.22. This seems to have formed as a result of fracturing of some underlying, more ancient landform, followed by subsidence or erosion of the fields contained inside the walls formed by the ridges. Numerous small dark features are scattered around, the nature of which is hard to guess, except that they seem to have shed their covering of carbon dioxide frost sooner than their surroundings.

The existence of layered terrain near the polar caps, which can sublime and re-form, trapping large amounts of airborne dust as the atmosphere snows out onto the pole in the winter, is less surprising than the fact that thick layers are also found at more temperate latitudes (for example in Meridiani, almost on the equator, see Figure 6.23). This has led to speculation that, rather than having simply evolved from a warmer to a colder climate, Mars may actually alternate between having a cold thin atmosphere (as at present) and a warmer thicker atmosphere able to support precipitation from water clouds. This could conceivably happen if additional atmospheric carbon dioxide is released from the 'permanent' polar caps at times when the orbital characteristics of the planet change, as they are known to do on very long time scales. Such periodic orbital changes are thought to explain the terrestrial ice ages, and are more extreme for Mars because of its more eccentric orbit. On the other hand, it now appears

Figure 6.23 Layered terrain in the Meridiani region, imaged by the *HiRISE* camera on *Mars Reconnaissance Orbiter*.

that the polar caps are not massive enough to supply the necessary material to create an atmosphere thick enough to induce a climate warm enough for liquid water. This would be the case even if the caps consisted entirely of frozen carbon dioxide and were completely vaporised. In fact, the caps at both poles that remain in the summer, after the carbon dioxide that freezes out during the winter has evaporated, are now known to consist almost entirely of water ice.

The most recent estimates of the volume of ice contained in the polar caps come from the laser altimeter on *Global Surveyor*. Elevation measurements of the north polar cap, which is about twelve hundred kilometres across, show it has a maximum depth of three kilometres and an average thickness of approximately one kilometre, corresponding to a total volume of about a million cubic kilometres. This is less than half that of the Greenland ice cap, and only about four per cent of the Antarctic ice sheet. It is also at least ten times less than the volume of the ancient northern ocean that would have filled the low-lying regions up to the apparent coastlines on Mars. Assuming this interpretation of the surface record is correct, the rest of the water must now be stored below the surface and in the (much smaller) south polar cap, or have been lost to space.

6.7.4 Clues from atmospheric composition

Most elements exist as a range of isotopes, each with a slightly different mass. Where the ratio of the abundances of these has been measured in the Martian atmosphere, evidence of fractionation (relative depletion of the lighter isotopes) is found that suggests a major portion of the atmosphere has been lost over time. Loss certainly occurs in the upper atmosphere, where the concentration of gases with altitude is determined by diffusion, producing an increasing proportion of the lighter gases at greater heights. Thus, if gases escape from the top of the atmosphere in significant amounts, by virtue of their thermal motions or in some other way, over long periods of time the heavier isotopes can become enriched.

Attempts have been made to estimate the total amount of atmosphere originally present on Mars based on the abundance of the two nitrogen isotopes, ^{14}N and ^{15}N. Measurements show that the heavier isotope ^{15}N is enriched by nearly a factor of two relative to ^{14}N when compared with the terrestrial case. It is unlikely that the ^{15}N/^{14}N ratio differed from that of the Earth at the time the planets formed, and the enrichment on Mars is probably the result of the more rapid loss of the lighter isotope. An analysis indicates that between twenty and three hundred times the present amount of nitrogen would have to be present originally to produce the observed enrichment in this way. The amount of water that would have been lost at the same time is estimated to be equivalent to a depth of more than a hundred metres averaged over the planet.

There are problems with this, however. The ratio of the heavy to light isotopes of hydrogen, written D/H, has been measured spectroscopically in the Martian atmosphere, and in water contained in Martian meteorites, to be five times that of the Earth. Similarly, the argon ^{38}Ar/^{36}Ar ratio is thirty per cent enriched over Earth's while the ^{136}Xe/^{130}Xe ratio is about twenty per cent greater than that in the Sun and in meteorites. Krypton and xenon isotopes are close to the solar values, however, and some of the nitrogen trapped in Martian meteorites has isotopic ratios in nitrogen, oxygen and carbon that are similar to Earth's and not to Mars' present atmosphere.

It is hard to reconcile these findings without assuming that there are at least two reservoirs of these gases on Mars, only one of which was fractionated during the loss of the early atmosphere. The other may have been released from the interior at a later time. At present, there are too few measurements, and too little knowledge of where on the planet the samples came from, to make full use of the information that these kinds of data obviously contain. However, the analysis of future returned surface materials containing a wider variety and abundance of volatile species (including compounds of interesting elements such as sulphur and chlorine, as well as carbon, oxygen and noble gases) clearly offers great potential for deciphering the long-term volatile evolution of Mars.

6.7.5 Aeolian deposits and surface modification

Winds, and wind-blown dust, have been significant forces in shaping the surface of Mars, with huge amounts of material having moved around the planet. Deposition features such as dunes are seen almost everywhere in a variety of

Figure 6.24 Sand dunes of all shapes and sizes, built up by wind-blown dust, are present all around Mars. These versions, one to two metres high, were photographed at close range in Endurance Crater by the *Opportunity* rover.

sizes and styles, like those in Figure 6.24. The elongated, wind-eroded ridges known as yardangs are also common, and extensive regions of etched and eroded terrain are found. Major changes in surface brightness are sometimes seen following dust storms, providing evidence that deposition and removal of material by aeolian processes are modifying the surface on a large scale even today. The estimates of erosion rates from materials seen at the *Viking* and *Pathfinder* sites would be too low to explain the transport that has apparently occurred unless most of the latter occurs during relatively infrequent, energetic events, presumably the approximately biannual global dust storms.

Remarkably, there are wind-blown dunes on the slopes and in the calderas of the highest volcanic mountains, where the atmospheric pressure is less than a millibar, one-thousandth of the pressure on Earth. Saltation – the movement of particles on the surface by wind – is not expected to occur at such low pressures, according to theory, unless the winds are much higher than the current models show at these locations. Either they were formed in mighty storms, by some other process, or long ago when the pressure was higher. This is another puzzle to add to the list for Mars explorers of the future.

Figure 6.25 The older lava flow on Mars' Elysium Plains (left) is more eroded and heavily cratered than the geologically recent flows on the face of Olympus Mons (right).

6.7.6 Evidence for modern volcanism

Most models of the internal structure of Mars suggest that the interior should have cooled and solidified in the distant past, perhaps as long as two billion years ago. The main difference between Mars and the Earth in this respect is size: smaller bodies cool much faster. It also depends on the composition of the interior, which affects the conductivity and melting point of the lithosphere, and the release of chemical, radioactive or potential energy.[9] A cold, solid core appeared until recently to be supported by convincing evidence, principally the absence of a significant magnetic field. Since the Earth's field is apparently produced by dynamo action in its molten iron core, the absence of similar activity on Mars seemed clearly to indicate that the core had frozen. In addition, although there was plenty of evidence for volcanoes on Mars, they all seemed to have run out of molten lava long ago, further evidence that the interior was no longer hot enough to contain liquid magma. More recent evidence has led to doubts about this, however. As the images of the surface get better, it begins to look as if some Martian volcanoes have produced fresh lava fields fairly recently, perhaps within the last few million years. Figure 6.25 shows some of these young lava flows on the slopes of the largest Martian volcano, Olympus Mons.

---

[9] Potential energy is released as heat if relatively heavy substances are able to migrate closer to the centre of the planet. This is an important process soon after a planet forms, but it is not known whether it still operates inside Mars to any significant degree.

The most recent evidence comes from studies of the frequency with which craters and sand dunes occur on lava beds. While younger flows will generally have fewer of these features than older ones, putting precise dates on any particular flow is something of an art. According to the best estimates to date, a number of flows are as little as ten million to a hundred million years old, and some, in an area the size of Canada covering part of the Elysium and Amazonis plains, may have occurred in the last few million years. Supporting these age estimates is the fact that some of the Martian meteorites found on Earth are made of volcanic rock formed only a hundred and fifty million years ago. The age estimates from the meteorites, unlike those from photographs of flows, are based on laboratory analysis of radioactive isotopes in samples of the meteor and are considered fairly reliable.

The existence of these apparently young flows show Mars may have experienced volcanism and subsurface heating that is so recent in geological terms that it is unlikely to have stopped entirely. Major eruptions could be rare, thousands of years apart, which would explain why none has been observed in recent times. Less energetic heating could be continuously melting subsurface ice to create groundwater that escapes on the surface in a less dramatic fashion than full-scale volcanoes, an intriguing possibility for which recent evidence has also been accumulating.

6.7.7 Evidence for recent surface liquid water

High-resolution photography from orbit reveals features on Mars that can be best explained by the presence of liquid water flowing relatively recently where geothermal activity – residual volcanism – melted the underground ice and caused the flows. However, these features are found in some of the coldest regions of Mars, where models predict thick layers of permafrost.

Perhaps the most intriguing feature of these images is that they suggest the presence of water on Mars not hundreds of millions or billions of years ago, but rather, within thousands of years or even more recently. Typically, they show landslides where material has moved down a slope, forming an apron of debris at the bottom, as in Figure 6.26. They resemble so-called 'weeping' features on Earth, produced by water percolating through the ground and causing material to fall as it emerges out of a hill or cliff. The features are rare and mostly occur in the southern highlands polewards of thirty degrees south or thirty degrees north. They were originally thought to be uniquely oriented facing towards one pole or the other, always facing away from the Sun, although some more recently identified examples do not follow this rule. The features are young, free of craters, dust deposits or signs of erosion. In one remarkable case, the debris extends down a crater wall and across a dune field. Dune fields usually form and change quite rapidly, suggesting this collapse was very recent, possibly only a few years old.

The temptation to identify the gullies on Mars with flowing water is almost irresistible. Not only do they closely resemble features on Earth that are definitely

Figure 6.26 A high-resolution image about five hundred metres across by *HiRISE* of a range of gullies in the walls of a small crater that lies at low southerly latitudes about sixty degrees from the pole. The gullies appear to be of various ages, with younger ones cutting across older, more eroded ones, and have the appearance, at least, of having been created by short-lived eruptions at various times of liquid, possibly a very salty brine, from the subsurface.

produced by water erosion, but they fit with the evidence for a warmer, wetter ancient Mars, with the water now trapped below the surface as ice that can be melted in local volcanic hot spots. There is also the enticing possibility that life survives in the warm subsurface water whose presence is signalled by the flow into the gullies, where it can be found and sampled, perhaps quite soon.

However, it is first necessary to explain how liquid water could possibly survive under present conditions on the surface of Mars. Any water below the surface should be frozen to a depth of several kilometres, and even if geothermal heating drove liquid water to the apparent flow depths of only about a hundred metres, it should evaporate quickly in the thin Martian atmosphere instead of dripping down the crater wall. Some researchers have tried to create alternative explanations for the phenomena, mostly centred on carbon dioxide, perhaps compressed and liquefied, or stored as a clathrate, as the working fluid. A clathrate is a compound in which molecules of a gas are physically trapped inside the crystalline structure of frozen material, in this case carbon dioxide gas trapped in the water ice lattice. Clathrates can be stable underground but not on the surface, so if some disturbance – a marsquake for instance – suddenly exposed a deposit to the atmosphere it is possible that this would lead to an eruption of gas that could produce the landslide features. Similarly, pockets of liquid carbon dioxide could be trapped below ground if the pressure were high enough (greater than about five bars) and the temperature low enough (about minus seventy degrees centigrade, which is close to the temperatures found on

Mars at the latitude where the features are most common). Again, this reservoir would erupt if exposed to the pressure of the atmosphere at the surface, although as with the release of gas from a clathrate, it is difficult to picture this explosive process forming the gully features that look as if they were formed by the gradual seepage of a liquid. Also, it is difficult to imagine the mechanism that compressed the gas until it liquefied, or that trapped it in icy clathrates, and then released it, especially in the large amounts required to make the observed gullies. Each of them would require at least a ton of carbon dioxide, according to one estimate.

So, water still remains the most likely cause. It was once present on the surface in large amounts, and this and its isotopic ratios suggest an efficient exchange of water between the planetary interior and the surface in the past. It remains conceivable, given a source of heat and perhaps a large admixture of brine in the water to lower the freezing point, that it could still occur in some locations. The subsurface water is unlikely to be pure, and water with high concentrations of certain soluble salts can have a freezing point low enough to make a plausible case for survival as liquid on the surface, at least for a short time.

Alternatively, the gully features may not be as young as they look. The erosion rate on Mars can be slow under some conditions, and some geologists think the gullies may be as old as fifteen million years. Even the dunes may not be recent, and at least one case has been found where a dune seems to have moved over the apron, rather than the reverse, suggesting that the apron may not be recent. If the 'weeping' features are in fact at least five million years old, then there may be a simpler explanation for their formation, involving the cyclical changes in Mars' orbit. As further discussed later, these occur on a range of time scales and produce different effects on the climate. Very large, very slow changes have been invoked to account for the ancient floods that the geological record shows occurred billions of years ago. But some of the cycles act on relatively short time scales, and one result of these is to alter the tilt of Mars' axis of rotation. At times when the poles pointed more towards the Sun in summer than they do now, the poleward-facing slopes in some canyons and craters could have experienced exceptional levels of warming by solar radiation. Then cliff faces that had been frozen hard could soften to the point where springs could break out from subterranean pools of geothermally heated water or brine, particularly were these happen to be just below the surface.

It still remains to be explained how the liquid survived on the surface long enough to produce the flows. However, the time scale of the flows that produced the gully features, which are only tens or at most hundreds of metres in extent, could be quite short, perhaps less than an hour. During that time, ice might form a layer on top that insulates the fluid beneath, allowing it to continue to flow long enough to produce the observed features, especially if the water is very salty so it has a low freezing point. The gradual decline in Mars' internal heat over the ages, possibly accompanied by a simultaneous decline in the thickness of the atmosphere and hence the surface temperature, would be consistent with the fact that the large water-produced features, the river valleys and shorelines, appear to be ancient, while the small features like gullies look very young. The latter are also more localised, perhaps because only those regions with good connections to the warm interior remain geothermally active as the Martian interior continues to cool.

All of this involves a large number of unproven assumptions – briny conditions, relatively recent fluctuations in axial tilt, localised geothermal effects, ice-cover, and favourable heat balances. Obviously, the gully features are promising sites for future exploration, when all of these can be checked out. Drilling at such a site might even give future Mars astronauts access to liquid water, which they could check for signs of microbial life, and of course utilise as a key resource for future Mars bases. The *Global Surveyor* team that first saw the gully features has estimated that the water necessary for their formation would be sufficient to supply one hundred people for at least twenty years, even without recycling.

6.8   Mechanisms for climate change

The idea that fluctuations of climate have taken place on Mars recently, and that they may still be going on, now joins the widely held conviction that larger changes have certainly taken place over very long geological time scales. The most dramatic and large-scale evidence etched in the surface, first identified by *Viking* more than thirty years ago, corresponds to climate change around four billion years ago, while the weeping gullies and the polar ice caps show evidence for change that is very recent and possibly even current. Thus, all time scales are evident, covering most of the history of the Solar System and all of the time that Mars, and Earth, were recognisably planets as we now know them.

The task now is to make sense of all the different mechanisms that may have contributed to this vast tapestry. They are the same for Mars and Earth, the differences only a matter of degree.

6.8.1   Changes in the output of the Sun

Like any star, the thermonuclear reactions that produce the energy given off by the Sun gradually change the composition of the core and alter its physical structure. As a result, according to most current theories, the Sun gradually has got hotter over its lifetime of about four and a half billion years, so that, at the time of the formation of the water-related features on Mars, it is thought to have been twenty-five to thirty per cent less luminous than today. Since the geological and biological evidence suggests that not only Mars, but also the Earth, were warmer than at present, this presents a problem known as the Faint Young Sun Paradox. If Mars received a third less solar energy today, its mean surface temperature would fall below the freezing point of carbon dioxide and most of the atmosphere would condense on the surface, a global version of the phenomenon that presently occurs only in the polar night in the winter. The mean temperature of the whole Earth would also drop to Arctic levels, around three degrees below freezing, and our planet would become icebound.

It has generally been assumed that the early weak Sun was compensated on both planets by an enhanced greenhouse effect, due to thicker atmospheres perhaps containing larger concentrations of ammonia and other strong greenhouse gases that have subsequently vanished. Additional carbon dioxide and water vapour may have been vented from the hot interior of the planet, and this could have been supplemented by impacting cometary material on a larger scale than happens now. However, there are difficulties with this, as we will discuss later.

Perhaps the early Sun was not as dim as most current astrophysical theories suggests, or possibly the fluvial features on Mars somehow formed in colder conditions, as might be possible if the water was very salty and had a low freezing point (although not, obviously, if the atmosphere was frozen!). Recent observations of young stars elsewhere in the Universe cast doubt on whether there really is a definite progression in heat output with age, while some models find that the Sun was actually warmer in the past, which would have great significance for Mars if true. Finally, ingenious models have been produced that circumvent the whole problem by postulating that the channels on Mars could have been carved by volcanically heated water or mud flowing underneath a thick layer of ice.

6.8.2 Orbit and spin axis variations

Mars' equinoxes precess just as Earth's do, but much faster and with greater effect because Mars' orbit is so non-circular. Carl Sagan proposed in 1971 that Mars' past warm climate occurred when aphelion and perihelion fell during spring and fall respectively, so that neither pole was cold enough to freeze carbon dioxide. The thicker atmosphere would then create enhanced greenhouse warming, resulting in conditions in which liquid water can exist. In fact, recent measurements show that there is insufficient carbon dioxide in the caps to achieve this; the actual pressure rise would be less than two millibars, an increase of only about one-third over the current value. Also, the cycling of the equinoxes takes only about fifty thousand years, and Mars' current cold state has persisted for far longer than that.

However, other changes in Mars' orbit and inclination occur on longer time scales, and it remains likely that a version of the Milanković model[10] used to explain Earth's ice ages will contribute to understanding the Martian climate and its changes through time. Recent calculations have fairly reliably quantified the

---

[10] Milutin Milanković worked on the effects of orbital variations on climate in the 1930s when he was Professor of Applied Mathematics at the University of Belgrade. Much more sophisticated calculations are possible now with modern computers and better data about the Solar System, but the problem is so complex that there are still many uncertainties.

changes in insolation (energy per unit area arriving from the Sun) that occurred in the past, and those that will occur in the future, as a result of variations in the planet's orbit. These variations result from not just the changing eccentricity of the orbit, but also from the effects of the secular perturbations of the other planets in the Solar System, and from the precession and obliquity variations of the spin axis. The latest calculations include the gravitational effects of the whole Solar System, including all eight planets and the Moon, on the orbit of Mars. Even minor effects due to the polar flattening of the Earth and Sun and tidal dissipation in the Earth–Moon system are taken into account, as are the effects of general relativity.

The most striking feature of the resulting plots is an increase about five million years ago in the axial tilt (obliquity) of Mars to more than thirty-five degrees, producing a large increase in insolation at the summer pole, as shown in Figure 6.27. Some recent calculations suggest that obliquities as high as

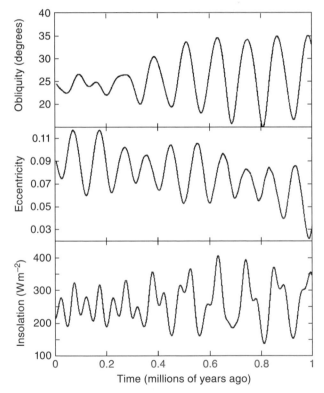

Figure 6.27 Calculations of (a) the obliquity (axial tilt), (b) the eccentricity, and (c) the insolation at the north pole of Mars at the summer equinox over the last million years, allowing for the gravitational effects of the other bodies in the Solar System, including the Moon, Pluto, tidal dissipation terms, and the oblateness of the Sun and Earth. In the first half a million years, the obliquity variations are small and the insolation is dominated by precession, while later changes in obliquity become more important. (After Laskar *et al.*, 2002)

eighty-two degrees can occur, in which case the poles would actually point nearly straight towards the Sun, and the polar caps would melt or sublime. This phenomenon was invoked above as a possible explanation for the release of subsurface water to produce the gullies seen in some, mostly poleward-facing, slopes; how much of an effect it would have on the mean global climate is debatable, since there is still the problem that the amount of volatile material in the superficial caps is not large enough to make an enormous difference. Subsurface deposits could be released, however, and current thinking is that these are probably much larger than those forming the polar caps.

It would be nice to know what calculations of this kind have to say about the orbit and tilt of Mars in epochs more likely to be associated with the formation of the major fluvial features, that is to say, several billion years ago. Unfortunately, the calculations become unstable on such long time scales and the predicted variations in Mars' orbital parameters become unreliable. All that can be said with certainty is that orbital variations large enough to produce major climate change on Mars cannot be ruled out over periods comparable with the age of the Solar System.

6.8.3 Loss of a dense early atmosphere

Although it has often been questioned, and there are other possibilities, the most popular possible explanation for the fluvial features preserved on the Martian surface remains the idea that the planet used to have a much thicker atmosphere, which kept the climate warm and wet through the greenhouse effect. If this was so, then what was this early atmosphere like? What was its composition, and how high was the surface pressure then?

Since the similarly sized planets Earth and Venus originally had something like one hundred bars of carbon dioxide in their early atmospheres, which Venus retains and Earth has mostly lost through weathering into carbonate rocks and sediments, simple scaling suggest that Mars, with its smaller mass, would have started out with about one tenth this amount. Studies that have been carried out of isotopic ratios in the noble gases tend to support this estimate. Currently, there is of the order of one-hundredth of a bar in the Martian atmosphere and probably something between ten and a hundred times as much in the polar caps, regolith and subsurface ices.

The question is then to ask what happened to the early thick atmosphere. Planets, including Earth, lose the lighter components of their atmosphere all of the time, through escape into space as a result of the thermal motions of the molecules. They also gain some, from the interior via volcanoes, and from space as icy mini-comets and other incoming debris with a volatile component. The rates for all of these are quite small at the present time, however, leading to the observed (short-term, at least) stability. Could any known process have taken Mars from a thick atmosphere protecting a warm surface, to its present state?

6.8.4 Mechanisms of atmosphere removal

Atmospheric gas can be lost due to ejection by impacts by large objects, an occurrence that was much more common in the early days of the Solar System when there was a lot more randomly drifting debris than there is today. The efficiency of an impact obviously depends on the mass of the object coming in, but it also depends on the mass of the planet, with more material being removed from a smaller planet because of its smaller gravitational field. Thus impact erosion is relatively inefficient on Earth and Venus, while the impact of a sufficiently large and fast object (more than three kilometres across and travelling at least fourteen kilometres per second) will create a plume on Mars that expands faster than the escape velocity. This could sweep away a lot of the atmosphere, especially if the impact is oblique like a brush stroke, rather than a direct hit.

Very small impacts can also remove gas from the atmosphere, if they are very fast and frequent. All the planets are exposed to streams of protons and helium ions expelled from the Sun to form the solar wind. Since these are charged particles, a planet like the Earth, which has a magnetic field, deflects the solar wind around the planet, and this may also have been the case with early Mars. When Mars' magnetic field disappeared, as the core cooled and stopped acting as a generator, the solar wind could interact more directly with the atmosphere. This would alter the rate at which sputtering took place, a process in which the solar wind particles collide with upper atmosphere atoms or molecules and eject them into space.

Whether atmospheric loss by solar wind erosion is increased or decreased by the presence of a planetary magnetic field has become controversial. The conventional idea that a field deflects the solar wind particles and screens the planet from loss has been challenged by some leading magnetospheric scientists, who have obtained data that show *higher* rates of total mass loss at present from Earth than from either Mars or Venus. It has been suggested at meetings that the reason may be that Earth's field actually helps the planet to trap more energetic particles than it would otherwise, increasing the chances of removing atoms and ions from the upper atmosphere. In this scenario, the loss of a thick early atmosphere on Mars may in fact have been assisted by the presence of a powerful planetary magnetic field during the first billion years or so of the planet's history.

Atmospheric gases do not have to be lost from the planet in order to escape from the atmosphere. On the Earth, weathering is a very efficient process, which removes carbon dioxide by converting it into carbonate rock. The process requires standing water and rain, in which the carbon dioxide dissolves, forming carbonic acid, which then interacts with silicate rocks, producing carbonates. The crust and the polar caps are also storehouses for atmospheric gases, not only as ices of water and carbon dioxide but also by adsorption into minerals. Finally, photochemical mechanisms can turn stable gases into unstable ones, which are then removed by chemical reactions with the crust.

6.8.5 Evidence for atmospheric loss

Probably, all of the above processes contributed to the loss of some of the early Martian atmosphere. But is there any evidence for a large loss, as would have had to occur if an early thick atmosphere was responsible for the water features? Some comes from the isotopic ratios in Martian gases – the relative abundance of heavier and lighter forms of the same element, such as carbon and oxygen.

The removal of gases from the top of the atmosphere by solar wind sputtering and, in the case of the very light gases hydrogen and helium by thermal escape, leaves a signature in the present-day atmosphere, because the lighter isotopes of a gas such as nitrogen escape more easily, and enrichment of the heavy isotopes takes place. The current rates of atmospheric escape can be measured by remote sensing and *in situ* instruments carried on an orbiter. Clues to what escape has taken place in the past may be found from the composition of the remainder, for instance the abundances of the noble gases neon (Ne), krypton (Kr) and xenon (Xe). The ratios of common isotopes such as $^{36}Ar/^{38}Ar$, $^{12}C/^{13}C$, $^{16}O/^{17}O$, $^{16}O/^{18}O$, $^{14}N/^{15}N$ and $^{2}H/^{1}H$ provide quantitative information on the way the atmospheric composition has changed over the aeons, and have led to estimates that around ninety per cent of each gas has been lost by this process. There are already some partial data on these ratios from mass spectrometers on landers, from spectroscopy, and from laboratory analysis of atmospheric gases captured in the Martian meteorites. Although the answers these provide are valuable, they remain far from providing a complete picture of the history of the Martian atmosphere. This may have to wait for sample return from Mars, which will surely include a bottle of atmospheric gas.

Since impacts, unlike other forms of escape, do not strongly favour the removal of lighter isotopes, are not fended off by any early magnetic field no matter how strong, and were more common, and more violent, in the early history of the planet, it may be that most of the atmosphere was removed by impacts first. When all processes are taken into account, the estimates indicate that the Martian surface pressure could have been in the range one hundred to six hundred millibars at the time the geological record began. This is not high enough for liquid water to exist, however; a pressure at least two times greater would be needed for that. If the atmosphere had been as thick as three bars, sputtering and impact erosion would leave behind thirty to a hundred and fifty millibars. The regolith and polar caps could have taken up some of this remaining gas, with the rest converted to rock in the form of carbonates. For such a relatively small quantity of carbon dioxide, the carbonate reservoir is small enough that weathering by liquid water is not required and the fact that carbonate features are in short supply on Mars is less of a problem.

However, even if much higher pressures did occur, there are still problems with the theory that an atmosphere consisting entirely, or mainly, of carbon dioxide warmed Mars during its early history. When the calculations are done, the temperature does not rise high enough unless other greenhouse gases such as methane and ammonia are also present, and probably thick clouds, too. Even then, it is far from certain that Mars ever had an atmosphere with adequate heat-trapping properties – it is just too small and too far from the Sun for any combination of plausible pressures and compositions to warm the planet as much as greenhouse gases did on Earth and Venus in their early histories.

This dilemma is particularly serious if the Sun was dimmer at the time when the Martian climate appears to have been warmer, as we have seen. Recent suggestions that models of the history of the output from the Sun are more uncertain than previously thought, such that it now seems possible that the sunfall on Mars (and Earth) was actually *greater* four billion years ago than it is now, may provide the solution.

6.8.6 Flash flooding by impacts?

The large craters and the river valleys on Mars appear to be about the same age, so it is possible that the warming needed to produce the rivers came from the impacts of giant bodies like asteroids and comets. In between such infrequent, catastrophic events, the planet could have cooled to something like its present austere condition. The fact that the river valleys show few tributaries could be explained by flash flooding rather than slow erosion of the kind familiar on Earth.

This scenario is obviously far less favourable to the chances for early life on Mars. The temperatures in the water and soil would have exceeded the boiling point of water at first, and then declined in a few years or decades after the impact to levels too cold for life to become established. Proponents of this scenario describe Mars throughout history as 'a cold and dry planet, an almost endless winter broken by episodes of scalding rains followed by flash floods'.[11]

6.9 The Martian climate: a current paradigm

While the story so far on the Martian climate includes increasing perplexity as more detailed information has become available, and no two Mars researchers would agree about the details, an overall picture has begun to emerge. The time scale is still very vague, but many of us are more confident than before that we know the basic causes of massive climate change in the past, and present-day trends, including the role of internal activity affecting the availability of water on the surface and in the atmosphere. I can try to summarise how it looks to me at present, no doubt with surprises still to come.

Everyone agrees that, like the other terrestrial planets, Mars has a secondary atmosphere. That means that any atmosphere that surrounded the planet when it formed, which would have been made up mostly of light gases hydrogen and

---

[11] Quote from Segura *et al.*, 2002.

helium, was lost to space, and later replaced by outgassing from the interior of the planet, a process that is still moderately active on Earth, probably very active on Venus, and maybe just ticking over on Mars. During the first billion years or so of its existence, Mars was volcanically active and large quantities of carbon dioxide, water vapour and other gases were pumped continuously into the air; at this stage the Martian atmosphere was much thicker than it is now and the surface pressure was more like Earth's. The composition of the atmosphere was such that the greenhouse effect associated with carbon dioxide, water vapour, clouds and minor constituents such as methane was sufficient to maintain the surface temperature above the freezing point of liquid water. Mars' inventory of water was also not too different from Earth's and in this early epoch large open bodies of liquid water were present on the surface, fed by rain and runoff from the highlands via an extensive network of river valleys. The standing water was very salty, using salt in its broader sense to mean any soluble compound and not just sodium chloride, which helped it to stay liquid when cold.

The climate changed dramatically when the interior of Mars began to cool. This had the two-fold effect of first reducing and then eliminating the planetary magnetic field, and a declining rate of volcanism. Without its magnetic shield, the rate at which the upper atmosphere lost gases to space as a result of the action of the solar wind increased, while at the same time the lost gas was less rapidly replaced by the increasingly torpid volcanoes below. Large impacts by stray bodies colliding with Mars were still a common enough occurrence to have a role in further depleting the atmosphere.

As the atmosphere got thinner, the surface got cooler and fell below the freezing point of water. When this happened, the oceans froze and the amount of water vapour in the air dropped dramatically, further lowering the mean temperature. The entire surface of Mars froze. Impacts continued, as did volcanic eruptions, although neither was as frequent as before. The occasional very large impacts and sporadic eruptions both could melt the ice on and under the surface, and cause a flood of liquid water, lava and ejecta across the surface that would soon, in geological terms, subside. The thin, dry air eroded the exposed rock on the higher mountains and plateaux, producing dust that blew around the planet and gradually covered over the frozen seas. Volcanic eruptions then ceased entirely, and the source of atmospheric gas to balance the losses was reduced to low-level seepage through a few remaining active vents. Atmospheric loss due to impacts also effectively ceased, reducing the rate of decline in the total mass of air on the planet. At some point the air became so cold that carbon dioxide, its main constituent, also began to freeze out at high altitudes and in the polar winters, further reducing the pressure and the temperature.

This is the state in which we find Mars today. The greenhouse effect due to the present thin atmosphere is less than ten degrees centigrade and so, while it is likely that gases are still being lost and so this will fall still further, the rate is slow now and a relatively stable state has been reached. The sulphurous volcanic gases of aeons gone by are now locked in the sedimentary rocks, rich in sulphates, that the exploration rovers recently stumbled across. Most of the water lies not far below the surface, frozen solid. Relics of anything that might have lived in those warm oceans, billions of years ago, are down there too, a fossilised history book that has not yet been opened. Some of their hardier, microscopic descendants may still swim in subarean pools heated by the last flickers of geothermal activity that once powered the mighty volcanoes, also undetected by us, for now.

Further reading

The Martian Climate Revisited: Atmosphere and Environment of a Desert Planet, by P. L. Read and S. R. Lewis, Springer-Praxis Books, 2004.

Elementary Climate Physics, by F. W. Taylor, Oxford University Press, 2005.

Possible ancient oceans on Mars: evidence from Mars Orbiter Laser Altimeter data, by J. W. Head, H. Heisinger, M. A. Ivanov, M. A. Kreslavsky, S. Pratt, and B. J. Thomson, *Science*, 286, 2134–2137, 1999.

Orbital forcing of the martian polar layered deposits, by J. Laskar, B. Leverard, and J. Mustard, *Nature*, 419, 375–377, 2002.

Mars Climate Sounder: an investigation of thermal and water vapor structure, dust and condensate distributions in the atmosphere, and energy balance of the polar regions, by D. J. McCleese, J. Schofield, S. Calcutt, M. Foote, D. Kass, C. Leovy, D. Paige, P. Read, M. Richardson, F. W. Taylor, and R. Zurek. *Journal of Geophysical Research*, 112, 2007.

Intense polar temperature inversion in the middle atmosphere on Mars, by D. J. McCleese, J. T. Schofield, F. W. Taylor, W. A. Abdou, O. Aharonson, D. Banfield, S. B. Calcutt, N. G. Heavens, P. G. J. Irwin, D. M. Kass, A. Kleinbohl, W. G. Lawson, C. B. Leovy, S. R. Lewis, D. A. Paige, P. L. Read, M. I. Richardson, N. Teanby and R. W. Zurek, *Nature Geoscience*, 1, 745–749, 2008.

Environmental effects of large impacts on Mars by T. L. Segura, O. B. Toon, A. Colaprete, and K. Zahnle., *Science*, 298, 1977–1980, 2002.

Chapter 7
The search for life

7.1 Are we alone in the Universe?

It seems unthinkable that we should be alone, the whole thing is so vast and varied. There must be other stars like our Sun that have planets not much different from the Earth, and they must have had a history not too different from ours, including, surely, the rise of living things. Enough is known about the Solar System now to be virtually certain that there are no advanced forms of extraterrestrial life closer than the nearest stars. About three hundred extra-solar planetary systems have already been detected from ground-based observations, although the techniques are currently most sensitive to the detection of large, gas giant planets orbiting close to the star, the so called 'hot Jupiters'. Space-borne telescopes now in the planning cycles of the European and American space agencies will be capable of finding systems with terrestrial planets at distances where the population of Sun-like stars is sufficiently large that discoveries of really Earth-like planets are nearly certain. In a few years or decades after that, astronomers will be able to probe atmospheric conditions on these distant worlds. Some of them will have water, oxygen and equable temperatures and will be the targets for attempts to identify and even communicate with any intelligent species that may exist there.

Until then, most astrobiologists[1] have their focus nearer to home, on the possibility that the nearby and already accessible planets may have simpler

---

[1] The term *astrobiology* was used in the pioneering days of space flight to describe human biology in its ventures into a new environment beyond the Earth. Today the term is extended to describe the study of life that actually originated beyond Earth, also called exobiology. Astrobiology and exobiology are often used interchangeably, but it can be useful to distinguish between speculation about, and the search for, indigenous extraterrestrial life (exobiology) and the study of life in the context of the Universe (astrobiology). The latter definition includes the former, but also admits questions about the nature of life, its origins on the Earth, and the future for humans in the cosmos. Some say neither is a proper science, since there are no known examples to study beyond those already well catered for. Nevertheless, astrobiology journals and conferences, and even university departments, are springing up all over and take themselves very seriously.

forms of life, such as microbes. The other three terrestrial planets in our Solar System have problems when it comes to habitability: Mercury has huge temperature extremes and almost no atmosphere, Venus is dry and very hot, and Mars is cold and frozen. Mars is less different from Earth than the other two in climate terms, and as we have seen it may have been warm and wet at the same time as life was developing on the Earth. Recent exploration suggests that Venus was probably cooler and wetter then, and also more like the Earth, but if it did have life it is going to be much harder to find any trace of it there, with the ancient ocean basins (if indeed that is what they are) buried under layer upon layer of searing volcanic lava, at the bottom of a hot and massive atmosphere. Mercury is generally assumed to be a non-starter in the exobiology stakes, but it has been shown recently to have deep deposits of what must be water ice in craters near its polar regions and these may contain interesting organic material also.

Saturn's giant moon Titan is like another terrestrial planet, roughly Earth-sized and with a nitrogen atmosphere, clouds and a hydrological cycle.[2] Its surface pressure is closer to Earth than Mars or Venus, but the temperature is so low that all of the water is frozen hard. While the processes in the methane-rich atmosphere drive organic chemistry that is of huge exobiological interest, no one really expects to find extant life on Titan, and unlike Venus and Mars, it has no discernable history of a more benign climate in the past. Many of the other outer-planet moons have been shown with reasonable certainty to have liquid water layers, probably rich in ammonia and other soluble material, at some depth in their interiors, and on two of them, Jupiter's Europa and Saturn's Enceladus, they may be within a few kilometres of the surface. Still, they are dauntingly inaccessible, not only to the robot submarines that would nose about in the depths and to the sample return spacecraft that would bring back some of the solution for analysis (to mention two possibilities that have been studied), but also to the direct sunlight and lightning that deliver the 'high-grade' energy on which Earth-like biogenesis probably depends.

That leaves Mars as the best bet by far, for the time being, in the quest for life outside the Earth, not just in terms of its proximity and being able to search for life actually on the planet with the tools and budgets that exist now, but also in terms of the chances that there is something life-related to find. It is also the only place where there is much possibility, in the relatively near future, that robots or astronauts can actually hope to (literally and metaphorically) dig up any survivors or artefacts that may be there.

So, did life ever arise on Mars, and if so, what remains of it now? The 'quick-look' search is over, and the answer inconclusive. The question needs to be posed more carefully in order to make the more difficult and subtler investigations that must follow more focussed and as direct and efficient as possible. A great deal of thought has gone into how this could be done, and it is helpful to look at the

---

[2] See: *Titan: Exploring an Earthlike World* by A. Coustenis and F. W. Taylor, World Scientific Publishing, 2008.

question in several different ways, starting from first principles with a more detailed definition of what to search for and what techniques to use.

7.2 What is life?

A convenient and generally acceptable, to scientists at least, definition of life is: life is a system of materials and reactions that can make copies of itself via a succession of chemical reactions.[3] So far as is known, certain atoms and molecules are always involved, the most fundamental being the carbon atom. Some pre-biotic molecules are assembled from these materials fairly easily given an energy source. For instance, we now know that amino acids are synthesised in the atmosphere of Titan by the action of solar ultraviolet radiation on the nitrogen and methane that it contains. The crucial step is how these go on to make up more complex systems that use energy and nutrients from the environment to drive the reactions that lead to the growth and reproduction of larger, more complex molecules. Enough variants exist of these large molecules – nucleic acids, proteins, sugars and lipids – in Earth-based life to suggest that there could be a variety of ways in which this might happen. If that were true, then life that started somewhere totally isolated from the Earth might conceivably evolve along a different pathway, even under quite similar conditions. It is important to allow for the possibility that there is some broader, universal set of principles presently hidden from us that underlie not only the origins of life on Earth, but also the possible origins of life elsewhere.

Some substances that may have been around from early times on the Earth have the capacity to convert light energy into chemical energy, the familiar modern example being chlorophyll, the pigment responsible for the green colour of plants and leaves. The so-called redox reactions,[4] in which hydrogen serves as a source of free energy, are important for some primitive organisms and could also have been available for early forms of life. The key seems to be the organisation of these materials and processes into enclosed systems with boundaries that can set up gradients between reactive substances in solution. It is these gradients that start the synthesis of larger molecules and high-energy compounds. In the known examples, the containment is achieved by a metabolism based on cells with semi-permeable walls. The solvent they contain has to be liquid water; nothing else has the properties that allow the substances essential for biochemistry to stay in solution together. The physical properties of water dictate a fairly narrow range of temperatures for the genesis, and the survival, of life. Thus any search for life has to 'follow the water': liquid water for extant life, frozen water and clay-like minerals as indicators of once-wet environments,

---

[3] Erwin Schrödinger proposed a different definition in 1944: 'Life consumes and transforms the energy of its surroundings'.

[4] Redox reactions are those that involve the exchange of electrons, achieving both reduction and oxidation. A well-known example is the electrolysis of water ($2H_2O \rightarrow 2H_2 + O_2$).

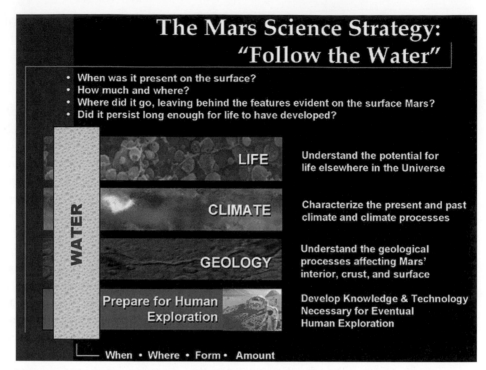

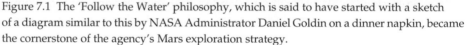

Figure 7.1 The 'Follow the Water' philosophy, which is said to have started with a sketch of a diagram similar to this by NASA Administrator Daniel Goldin on a dinner napkin, became the cornerstone of the agency's Mars exploration strategy.

and layered landforms that allow the history of water in an area to be reconstructed. Figure 7.1 shows this symbolically in a diagram that became a sort of coat of arms for NASA in the 1990s.

Water may be essential for life, but by no means does it guarantee it. Mars may have had oceans, rivers and rain long ago and still stayed lifeless. Searching for microbes or fossilised remains of more advanced species may turn out to be nugatory if life never existed, even in conditions similar to those on Earth where it obviously did evolve. The search continues to makes sense so long as the possibility has not been ruled out: for Mars, the consensus among scientists is probably close to fifty-fifty. The immediate priority is the less ambitious but considerably easier task of searching for the precursors of life by establishing what organic molecules are to be found in atmosphere, ices, rocks and soils. Studying these and the range of complexity they represent should, in principle at least, reveal how far down the path to life Martian chemistry went, according to the definition that life is a sophisticated form of carbon chemistry that involves self-replicating molecules.

It is now generally accepted that carbon is the fundamental building block for life, not only on Earth, but, so far as anyone can see, in all of life's possible forms. No theoretical model for a metabolism that is not based on carbon chemistry has been devised, and most molecular biologists now believe that none is conceivable. On Earth carbon is found in the atmosphere, mainly as carbon dioxide, CO_2, carbon monoxide, CO, and methane, CH_4; and within soils and ices in a variety of forms, some life-related and others not. *In situ* exploration and sample return missions[5] can determine the nature and amount of organic material in representative soils and ices of the Martian crust and search for the more complex organic molecules that are life-related, i.e. amino acids, proteins, carbohydrates, lipids and nucleic acids. Once found, the next step will be to measure their distributions and concentration gradients, and to relate these to the geological and isotopic evidence for the dates and relationships between the regions where they occur. Then the complex task of working out how they originated can begin.

7.3 Ancient and modern life

Obviously, there is no longer any hope of encountering the race of intelligent canal-building Martians that once inspired the imagination of Percival Lowell and a great many others a century or so ago. As little as fifty years ago, many would still have bet on lichens and mosses as a feature of life on the surface of Mars, but now that too is ruled out. Today, the talk is more in terms of bacteria clinging on in underground aquifers, warmed by the residuals of the once-active volcanoes that stud the surface to this day. If astrobiologists are really lucky, they may be accompanied by larger creatures, worms of some kind perhaps, that burrow in the damp soil near these subterranean warm springs. Or there may be nothing at all.

Whether or not modern Martian microorganisms do show up, it is possible that there were larger and more advanced life forms in the warmer wetter era, which have died out and left traces, rather as the dinosaurs did on Earth (although no one is suggesting there was ever anything that large on Mars). It is reasonably certain now that Mars was not always such a hostile environment for biology as we know it on Earth. It seems likely that rain fell from a cloudy atmosphere that was much thicker than the present one, and collected on the surface in warm seas. Life arose on Earth under such conditions, and it does not seem unreasonable to assume that it could have done so on Mars, as well. If the climate on Mars was Earth-like for long enough, the life forms may have been quite advanced – plants and fish of some kind, say – but these would have found it difficult to survive the dramatic change in climate that manifests itself in the more obvious aspects of the geological record inspected to date.

The chances of finding traces of past life are probably higher than those of finding extant organisms. Anything that evolved on early Mars will have left a record that can be found, if mission planners know where to look. Some notorious but implausible examples are illustrated in Figure 7.2; studies on a smaller

---

[5] In chemistry, 'organic' just means carbon-containing, whether life-related or not.

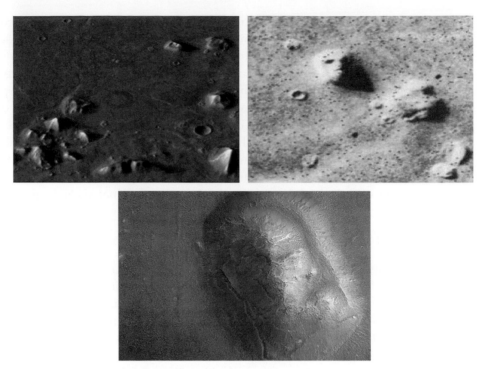

Figure 7.2 Above left, the 'Aztec city', and above right, the giant 'face' on Mars, as seen by the cameras on the *Viking* orbiters. If these really were what their popular names imply, then they would be examples of artefacts that can be studied to learn about past life on Mars. The image below shows what the face looks like under different lighting conditions and at higher resolution, as seen thirty years later by the *HiRISE* camera on the *Mars Reconnaissance Orbiter*. The real artefacts, if any, will be more subtle.

scale will be more likely to bear fruit. Martian fossils may be morphological, like the shells and bones so familiar on Earth, although probably much smaller, or just chemical biosignatures of ancient life such as complex organic compounds, perhaps even beds of coal or oil. More subtle clues are found in isotopic and trace element signatures, and in the presence of biominerals, which are substances that are associated with life and its habitats although not themselves ever alive. Examples of Martian biominerals already found would be the magnetite found in the meteorite ALH 84001, or the silica deposits uncovered by the *Spirit* rover. The latter must have required large amounts of hot water to form, and is like the silica found on Earth in hydrothermal vents or geysers that also contain microbial life.

7.4 Terrestrial vs. Martian life

A fundamental problem for the astrobiologist is that no one yet understands how life formed here on Earth. The manner in which the convergence of organic chemistry, water and energy leads to living organisms remains exquisitely obscure. Persevering in the face of these hurdles, he or she must follow the same paths of study as those employed by the Earth biologist and paleobiologist, with an even greater emphasis on pre-biotic chemistry and habitable environments.

Any view of possible extraterrestrial life is always going to be based on experience of life on Earth. This was implicit in the previous section, where we took the current paradigm developed in an earlier chapter, that conditions on the surface of Mars were once more Earth-like, and asked whether Mars then behaved for a while like a second Earth and followed more or less the same evolutionary track. Another way of putting this is to ask whether Mars was habitable by any living organism of the type found on Earth. In other words, if specimens of various terrestrial life forms had somehow been delivered to Mars back then, could any of them have survived, or mutated to survive? This is a difficult but straightforward question to which exobiologists feel comfortable in being able to find an answer, given enough time and effort.

Suppose the answer is yes, and Mars could have supported Earth-like life. The exobiologist can then take a step back and ask, did any then arise there without being introduced from outside? Again, it may take a mind-boggling number of expensive expeditions to search, but still, it is possible to feel comfortable with the question in principle, since biologists know how to recognise terrestrial organisms when they see them. This assumes, of course, that fossils or biosignatures of an Earth-like species, but one that arose on Mars, have survived or evolved to exist in today's cold, dry climate. Again, given enough hard work it is possible in principle to establish whether that is the case, although of course it would take a very long time to exhaust all of the possible types of relict and all of the locations that might need to be searched.

Suppose no evidence can be found for Earth-like life, whether extinct or extant. Could there be some sort of life on Mars that is unfamiliar to Earthlings, as a result of following a very different evolutionary path? Things get much more difficult now. While most scientists have trouble reconciling the silicon-based life forms and electromagnetic or fifth-dimensional wraiths of science fantasy with the current scientific oeuvre, they have no problem admitting that the latter may be incomplete in some way. Perhaps the only hope is that serendipity will provide the clues that will find it and enable it to be recognised when seen. Science has worked that way many times before.

Again it may be reminiscent of science fiction sagas, but it is important to ask about any new life that manifests itself, Earth-like or not, whether it represents a threat. While it will probably be necessary to go much further than Mars to encounter the likes of Ming the Merciless or similar characters, it is not too difficult to imagine a virus-like organism on Mars that is not found on Earth, but which is compatible with it, and which would spread alarmingly upon its return to a benign environment like Cape Canaveral, Florida or Houston, Texas.

7.5   Habitats

The search for life begins with the search for habitats – places that look like they could harbour life now, or could have done so in the past. That is where the life will be, if it exists; if a benign habitat is found that shows no sign of life, this may be as close as anyone can get to showing that there is no life on Mars. (This is setting aside for the moment the possibility that life on Mars is utterly bizarre and lives in places quite inconsistent with any current concept of the biological sciences.)

A convenient and simple definition of a habitat for extant life that is consistent with most current thinking would say that it is anywhere where liquid water can be found. The recent evidence for seepage from cliffs and for the escape of methane into the atmosphere, for example, lends hope to the idea that there are such places on Mars that can be reached with rovers and drills, or at least with persistent teams of astrogeologists camped on the surface, also equipped with drills. Ancient potential habitats are where sedimentary deposits formed in surface bodies of water, such as the fine-grained, clay-rich deposits that precipitated rapidly from water at the bottom of seas, lakes and streams. In this case the water is no longer present, but if organisms were present when it was, then their fossil biosignatures should be preserved in strata, just as they are on Earth. It should be possible to date these strata from other evidence that they contain – the composition of the deposited material, the composition and isotopic ratios of trapped atmospheric gases, the relative amounts of different radioactive materials, and so on.

One of the major reasons that the steps in pre-biotic chemistry that led to life on Earth are unknown is that the long-term recycling of the surface by the overturning associated with plate tectonics has largely destroyed the record of those early events. The exciting possibility exists that such motions stopped early enough on Mars to have left a record of early pre-biotic chemical events in an Earth-like setting. This will be fascinating even if life never existed on Mars, since it will help us to understand the early climate and the chemical and geological steps that produced an environment similar to the one that preceded the appearance of life on Earth.

Looking for pre-biotic chemistry requires a different approach from the search for extant biochemistry, although the initial steps are similar. First comes the search for modern aqueous environments (groundwater, ice–brine transitions, hydrothermal systems etc.) and ancient sedimentary rocks. The process of identifying these by remote sensing, and then landing rovers to determine their mineralogy, geochemistry and inventory of organic materials, is already underway as we have seen. Future generations of unmanned missions will gather this information as a function of time, including radiometric dating and drilling to get samples from strata laid down during different epochs of Martian history at various depths in a number of locations scattered around the planet. From the amount and type of carbon compounds in the atmosphere and crust at different geological times, the pre-biotic chemistry during different climate epochs can be extracted and related to any tendency for life to develop.

At some stage, as the history of the Martian biosphere is unfolded, it is likely that the measurements required will exceed the capabilities of robot explorers. In the life sciences, even more than in geological or atmospheric studies, the sophistication of the apparatus and methods potentially required is almost unbounded. It will then be necessary to return samples to the laboratory on Earth. Particularly if the trail uncovered is an interesting one, it could make sense to plan the return of biological samples well before the robots have completely exhausted their possibilities. Proceeding to obtain samples from at least six well-dated sites, that have different geological ages and rock types, would certainly save time and possibly even money, depending how the robotic technology evolves and the follow-on experiments implied by earlier discoveries. It certainly seems inconceivable that a manned expedition could be planned without extensive experience with sample return techniques, and without the clues they would provide about the opportunities and possible threats contained in the Martian environment.

7.6 Biosignatures

Having identified the most promising habitats, the search for life can begin in earnest. As always, some careful thought has to go into planning exactly what will be measured, and what tools will be needed. It would be rash to design an experimental payload for a mission, especially a robotic one with very limited capability to adapt to what it finds, if it was only capable of detecting living things and not sensitive to signs of pre-biotic chemistry as well. The instruments must be able to detect and measure the materials and processes that are involved in the entire sequence that transforms simple compounds to complex biomolecules and on to cells and microbes. Figure 7.3 shows this sequence as laid out by the Center for Life Detection at the Jet Propulsion Laboratory, identifying important examples of the various 'biosignatures' – signs of extant or dormant life – that result at each stage.

The crucial importance of liquid water has already been emphasised. The other *sine qua non* so far as metabolism is concerned is energy in a suitable form: solar radiation, electricity, radioactive decay, or some form of chemical source like a redox reaction. Sunlight is a mixed blessing because the ultraviolet photons and subatomic particles that it contains are energetic enough to drive a vast range of chemical reactions, but they can overdo it to the point where they damage living cells. On Mars, there is no protective ozone layer to absorb the ultraviolet rays, nor enough mass in the atmosphere to attenuate the particle flux from space to a safe level. Also, as we have seen, liquid water is not stable on the surface of Mars but only underneath it at depths where sunlight does not reach. Lightning, an intense source of energy that may have been the key to life on Earth, is mostly absent on Mars although it may occur in the more vigorous dust storms. Other energy sources might be available deep below the surface, for instance certain oxidising chemical reactions, geothermal heat from residual

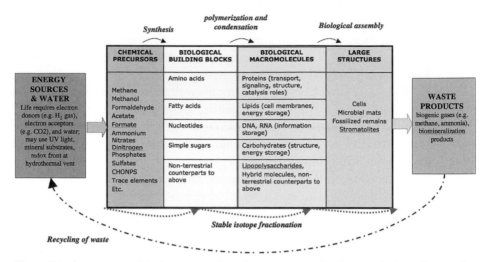

Metabolic processes transform chemical precursors into biomolecules ...

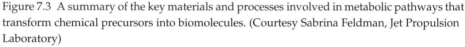

Figure 7.3 A summary of the key materials and processes involved in metabolic pathways that transform chemical precursors into biomolecules. (Courtesy Sabrina Feldman, Jet Propulsion Laboratory)

volcanism, and heat from the decay of radioactive elements. The extent to which these contribute inside Mars is unknown, but in any case they tend to be the wrong sort of energy, too low-grade to create life, although possibly capable of sustaining it, with minimal evolution, since the early days of an Earth-like climate on Mars.[6] Life on Earth probably needed solar ultraviolet light and perhaps also lightning to drive the early biochemical processes that led to life. If Mars had Earth-like conditions on its surface for a long period in the distant past then there could have been water, sunlight and possibly abundant lightning there too. This may reduce the problem to the much easier one of adapting and surviving underground on low-grade sources of energy such as chemical conversion of some kind of food, and geothermal or radiogenic heating.

Finding water and evaluating energy supplies is relatively easy. The next step is to do the same for the basic precursors of life, molecules such as carbon dioxide and carbon monoxide (the commonest oxycarbons, known to be abundant on Mars) methane (the commonest hydrocarbon, recently detected on Mars), ammonia or ammonium salts (the commonest source of chemically available nitrogen), simple phosphorus- and sulphur-containing gases or salts, and so on. Again, these are relatively easy to detect and measure and much progress has already been made with more to come soon.

[6] Whether life could get started with only the low-grade forms of energy that might be available deep underground is a key issue for Jupiter's satellite Europa, where there is almost certainly a lot of liquid water below an icy crust a few kilometres thick. There may be warm water and nutrients, but if the habitat has never been exposed to energetic particles from space, ultraviolet photons from the Sun, or lightning in an atmosphere, should we expect to find life?

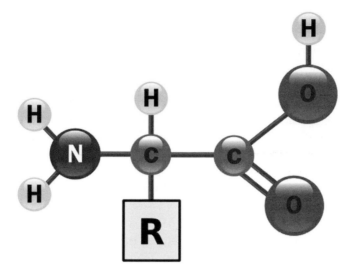

Figure 7.4 The general structure of an amino acid, with the central carbon atom attached to the 'amino' group on the left, the carboxyl group on the right, and another group R, which is different for each amino acid. For the simplest, glycine, R is just a hydrogen atom.

Next comes the 'building blocks' of life – molecules of intermediate complexity that can be synthesised from the simple precursors, for instance by the action of solar ultraviolet photons on an atmosphere containing methane and ammonia (like present-day Jupiter, for instance, and probably the early Earth and Mars). The best-known examples of these are the amino acids, which consist of a central carbon atom with three groups attached, as shown in Figure 7.4.

Amino acids combine into groups to form proteins; similar polymerisation reactions build up other large molecules such as carbohydrates, lipids and nucleic acids. With a wave of life's magic wand, these assemble into cells, amoebae and bacteria. Living things produce waste, and some of this can become the input at the beginning of the same or a different chain, just as the oxygen given off from plants is used by animals to live. Another important point made in Figure 7.3 is that the processes making up the chain can involve isotope fractionation. Since for example they make different use of atoms of carbon 12, 13 and 14, it is possible in principle to tell whether a smear of carbon-containing matter that might be found on Mars was once alive, or not. Of course, reality is rather more complicated than that, nevertheless this is another way in which isotope ratio determination is a valuable part of any suite of measurement objectives directed towards the time-honoured and elusive question of finding life on Mars.

7.7 Astrobiological experiments on Mars

The interpretations placed on colours, markings and 'canali' on Mars by early astronomers, however misguided, laid the foundations for Martian astrobiology more than a hundred years ago. However, experimental life science on Mars really began with *Viking*. In a sense it also ended then, since except for the maverick *Beagle 2*, none of the missions flow to Mars since *Viking* has had the detection of life as their declared goal. *Viking* was a technical success, marred by the fact that it did not come up with indisputable evidence – most would say any evidence at all – for even microscopic organisms on the surface of Mars. The saga of whether *Viking* made the wrong measurements, in the wrong place, and whether the data it did get have been correctly interpreted continues to the present day. *Beagle 2*, as we saw in Chapter 3, crashed on landing and did not return data of any kind. Its promoters and supporters – who are numerous – remain keen to try again at the earliest opportunity.

As described in Chapter 2, *Viking* evolved from a project called *Voyager*, intended to carry detection experiments to Mars, which was much studied at the Jet Propulsion Laboratory and elsewhere in the early days of planetary exploration. In the span of only fifteen years from the first launch into low Earth orbit to the first landing on another planet, astrobiologists found themselves with the access needed to apply instruments with some resemblance to those used in their laboratories. This became the objective of the *Viking* missions, whose two life-seeking landers, together with their companion water-seeking orbiters, were the first off-planet contributions to astrobiology. Fired by the confidence gained during the era of *Apollo* missions to the Moon, *Viking* cut straight to the chase in the direct search for life beyond Earth. Its official objectives were slightly more muted: to obtain high-resolution images of the Martian surface, characterise the structure and composition of the atmosphere and surface, and search for evidence of life.

Each *Viking* lander carried instruments weighing a total of just under a hundred kilograms to study the biology, chemical composition (organic and inorganic), meteorology, seismology, magnetic properties, appearance and physical properties of the Martian surface and atmosphere. Two cameras capable of scanning over all directions were mounted near the sampler arm, which had a collector head, temperature sensor and magnet on the end. A meteorology boom, holding temperature, wind direction and wind velocity sensors extended out and up from the top of one of the lander legs; a pressure sensor was attached under the body of the lander. A seismometer, magnet and camera test targets, and magnifying mirror were mounted opposite the cameras, near the communications antenna. The lander had an environmentally controlled compartment that held the 'labelled release' experiment shown in Figure 7.5, intended to detect the uptake of a radioactively tagged liquid nutrient by microbes. The pyrolytic release experiment illustrated in Figure 7.6 involved heating soil samples that had been exposed to radioactively tagged carbon dioxide to see if the gas had been taken up by organisms. The gas exchange experiment in Figure 7.7 incubated a Martian soil sample in a nutrient solution for twelve days and analysed the gases that were released. The total mass of all three *Viking* biology experiments was fifteen kilograms, with a peak

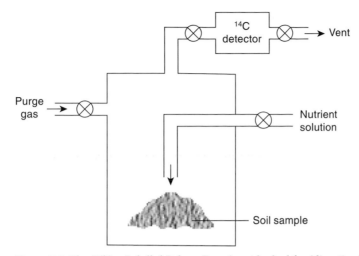

Figure 7.5 The *Viking Labelled Release Experiment* looked for 'digestion' of a nutrient broth by putative microorganisms on Mars. An aqueous solution containing many simple ^{14}C-labelled carbon compounds (formate, glycine, alanine, lactate and glycolactate) was injected into a Martian soil sample, and the gases released analysed for ^{14}C. Nine experiments were performed, including three controls in which the specimen was heat-sterilised first; only the unsterilised samples produced labelled gases.

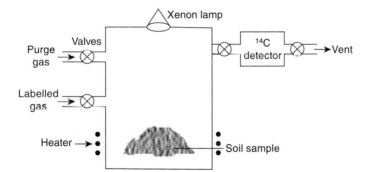

Figure 7.6 The *Viking Pyrolytic Release Experiment* placed a Martian soil sample in an atmosphere of ^{14}C-labelled carbon monoxide and carbon dioxide and shone a xenon lamp on it for five days. To see if any of the carbon had been taken up by microbial life, the gas was purged and then the sample heated to six hundred and twenty-five degrees centigrade. The gases released at this high temperature were analysed for ^{14}C, finding that both the sterilised and unsterilised samples had produced labelled gases, suggesting an inorganic process was responsible.

power consumption of only fifteen watts, the same as a low-power fluorescent light bulb.

Viking was a great scientific undertaking, boldly adopting the extraordinarily difficult objective of seeking evidence of pre-biotic chemistry and living organisms in a largely unknown environment. Perhaps it was unfortunate that the

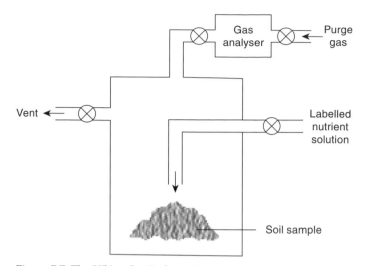

Figure 7.7 The *Viking Gas Exchange Experiment* incubated a Martian soil sample in a nutrient solution for twelve days and analysed the gases released. All of the samples, including those that had been sterilised first, gave off copious amounts of oxygen, suggesting an inorganic process was responsible, probably involving peroxides in the soil.

mission was caught up in the public and governmental expectation that life on another world would be discovered. Today, it is possible to look back on the many NASA press briefings and written and filmed presentations by *Viking's* investigators as they speculate on the forms of life that might be seen on Mars.[7] The scientists used cautious wording to describe to the public and government the expected results of their exploration; everyone directly involved knew that the probability of finding life was extremely small. However, the fact that they represented that it could happen at all gave licence to the media to advertise the mission as one that could come back with the news that Mars is inhabited.

In the event, both landers used their extendable arms to feed soil into their instruments, and both found Mars to be sterile. Was this first direct search for life too simplistic? Were the *Viking* biology investigations attempted before sufficient information was known about the planet? Certainly, knowledge was limited, and yet the investigations performed by the landers seem well considered and executed even from our perspective thirty years on. The conclusions reached, based upon the measurements made by *Viking*, also remain credible.

The possibility that *Viking* did look in the wrong places or perhaps carried instruments that were incapable of making an unambiguous detection of life is, for some, an ongoing controversy. Gilbert V. Levin, who designed the labelled release biology experiment on both landers has consistently claimed his results were compatible with biological activity, and in 1997 he went on record as saying

---

[7] Some of them are on the Web, see for instance http://mars.jpl.nasa.gov/gallery/video/viking30/index.html

he believed that both *Vikings* had detected living organisms. In March 2000, Steven Benner, Professor of Chemistry at the University of Florida, published a paper about the sensitivity of the *Viking Gas Chromatograph–Mass Spectrometer* experiment entitled *The Missing Organic Molecules on Mars*. In this, he concluded that the instrument was insensitive to some kinds of organic molecules, including those he would expect as relicts of the microbial life that might have been on Mars. Most scientists, however, feel that Gerry Soffen, *Viking*'s lead scientist, summed it up realistically when he concluded at the end of the mission that it found 'no organics on Mars, no life on Mars'.

So profound was the disappointment over *Viking*'s nugatory search for life that NASA did not return to Mars for twenty years. Another factor in this was the concurrent disenchantment with lunar exploration by *Apollo*, which affected the public enthusiasm for robot expeditions to Mars just as it helped to kill von Braun's dreams of an early manned expedition. Although NASA supported numerous studies for rover and sample return mission concepts, those involved came to realise that the decision makers in Washington did not want to hear any proposals for any kind of real mission to Mars, especially an expensive one. Eventually, all but a few planetary biologists were lost from NASA's space science programme or made the transition to another field of space research. Early attempts by space scientists in Europe to mount a modest Mars mission of their own, called *Kepler*, also failed, with ESA preferring comets and the outer planets as the objectives for its fledgling programme of Solar System exploration. At about the same time, the Russians had their spectacular failures with the *Phobos* and *Mars-96* missions, and then suffered internal political changes that meant they had to leave the game. The revival came with a refocussing on Mars geology and climatology in the USA, and with a desire to recover the goals of their experiments on *Mars-96* by the Europeans and the Russians. All were fuelled by the excitement generated, albeit probably erroneously, by the discovery of life signs in ALH 84001, and by the promise, also mostly erroneous as it turned out, of much cheaper missions by applying the principles of Faster, Better, Cheaper.

7.8 Renewing the search for life on Mars

As searching for life on Mars became fashionable again in the 1990s, and now on multiple continents, all concerned agreed that the sequence which exploration and life detection on Mars should now follow is:

firstly, orbital surveys to identify the best sites to land;
secondly, rovers equipped with astrobiological payloads to analyse accessible materials;
thirdly, deep drilling to find water and organics;
fourthly, robotic sample return of carefully selected and verified material;
and finally, laboratories on Mars staffed with humans.

The technology needs for such a programme include: very large, very high resolution cameras and spectrometers on low, polar-orbiting satellites; long-range rovers capable of surviving from months to years on the Martian surface and traversing distances measured in tens of kilometres; and lightweight, low-power tools for excavating and drilling. Even when sample return is being contemplated, it is obviously desirable to be able to identify aqueous minerals in rocks and estimate the relative ages of samples while actually on Mars, and ideally to develop advanced instrumentation capable of the *in situ* detection of life, as unambiguously as possible.

The simplest search for indicators of biological activity begins in the gases that make up the atmosphere, where they are most accessible. For instance, had it survived, the *Beagle 2* probe would have measured the carbon isotope ratios in atmospheric carbon dioxide for this purpose. It would also have measured the concentration of methane by sampling the atmosphere directly to confirm the remote spectroscopic detections from telescopic observations. Then it would have attempted the analysis of uncontaminated samples obtained from below the surface to search for organic molecules and signs of metabolic activity, isotopic fractionation and disequilibrium chemistry. This type of analysis is now a target for ESA's *ExoMars* and NASA's *Astrobiology Field Laboratory* missions, both under construction but with the launches still some years away (see Chapter 8).

Increasingly larger and more sophisticated robots performing drilling and biological analyses can go on making massive progress in the assessment of the biological history and potential of Mars, but at some point the emphasis will switch to samples returned to Earth for analysis in a full-scale laboratory staffed by humans. Back in the laboratory on Earth, they will be analysed to find the molecular structures of organic compounds, their isotopic compositions (D/H and $^{12}C/^{13}C$, for instance) and their oxidation states. The results from *Viking* that suggested oxidation by something in the Martian soil is responsible for the selective destruction of organic compounds still need to be comprehensively investigated, more than thirty years later. The theory developed after *Viking* was that atmospheric chemistry near the surface produces peroxides similar to those used in disinfectant solutions in the home. This comfortingly suggests that bacteria would avoid this problem by living below the mixed part of the regolith, one or two metres down where they can be fairly easily reached by the drills on the next generation of rovers. But this remains unproven, and the reality is likely to depend very much on terrain type and locality on the planet.

A sensible strategy to pick up the search for life where *Viking* left off might start by measuring the distribution of oxidants on Mars before anything else, since (assuming problems of access to samples can be solved) this is relatively easy and is likely to have been a controlling factor in determining where, when and how life might have developed. It is already part of the strategy for the missions already under development to conduct *in situ* experiments at a well-targeted low-latitude site down to at least one metre depth to determine the distribution of oxidising compounds like peroxides, and gradients in the concentration of the electrochemically active species like oxygen and hydrogen that they produce. The challenge here is to conduct sufficiently sensitive measurements, as the interesting species are likely to be present in very small amounts, parts per million or less.

7.9 Zeroing in on habitats from orbit

The precursor orbital studies, including high spatial resolution images, maps of surface mineralogy, and measurements identifying near-surface water, water bound in rocks, and subsurface ice, are all underway. Global imaging from orbit at a spatial resolution of better than one metre with the high-resolution camera on *Mars Reconnaissance Orbiter* is finding geomorphic features like ancient shorelines, flow patterns of various kinds including dry river beds and river delta deposits, and more recent-looking flows apparently from hydrothermal springs, all indicative of aqueous processes. Infrared spectral mapping of sedimentary deposits, including the various types of layered terrain now found all over the planet, is used to identify the mineral types that are the best candidates as potential repositories for fossils. These include carbonates, phosphates, hydrated silica and metallic oxides such as haematite, sulphides, sulphates, borates, halides and clays. Attempts are being made, using the *Mars Climate Sounder* that normally measures temperature and humidity profiles in the atmosphere, to detect thermal anomalies on the surface that might reveal regions of geothermal activity. This is difficult, however, and regarded as a long shot; such regions, if indeed they exist at all, are probably too small in size for the temperature contrast to be detected from orbit, especially through the shifting atmosphere. This sort of survey would best be done by an aeroplane or a balloon, flying low over the surface. Plans for such a mission have come and gone, and currently remain out of favour.

Liquid water in (so far hypothetical) geothermally heated reservoirs below the surface may be detected directly by advanced versions of the ground-penetrating radar now being employed by *Mars Express*. High-resolution thermal infrared remote sensing, whether from orbit or closer to the surface, may eventually detect the long-suspected local geothermal 'hot spots' in areas of previous volcanic activity. Such devices can also search for local concentrations of gases, including water vapour, emitted from the vents in the crust that volcanologists call fumaroles. Most of the gas emitted by volcanoes, on Earth at least, is carbon dioxide, which is the same as most of the atmosphere on Mars and therefore difficult to detect. Water vapour is also emitted, giving rise to the exciting possibility that reservoirs of warm water are present just below the surface, heated by the residual volcanic effect, the Martian equivalents of the hot springs and geysers found in Yellowstone or Rotorua. Small anomalies in the water vapour distribution around the planet that could be associated with such an emission have been seen, but the atmosphere is too cold to retain large amounts for long, and so the results remain inconclusive.

The methane signature may be the key that will lead us to the most likely habitats for life on modern Mars. Most living things emit methane as a by-product of their metabolism: lots of it, in the case of humans or cows, but decaying plants and microbial life forms emit it, too. So, whether the methane is volcanic or organic in origin, or both, it is bound to give a valuable clue where to search for life. Small traces of methane, only about fifteen parts per billion or 0.0000015%, were detected in the Martian atmosphere in 2004, by ground-based spectroscopic observations and, almost simultaneously, by the same technique from orbit by *Mars Express*. More recent data have provided signs that it may be variable across the planet and with time, possibly with plumes from localised sources producing high concentrations in specific locations. Methane lasts only a relatively short time under Martian conditions before it is dissociated by solar ultraviolet rays and turned into water and carbon dioxide, so if it is present in any significant amount there must be a current source, estimated at around two hundred tons per year. If the variability of the atmospheric methane is confirmed, then mapping the distribution with dedicated orbiting spectrometers, more sensitive than those available at present, would be able to pin down where the largest sources are with considerable precision. These may well be places where residual volcanism is leaking gases into the air through the fumaroles, which are more subtle, near-dormant versions of the active volcanic craters familiar to us on the Earth. This would be a great place to land a sophisticated robotic explorer, to investigate the source in detail.

7.10 Astrobiological rovers

The instruments and tools that are designed to search for relics of past life are not the same as those used to look for evidence of present life, and a single payload may not be able to accommodate both. On present evidence, the former are probably more likely to succeed, given the climatic changes that evidently have occurred, and finding layers containing ancient organic material in exposed cliff faces and canyon walls is probably easier than drilling into geothermal vents. Any search for fossils or life either succeeds or it fails; the community of Mars scientists is understandably determined, so far as humanly possible, to avoid any more 'all or nothing' *Viking*-type missions. If they were not to show this kind of prudence, how many attempts would the public support before they lose interest and the politicians withdraw funding? So, at planning meetings, the strategy is to proceed to the direct search for life on Mars by first locating and investigating those environments on the planet that were potentially most favourable to the emergence, persistence and survival of living organisms.

These environments fall into three main categories, starting with the ancient terrain found in the heavily cratered Martian highlands, which by definition were exposed to the atmosphere early in the planet's history when the pressure was high and rainfall and other long-departed phenomena may have left their mark. The second set of targets consists of the water-laid sediments in the valley systems and deltas, and the ancient ocean basins, particularly where spectral mapping shows hydrated minerals that are still exposed, and not covered over by wind-blown deposits of dust composed of relatively boring volcanic rock. Finally of course, there are the modern groundwater environments, such as craters with signs of seepage in their walls, the polar cap edges with exposed icy layers, and the fumaroles emitting water vapour and methane.

This still leaves a lot of scope, and a lot of observing, thinking and arguing to be done in order to make sure rare and expensive missions have the best chance of hitting the jackpot. However many reasonable assumptions are made about the environmental conditions necessary for the origin of life, the fact that they are necessarily 'Earth-centric' and based on a limited knowledge of the geological history of Mars demands some manner of progressive approach. The investigation strategy must cover a fairly extensive and diverse range of ancient and modern aqueous environments, where assumptions can be tested and new priorities set. For instance, digging and drilling will certainly be required, although it is not yet clear to what extent. The demanding requirements for doing either on a large scale, or to any great depth, may to some extent be reduced or even avoided altogether by taking advantage of locations where material of different ages has been exposed by nature, such as the walls of the great canyons, the ejecta of young craters, and the material accumulated in outflow channels. These in turn place their demands on the technical teams working on advanced designs, like the cliff-climbing rover in Figure 7.8. Sooner or later, either *in situ* techniques will be exhausted or something really exciting will be found.

Figure 7.8 A prototype for a rover that can explore and sample near-vertical cliff faces, where interesting strata may be exposed, under test at the Jet Propulsion Laboratory. The 'cliffbot' is held by a cable attached to two other rovers that remain anchored on the flat surface above.

7.11 Advanced tools in the search for life

The robots that explore and characterise potential biological sites will need to carry a range of instruments that have much greater sensitivity and versatility than those deployed on even the most recent lander and sub-soil sampler, *Phoenix*. Very capable *in situ* sampling and analysis will still be important even when the stage is reached where samples will be routinely returned to Earth and subject to detailed microscopic, geochemical and mineralogical characterisation in the laboratory. The quantities returned will never be large; currently, for mission planning purposes, NASA has decided that around one kilogram of soil and rock cores from each of three diverse sites can be taken to be the goal. It is important that the samples are the best and most interesting available, according to a selection process based on carrying out as much testing and pre-analysis as possible while still on Mars.

The tools that are needed for this job are many and varied. Some exist already and others are yet to be invented, perhaps in response to unexpected discoveries or mysteries uncovered as exploration progresses. Even old workhorses like mass spectrometers can be made smaller, faster and more versatile, and are the subject of ongoing research and development. The diagram in Figure 7.9 shows in very simplified form the range of techniques that the NASA Center for Life Detection has identified as needed to analyse biological material in each of the six levels of development discussed above.

Top of the list is the trusty microscope. Martian life is unlikely to manifest itself in any large-scale, conspicuous form, but there may be interesting evidence on smaller scales, analogous to the fossils that abound in some terrestrial rock

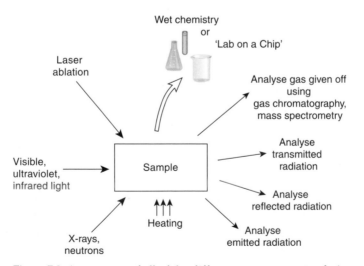

Figure 7.9 A summary of all of the different measurement techniques that can be combined into an integrated instrument suite capable of determining whether planetary materials show signs of past life.

samples of sedimentary origin. These were deposited out of water, initially as relatively soft clays, which later may have hardened into rocks. The rovers already on Mars carry devices for cutting away the surfaces of rocks and examining them at high magnification with a small camera to look at their microscopic structure (their *geomorphology*). Later versions will need to break rocks open, and drill through them to access specimens from depths of several metres below the surface, just as in softer soils. In fact, the most promising targets may be in rocky layers that are tough to penetrate, since they could be better preserved.

Optical, ultraviolet and infrared spectrometers are already well-known analytical devices that have been in use for more than a century. Spectroscopists continue to innovate, using wider ranges of wavelengths, higher spectral resolution, and higher sensitivity. Laser sources, including tunable lasers that do away with the old grating or interferometric device for wavelength selection, and micro-machined quantum detectors, increase sensitivity and reduce size and mass. Used with transmitted, emitted or reflected light, spectroscopy is a powerful and versatile tool for obtaining composition along with physical properties such as temperature and pressure of gases.

The addition of a polarimetric capability to a spectrometer allows it to measure the *chirality*, that is, the 'handedness' of the molecules, which on Earth is different for inert versus life-related molecules. Light that passes through them, usually in solution, is filtered in a preferred direction just as it is in passing through the lenses of Polaroid sunglasses. The effect is fairly easily measured by instruments fitted with polarising filters, so that the chirality of the molecules in the solution can be determined. Many life-related molecules, amino acids and proteins for instance, come in right- and left-handed versions, made up of the same atoms arranged differently. If produced via biological processes, they are nearly all right-handed, having evolved that way possibly because this form is slightly more stable. The ratio between the two kinds in a sample (or even a remote object) can be estimated by measuring the optical polarisation of transmitted or reflected light, providing a relatively easy and, in principle, quite robust test for the signature of past or present life. Advanced versions that work in combination with fluorescent dyes and microscopy can zero in on specific molecules, microbes and metabolic products, and introduce reactants and nutrients methods to help detect growth and metabolism.

Mössbauer spectroscopy is named after a German scientist who discovered in the 1950s a way of determining the amount of different iron-bearing minerals in a sample by irradiating it with very short wavelength electromagnetic radiation (gamma rays) and looking at the spectrum of the reflected or transmitted beam. A source such as radioactive cobalt is used to irradiate the target, and the backscattered radiation analysed by a detector in the instrument that physically oscillates to use the resulting Doppler shift to scan the spectrum. This technique was used by the *Opportunity* rover to show that the 'blueberries' that are abundant in Eagle crater contain large amounts of haematite, an iron-bearing mineral that on Earth is often formed in a watery environment. By extending the technique to obtain the abundance and oxidation state not only of iron but also of manganese and other redox-active metals, Mössbauer spectroscopy can in principle establish the potential of this class of energy sources for supporting biological activity and geochemical cycles.

X-rays are less energetic than gamma rays, but still powerful enough to induce fluorescence from a sample and relatively easy to produce and detect using a discharge tube or radioactive source. A solid-state detector records the fluorescent and diffracted X-rays that are emitted and scattered from the sample. If, additionally, a source of alpha particles (helium nuclei) is used to excite the emission of characteristic X-rays, the sensitivity to lighter elements such as carbon and oxygen is increased. Analysis of these yields chemical information about the elements present, including virtually all of those involved in rock formation, and mineralogical information about the crystal structure of rocks and soils. The method is not sensitive to specific minerals, although these can be inferred from the ratios of the elemental abundances, and it is not very sensitive to iron, making it complementary to the Mössbauer approach.

Another kind of spectroscopy is named after the noted Indian scientist, Sir C. V. Raman. In 1928 he was the first to observe the inelastic scattering of light, resulting in a shift in its frequency that is characteristic of the material doing the scattering. The resulting *Raman spectrum* is a unique fingerprint of not only the composition of the atoms in a sample, but also the structure and chemical bonding. Every molecule, in solid, liquid or gaseous states, has its own distinctive Raman spectrum, making the technique particularly useful for analysing specimens of organic material. In Raman's day the spectra were faint and hard to observe, but today much more powerful sources are available by shining a focused laser beam onto the sample. The scattered light from the sample is analysed conventionally by passing it through a diffraction grating and onto the pixels of a charge-coupled device (CCD) detector to obtain the distribution of intensity as a function of wavelength.

Lasers are also useful for vaporising solid or liquid samples so that their composition can be analysed in the gaseous or plasma state. Obviously, this is hard on delicate organic materials, and not recommended for use on living things, but it can be a powerful way of measuring the trace elements in a piece of rock, for example. Where ordinary spectroscopy would only reveal the bulk composition, traces of rare metals such as gold might be detected in this way. With a sufficiently powerful laser, a rover could analyse the composition of the different layers in an exposed cliff face remotely, positioned at the base, without the need to climb up or abseil down the cliff.

Gas chromatography and mass spectrometry, direct sampling methods that are usually coupled together to measure the composition of atmospheric, rock and soil samples, were first used on Mars on the *Viking* landers, and have been steadily refined since then. If the sample is of solid or liquid material it is baked in a small oven inside the instrument, or vaporised using a laser. The vapour is carried into a long tube containing material designed to selectively delay the passage of gases according to their adsorptive qualities. This is the 'gas chromatography' part. Increasing the temperature in steps will successively release different organic molecules into the mass spectrometer. There they are ionised by an ion source, electromagnetically separated by a mass analyser, and focussed in the electrostatic and magnetic sectors of the device onto an ion detector to produce a spectrum of mass versus the abundance of the molecules. Recent advances have improved the range, in particular the capability to measure very large molecules, including those of biological importance like amino acids and sugars,[8] and the sensitivity. They are also getting smaller, more robust, and less power-hungry, even minuscule in some cases where the instrument is focussed on a few target molecules rather than the more usual wide-ranging approach.

The ultimate evolution of tiny mass spectrometers, ion chromatographs, liquid chromatography and fluorescent imagers is the 'lab on a chip', devices miniaturised using technology similar to that used to produce modern computer processors. Some of the latest of these carry out biochemical assays to detect bacteria, viruses and cancers based on antigen–antibody reactions, and chemical analyses for fields like forensics and pharmacology, all in a device the size of a thumbnail. For instance, the existence of cell membranes might be inferred from the measurement and analysis of fatty acids, and small, robust instruments are being tested that could perform this measurement on Mars. As the technology advances, the potential for astrobiological studies is obvious.

7.12 Deep drilling to follow the water

It is one thing to have sophisticated analysis tools and quite another to have the means to collect and prepare the samples, and place them in the automated laboratory. *Phoenix* used a simple scoop to dig a few inches down, collect soil, and sprinkle it through a filter grid into an oven. Even this simple action, which had been extensively tested in the laboratory, was, in the event, fraught with difficulties. On Mars, lumpy, frozen material tended to get stuck in the grid, requiring a lot of pushing and tapping before analysis could begin. *Phoenix* was satisfied with a shallow sample because it was the first time the soil in the polar region had been analysed; for its successors, collecting the most telling samples for analysis will require drilling, digging or accessing exposed strata in valley walls to get below the weathered and dust-deposited layer on the surface to the more interesting pristine material below. The chances of finding liquid water increase with increasing depth, possibly requiring access down to a level of several kilometres below the surface. However, so of course does the cost and

---

[8] An application nearer to home is to screen material for infectious diseases. One device, small enough to fly to Mars, is currently being used to check white powder in politicians' mail for anthrax and other unpleasant biological agents.

complexity of carrying out the experiment, and really deep drilling may not be practical at all without the hands-on involvement of human engineers.

In current planning, the first drilling, rather than digging, experiment will be carried out by ESA's *ExoMars* rover in 2016. This will reach a relatively modest two metres below the surface, which is enough, probably, to get below the highly oxidised surface region to discover the nature of the more permanent regolith below. Some estimates, however, place the bottom of the oxidised layer as deep as ten metres, and obviously this will vary with location. Crucially, the *ExoMars* drill is attached to the rover so it can be moved about to drill a number of holes in different places, not just to examine the variability of the subsurface properties, but also to increase the chances of finding a layer that has not been sterilised by the peroxides produced at the surface and mixed or diffused downwards.

However, even pristine material will not be really exciting if it is bone dry. At the latitudes to be explored by *ExoMars* it may be hard to get deep enough to find significant amounts of water, even in the frozen state. In most places the ice-saturated layer will be twenty metres or more below the topsoil and will require a large, fixed drill. This is probably not beyond the power of future missions, as robotics continue to make remarkable strides, and is likely to be a task that has to be completed before the arrival of human explorers on Mars, who will have a practical interest in being able to access usable supplies of water. Assisted by machines, the astronauts will be able to dismantle the drill and take it elsewhere, and also to improve it to go deeper and deeper. The goal will be not just to find liquid water, and possibly fossils or even lifeforms; the non-biological goals will include acquiring core samples for climate change studies, and seismology to investigate the deep interior of Mars. When the drilling reaches bedrock, it will achieve improved mechanical coupling with the solid body of Mars for the detection of natural marsquakes and the use of sonar and small explosions to probe deep into the interior. These work best when multiple sites can be operated at different locations around the planet at the same time so the paths followed by the sound waves can be traced through the interior.

The gullies where liquid water apparently flowed recently and therefore geo-thermal sources may be active today present their own challenges. A rover at the top of the cliff drilling vertically would need a reach of at least a hundred metres to have much chance of breaking into the source region. However, if it were capable of descending or circumventing the cliff, to reach the bottom of the apparent flow, then the detritus, presumably rich in ice, could be probed with a much less capable drill, perhaps only ten metres long. It remains to be seen whether the robots that are being developed to scale cliffs and inspect the exposed layers will also be able to drill horizontally into the face of a cliff or gully. Like very deep vertical drilling, this may need human operators on the spot before it makes much sense.

Regardless of how they are acquired, the first cores of biologically interesting material will almost certainly need to be analysed in more detail than is possible by the instruments on the rover or those carried by early expeditions. Instead, the core will be carried to a cache at a location where it can be placed in an Earth-return vehicle, or stored until one comes along later. This will remain the case until human explorers set up full-scale laboratories on Mars, and even then, the best facilities (and most of the eager exobiologists) will always be back on terra firma. To the greatest extent possible, the robotic or human sampling teams will use their on-hand capability to examine the samples microscopically to detect the presence of any fossils, and organic chemical analysis to detect the simpler biomolecules. They will also make a compete visual and analytical record of the setting from which the sample was taken, so that the geological context, age and climatic environment in which the materials were formed can all be estimated as part of the analysis. All of this will form part of the selection process for the samples to be sent back to Earth.

7.13 Sample return vs. meteorites

The SNC meteorites first described in Chapter 2 provide 'free' samples from Mars. The famous ALH 84001 is unique among the SNCs, not only because some think it contains signs of biological activity but also in its geological properties. First among these is its great age, more comparable with 'ordinary' meteorites than with the other rocks from Mars, but it is also unusual in being classified mineralogically as an orthopyroxenite (similar in composition to a nahklite, but with a different crystal structure). Detailed studies of its structure and composition have shown that, after it initially cooled from lava about four billion years ago, it was reheated to a high temperature at least twice after that. These heating episodes were probably due to impacts, part of the meteoritic bombardment of Mars responsible for the heavily cratered surface, much of which is still apparent today. The second impact was presumably that which removed the rock from Mars and blasted it into space. Some time before that, the rock was in a wet environment and liquid seeped through its pores, depositing rounded globules of carbonate minerals. These contain the minute structures that have been interpreted by some as possible Martian fossils. Despite the excitement generated by their intriguing appearance, the objects are so small that the current consensus in most of the scientific community is that they probably have a non-biological origin after all.

Experts have estimated from radioisotope data that the carbonate globules formed three and a half billion years ago, around the time when Mars had a warmer and wetter climate, and the large-scale fluvial features seen today on the surface were formed. From examining isotopes formed by cosmic-ray exposure while it was in space, and then their subsequent decay, it was estimated that ALH 84001 left Mars about sixteen million years ago and joined countless others orbiting the Sun, until it collided with Earth around thirteen thousand years ago.

New SNCs continue to be discovered and new methods of extracting the information they contain continue to be developed. While the more recent studies remain sceptical about the evidence for life that they were initially thought to contain, they are still an ongoing source of information about the early history of Mars. It is often asked why anyone needs to resort to expensive projects such as sample return when Martian samples are available on Earth at virtually no cost. There are many answers to this, but the two main ones will be obvious from the discussions in earlier chapters, which we now recap.

The first is that all the SNCs consist of igneous rock – solidified lava – and this is not the rock of choice in which to search for fossils or biosignatures of any kind. It is not surprising that there are no meteoritic samples of Martian sedimentary rock, that is, of clays, sandstone or other solid material originally deposited by water, since these rocks tend to be soft and are probably not capable of surviving the explosive ejection from Mars, or entry into Earth's atmosphere. Now that it is certain that sedimentary rocks exist on Mars, by analogy with Earth they are the most likely repositories of any ancient biological remains. Thus the case for expeditions to bring them from Mars is stronger than ever.

A second reason for sample return is to obtain the provenance of the sample, that is, to understand where and if possible when it originated, and to know the properties of its surroundings. Even if Martian sedimentary rocks were to hand as meteorites, they would be far less useful than samples collected from known locations on Mars. In the latter case, there would also be information on the age of the sample (from its depth of burial, its position in strata, etc.) and on its history (from its location in ancient river beds or in one of the polar caps, for example) that might be vital for interpreting the evidence it contains.

7.14   Biological sample return

The 'lab-on-a-chip' experiments that exobiologists are busy designing for future life detection missions need to work on samples that are as small as possible. Experience with lunar samples and meteorites suggests that for most purposes a ten or twenty gram rock sample is sufficient to accomplish all of the important measurements. Unless there is a very obvious reason for returning a large object it will generally be most profitable to return a lot of small samples rather than a few large ones. This does not mean restricting the samples to small objects – the sample-gathering rovers will have to be able to carve up rocks, as well as extract them from under the ground or from exposed cliff strata.

The samples will need careful handling, to avoid mechanical damage and chemical or biological contamination. If they are obtained by coring, it is important that the stratification of the material is preserved, or at least recorded. For subsurface material that includes ices or brines, melting or evaporation must be controlled and temperature control, separate containment and hermetic sealing to prevent loss are likely to be difficult but important requirements. If extant life is found, the environmental controls that will be required stretch the imagination, especially if the organisms are to be brought in alive.

Coming back down to Earth, so to speak, the first material returned is likely to include samples of the Martian atmosphere for examination in the laboratory, as well as rocks, soil and ices. Although gases can be analysed *in situ* much more readily than rocks or organic materials, and *Viking* was able to make a significant start, a great deal of the information that is packed into the trace constituents and isotopes is below the limits of sensitivity for any reasonable flight instrument package. Any possible scenario for the development of life on Mars is fundamentally linked to the evolution of the atmosphere, through its control of the climate and the water budget, and its involvement in whatever biogenic reactions may have taken place.

7.15 Planetary protection

It is going to be extremely difficult to explore Mars without introducing biological contamination from Earth. While the Martian surface is quite effectively sterilised by chemistry and radiation, it is hard to be sure some terrestrial organisms will not find a niche where they can survive and mutate to adapt to their new surroundings. These Earthly bugs will soon become hard to distinguish from any native Martian life that still remains to be discovered, and they may be a threat to it as well. This could have happened already: the criterion adopted for *Beagle 2*, for example, was three hundred microorganisms per square metre, the same as for *Viking*. The experimenters compared this to a clean kitchen in a house on Earth, which has several thousands of millions of microorganisms on the same area. Still, three hundred is a long way from zero. The sterilisation process can harm delicate instruments and must be carried out thoughtfully, with a combination of heating to a hundred and twenty degrees centigrade (above the survival point for the most heat-tolerant bacteria on Earth, which is about a hundred and thirteen degrees) and chemical cleaning with alcohol and other solvents. The *Beagle* electronics were treated with hydrogen peroxide plasma and the airbags and the parachute system were irradiated with ultraviolet light, both intense forms of the process that makes Mars itself so hostile to microbial life, above ground at least.

Of course, the avoidance of contamination of Earth by anything that may be found on Mars is an even greater concern, and activities that risk either forward or backward contamination are governed by international treaties and agreements brokered by the United Nations in 1966 and 1979. Quarantine and sterilisation protocols are expensive and can never achieve perfect protection; also, they can be counterproductive with regard to the science goals. The exobiologists are unlikely to be enthusiastic about baking the samples prior to their return to Earth for analysis, or spraying them with disinfectant. This means isolation is the only real option until the risks are better understood. It will be very difficult to prevent human travellers from contaminating Mars when they arrive, and when they are brought home they will be even more difficult to sterilise or isolate than rock samples. The chance may be small for astronauts catching something nasty on Mars, or unwittingly bringing something home that then spreads, but few planners would wish to take the risk, and a lengthy quarantine can be anticipated. The development of efficient screening, cleaning and isolation techniques, and their certification, has to be a priority, along with a rapidly improving appreciation of the actual, as opposed to potential, risks, as exploration progresses.

7.16 The importance of human exploration

A recurring theme in all aspects of Mars exploration, particularly where the search for life is concerned, is the difficulty in knowing in advance when it will be necessary to progress beyond robotic exploration and remote analysis to sample return, and from there to manned expeditions. NASA has adopted a stepwise approach, using 'pathways', that will be discussed in the next chapter. This remains focussed on 'habitats' and will need radical modification if, at some point, robotic explorers and sample return missions show convincingly that life did exist on Mars at some time in the past. At that point, the questions that need to be asked will be more complex, and the observational and analytical capabilities that could be provided by humans and laboratories on Mars could be the more effective approach. The necessary political and popular support for ramping up the budget to cover this massive step might also be forthcoming, propelled by the promise of many more exciting discoveries, although this could depend on what else was happening on Earth at that time. Barring famine, flooding or war, support might be forthcoming even if the data were still ambiguous, so long as they were sufficiently intriguing and if robots seem to be reaching their limits. Finally, pragmatic issues might send humans to Mars even if the robotic programme was going well but before it had delivered any important conclusions, in other words social and political urges or strategic pressures may take over from science as the main motivator for going to Mars. Based on recent history, that does not seem terribly likely however, at least where NASA is concerned. Younger space-faring nations may be prepared to take greater risks than the seasoned bureaucracy in Washington, DC will presently contemplate.

Whether human missions become practical and desirable either from the scientific perspective, or from other rationales, the robotic orbital, surface and sample return programme will provide important information to support human missions, as discussed further below in Chapter 9. Apart from addressing practical questions such as the toxicity of materials like dust in the environment, the availability of water, the dangers of the radiation background, and the forward/back-contamination issues, these will include the development of technologies such as food and fuel production, communications and mobility.

7.17 Epilogue: the prospects for finding life on Mars

Even with a series of fairly optimistic assumptions about factors that are still uncertain – that Mars was warm and wet for a long time, that biology occurred, that it led to large life forms comparable to those on early Earth, and that some of these left fossilised remains (shells, bones etc.), the task of locating and retrieving these on Mars is clearly not going to be easy. A telling analogy that is sometimes heard in planetary exploration circles is with relatively desolate regions of the Earth, a planet that certainly has abundant life, primitive and advanced, and everything in between, but that does not always give up its secrets very readily.

Suppose one of the most advanced robot exploration vehicles, such as the mobile *Mars Science Laboratory* currently under development in the United States, or *ExoMars*, its European equivalent, was sent by rocket to Wyoming or Woomera instead of Mars, what would be its chances of finding life more advanced than bacteria? Colorado is full of fossils, but without advanced knowledge of where to look for them a Mars-style sample return mission to bring one back to California would have a very high statistical chance of failure. Even a human expedition, if confined to the conditions for living and travelling that they are likely to have on Mars, would be far from certain of finding much of exobiological interest if they landed in parts of the Sahara desert.

Of course, the robot or human explorers might set down in the most, rather than the least, promising locations, the Martian equivalents of downtown Denver or Hyde Park, whatever they may be. But this is unlikely and cannot be left to chance, not only because even robot sample return is expensive, but also because the disappointment of a boring result would kill stone dead the ambition of all but the dedicated enthusiasts on Earth. Governments and agencies would certainly back off until they got a new stimulus from somewhere, just as they did after *Viking* until the Allen Hills meteorite and its putative 'nanobacteria' came along.

One approach is to start with a modest search limited to bacteria that are likely to be ubiquitous, if they exist at all. This of course was the *Viking* approach, but (as the *Beagle 2* science team emphasised thirty years later) the surface and near-surface of Mars is efficiently sterilised by the ultraviolet rays of the Sun, and *Viking* might have found life had it been designed to dig deep below the surface for its samples. However, even if this is true, digging in a dust bowl, of which there are many on Mars, most of which present an enticingly smooth surface for a safe landing, would have to go very deep before it found material that had not been gardened by the wind and exposed to the deadly rays, possibly quite recently. Conversely, digging in a more difficult spot for landing, but one where there is a chance of finding a layer of clay, say, or even liquid water from a geothermally heated aquifer, is likely to increase the chances of an exciting result dramatically.

So, exploration and planning remains the name of the game for the time being. Finding the best current habitats requires not just extensive mobility, exotic tools and drills, and a lot of probes, but also continuing to develop a step-by-step understanding of what they are actually looking for. One thing that is certainly harder than finding a region of residual geothermal activity on Mars is proving that there aren't any, anywhere on the planet. Clues, like the possibility of

249

localised sources of water vapour or methane, detected spectroscopically from orbit, may be the way forward and must be followed up rigorously. Exposed strata of sedimentary rock are probably the best place to look for fossils of any hypothetical extinct creatures or plants, but no two regions are the same in terms of history, geology and geochemistry. Success is far more assured by looking carefully before each leap.

There are many different kinds of habitat for indigenous Martian life, depending, inter alia, on what kind of life it is. It is easier to look for the most suitable habitats for Earth-like life, assuming humans will go there in numbers reasonably soon and reside at least for a while. This motivation is likely to remain strong so long as we think there may be, or have been, life there. Suppose, however, all the exploration in the next few decades shows convincingly that Mars is completely and permanently dead. How much longer would it then be before Earthlings make the voyage to our planetary neighbour in person? We can only guess at the answer, but it is probably rational to assume, so long as an increasingly advanced civilisation endures on the Earth (which of course is itself not certain) that we will set foot on Mars, for cultural if not for scientific reasons, sometime within the next one hundred years.

If there are no biosignatures, scientists are left to contemplate whether Mars is of any exobiological significance at all (other than the profound philosophical questions raised by discovering a companion planet to the Earth that could have developed life, but didn't). It is known, for example, that Titan has organic chemistry in its atmosphere that leads to the formation of nitriles and amino acids and possibly a good deal more: much can be learned from this without meeting any Titanians or taking the risk of destroying life on Earth through bringing back the seeds of an epidemic. If in the end explorers find only a lifeless Mars, the corresponding question would be: what happened? What kind of chemical evolution took place on Mars, and how far did it get? Did it lead to the formation of pre-biotic organic molecules? Why did it not go on to the formation of self-replicating molecules, according to the essential definition of life?

The meaning of life is the evolution over a long period of time of molecules of continuously increasing complexity. Depending on what is found on Mars today, it is possible to try to determine how far this process progressed. If it stopped short of producing life, was that because it did not have enough time? Mars may have been warm and wet for more than a billion years, or possibly for only a few hundred million years, or even less. Perhaps that is not long enough for the complex molecules of life to develop (although evidence is gathering that it *was* long enough on the Earth). On the other hand, perhaps it reached some kind of less complex equilibrium state, and stopped there because conditions were never right for life. Perhaps Mars will tell us that life occurs rarely even under ideal conditions.

Part III
Plans and visions for the future

Having contemplated the history of Mars exploration to date in the first part of the book, and in Part II discussed the major goals that remain to be addressed, we come to the seriously difficult subject of trying to understand and predict the future of Mars exploration. Short-term plans are in place, of course, covering the next decade or so, but we have already seen how plans can change dramatically as a result of accidents, discoveries, politics and personalities. The professional planners generally restrict themselves to the next thirty years, with the implicit assumption that one generation is the maximum planning horizon for essentially every major human endeavour. Thirty years may also be the best estimate of how long it will be before the first human sets foot on Mars, although some would say this is optimistic while others assert that it could be sooner if only those responsible (meaning the public at large as well as those in the front line in government agencies) would set their minds to it.

In this final part of the book we will look at the many studies and plans drawn up by NASA and ESA teams, working mostly within the thirty-year horizon and with pragmatic ground-rules (and possessing ample resources of people and funds, for the studies at least). I will also stick my neck out and risk some speculation of my own as to how things will go, not just over the next thirty years but well beyond that. No one can really know what will happen fifty years from now, of course, or even in the next few years. There is no such thing as twenty-twenty foresight, and as recent history shows, where Mars is concerned the whole task is so difficult that we must hope new technology will come along and make reaching and living on Mars a lot easier and safer than it looks at the present time. But without making plans we will never begin.

Chapter 8
The future of the unmanned programme

Much remains unknown about Mars, and the urge to explore remains as strong as ever. The drive to understand mankind's origins, our ability to survive global change, and indications as to whether we are alone in the Universe, all find a focus on Mars. In addition, the cultural urge built into the DNA of our species to see far-off places, not only through telescopes and machines, but also in the flesh, remains a strong motivator. Without being able to explain why, except perhaps by vague reference to a nomadic past, we seem to need to travel ever further, and to find and overcome new frontiers. As a result, it is widely assumed that human travel to Mars is only a matter of time.

However, it is also clear that an expedition to Mars is a very difficult, dangerous and expensive undertaking with the technology presently available. Furthermore, it is likely to remain so, despite the innovations to be expected in any reasonable projection into the mid-term future. There are many possible approaches – *Apollo*-style direct missions, or via the *Space Station*, or via bases on the Moon or Phobos. A combination of these, or some other approach altogether perhaps using new propulsion systems, could be used, each with its own risks and potential setbacks. Progress will come a step at a time, following an agenda as unpredictable as that of the last fifty years, which started when von Braun first drew up his ambitious scheme for a human expedition to Mars. There are, of course, creditable but unofficial plans for getting on much faster, of which the best known is the 'Mars Direct' approach of Robert Zubrin and the Mars Society. Coupling the energy, ingenuity and ambition of the early pioneers to modern technological know-how far beyond that available to *Der Marsprojekt*, this would have men on Mars by 2025 or so at a cost of only a few times NASA's current annual budget.[1] However, although he had the ear of the Administrator, Daniel Goldin, for a while during the Faster, Cheaper, Better era in the 1990s at NASA, Dr Zubrin's success-oriented assumptions, which included manufacturing air and fuel on Mars from local material, involve serious risk-taking, and are

---

[1] NASA's budget was $17.138 billion in 2008.

unlikely to find favour in the current political climate, in the USA at least. The European Space Agency has been more adventurous in recent years, and makes no secret of its desire to put Europeans on Mars. China and India are headed for the Moon, and are unlikely to stop there.

Anyone without exposure to the real process of planning space missions might reasonably expect that the sequence of events is fairly straightforward. Scientists establish research goals and formulate the plans for exploration, and engineers work out how to make it all happen. Both remain aware that budgetary realities have a bearing on the pace of exploration, and that in this regard they are subject to political control. While all of this is broadly true, it is not the progressive, linear process that it would be in an ideal world. In practice there are powerful feedback loops that ensure space exploration remains an immensely complex interplay among scientific hypotheses, the practicalities of engineering the machines needed to get to Mars and then do the exploring, the programmatic drivers, especially budgets, and the domestic and international political elements of national ambitions in space. The level of popular support for any dangerous and costly venture, and the degree of enthusiasm expressed in the media, both of which wax and wane with each new discovery or technical failure, are also powerful regulating factors. Finally, large institutions, predominantly military, academic, and industrial, have important voices in setting directions and goals for the endeavours to which they make major contributions, and tend to have their own sub-agendas.

All present indications are that, barring global disaster, there will continue to be some kind of vigorous Mars exploration programme in the next few decades. Indeed, it could accelerate compared to current official expectations, as new space-faring nations like India and China join Europe and Japan in cooperating or competing with the original pioneers from the United States and Russia. It is also reasonably certain that the main motivation will continue to be the big three questions: to understand climate change on Mars, and relate it to the habitability of past and present Mars and global change on Earth; can life be found on Mars, or can it be proven that there never was any; and can humans be delivered safely to Mars and set up a base there? The last of these is deliberately posed in non-scientific terms: although the colonists will undoubtedly do scientific studies of various kinds, it is best to be realistic about the reasons for sending them to Mars, and these are primarily political and cultural, with science following some way behind.

8.1 Choosing the way forward

A problem that planetary scientists have in influencing the selection of exploration pathways is their difficulty establishing priorities between themselves. New planetary missions rarely come along at a rate large enough to satisfy everyone's diverse interests, so at budget time they have to decide which to favour among all of the potential targets in the Solar System. At a given planet, choices have to be made about the focus on objectives (such as geology versus atmospheric science) and the method of study (landers versus orbiters, for example). The destinations for exploration can be, and often are, selected by a reckoning of fairness in sharing the opportunities among science disciplines, assuming equivalent technical challenges. If clear priorities from scientists are lacking, other agendas of all sorts can take over the course of exploration.

Fortunately, for Mars scientists, the well-known public appeal of Mars attracts its share of government recognition and, consequently, sufficient funding to pursue a wide range of different projects, reducing the need for competition between them. Also drawn to Mars are the space architects and aerospace industrialists who are keen to contribute to the exploration strategy. Within NASA, the Mars programme has for many years had a separate budget line, good for about one mission per year on average (i.e. two launches in a typical launch window, twenty-six months apart). Occasionally, the influence exercised by individuals with oversight of NASA's budget can amplify or override all other priorities, as was the case when NASA Administrator Dan Goldin turned his attention towards Mars, and again, more recently, when President George W. Bush announced his *Vision for Mars Exploration*.

Like a game of chess, certain pieces always have to be present in top-level space programme planning:

1. A transportation system must be identified, planned, built and flown, with the entire infrastructure necessary to operate it.
2. 'Stepping Stones' must be identified and adopted or bypassed: do our goals for Mars require a space station in Earth orbit? In Mars orbit? With or without a man-made space station, how should the Moon figure in plans for Mars?
3. What level of robotic activity at the intended landing site is necessary prior to a human expedition?
4. What is to be the scope of human activity on Mars in the longer term, once initial exploration has taken place?

Human missions are the subject of the next chapter. Before getting to that, we look at plans for the unmanned missions that will fly to Mars before astronauts do. These have long featured two classes of mission very prominently – networks and sample return. There have been a number of false starts for each and no actual flight hardware has yet been built and flown. A third class of mission, the use of specially designed Mars aircraft for a global photographic survey, also comes up for consideration from time to time, but tends to be seen as impractical, and perhaps also now irrelevant since the advent of high-resolution photography from orbit with *Mars Reconnaissance Orbiter*.

The last serious proposal for an unmanned reconnaissance aircraft to explore Mars was *Kitty Hawk*, a hydrazine-powered airplane proposed in 1998 for deployment on Mars in 2003, marking the centenary of the historic first flight of its namesake by the Wright brothers (Plate 41). The plan, which was adopted initially but then cancelled in November 1999 for budgetary reasons, had the one hundred and thirty-five kilogram aircraft arriving folded into an aeroshell with a heat shield for the initial entry into Mars' atmosphere. After descending on parachutes to its operating level of about two kilometres above the surface, the wings would extend to their full ten metre span, and the modern *Kitty Hawk* would begin a three-hour flight, covering an estimated distance of nearly two hundred kilometres. To reduce the complexity, no attempt would be made to soft-land – the plane would crash at the end of its mission. The proposed scientific payload included: (1) a gravity gradiometer, for measuring the subsurface mass distribution; (2) a magnetometer, for measuring the magnetism of crustal rocks; (3) an electric field experiment, to measure signals from the Martian ionosphere and lightning; (4) a laser altimeter, for determining the topography beneath the aircraft; (5) an infrared imaging system, for determining rock composition; and (6) no less than six cameras, including a tail-mounted video camera to capture the aircraft and the surrounding terrain in-flight.

While aircraft will probably not feature again until they are needed for the transport of settlers and their goods between Martian colonies, it is still very clear at any recent meeting dealing with future Mars exploration that both networks and sample return remain very much on the agenda. In fact, the very complicated and often acrimonious discussions about how to move forward all seem to boil down to the rather simple logic shown in Figure 8.1. Networks first, then sample return, itself preceded by the development of the capability to drill deep for at least some of the samples.

In addition to advancing three key areas of Mars science – Meteorology, Seismology and Geology – networks provide the essential breadth of coverage that will help to understand the diversity of sites on Mars that are available for landings and bases. Still, it is widely held among the scientific community that rovers, networks and all other forms of *in situ* analyses including drilling are necessary but will not be sufficient, even with regular advances in robotic laboratory techniques, to resolve the climate and life issues, nor to prepare properly for human expeditions. For that, sample return is essential.

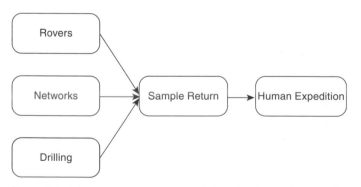

Figure 8.1 A 'common sense' flowchart showing how future missions will progress from essential precursors to sample return to human expeditions.

8.2 Towards Mars networks

As we have seen, networks have been planned before and come close to implementation, only to be discarded in favour of some other mission or plan.[2] When eventually implemented, the project will probably combine elements of all these shelved plans, including international cooperation for which it is ideal. Individual nations or agencies can build and deploy stations independently, and then share data to achieve the goal of a set of robotic observation posts deployed all over the globe. The individual stations are not very large and remain static once they are in place. They should be as long-lived and numerous as possible, and can be added to as resources permit. The question of how many stations make a valid network has been much discussed. Ideally, it takes at least twenty to provide reasonable coverage of a planet the size and complexity of Mars, but more than one mission planning exercise has concluded out of practical and economic necessity that as few as three, or even just one to test the technology, would be a very useful start.

As with much of recent Mars exploration strategy, parallel developments have taken place on both sides of the Atlantic. In the early 1990s, NASA initiated a study of a mission called *MESUR* (*Mars Environmental Survey*) that consisted of sixteen stations, while ESA embarked on *MarsNet*, which had four. The two teams worked closely together and exchanged representatives, speaking optimistically of flying both missions at the same time to make an international network of twenty stations.

In the event, only one station flew, the *MESUR Pathfinder* that carried the small *Sojourner* rover in 1997. By then the mission was just known as *Pathfinder*, since plans to follow it with sixteen copies to make a network had by then been shelved when the cost reached $1.5 billion, and kept rising. *MarsNet* failed to survive the ESA selection process, in part because the need for a complex delivery system and a communications orbiter to relay the data also pushed its price up too high. This was despite the cost-limited nature of the network (just four stations, and those were confined to low latitudes so they could be power by solar cells), which meant the scientific potential of the network concept could not be fully realised.

Bloodied but unbowed, the network enthusiasts, who were and are many in number, continued to look for a way to fly the mission. The Europeans tried *InterMarsNet*, a version of *MarsNet* in which ESA contributed the launch and the orbiter while NASA provided the landers and the science payload was shared. This also failed in the cut-throat competition with astronomers and solar physicists that generally characterises the ESA selection process. Eventually, *InterMarsNet* metamorphosed into *NetLander*, which nearly succeeded as a joint ESA–NASA project until the Americans pulled out in order to proceed directly

---

[2] A mission called simply *Mars Network* was proposed in the USA for flight in 2003, but this referred to a network of small orbiting satellites around Mars rather than a surface network. The main goal was to provide a communications and tracking infrastructure for rovers on the surface.

to a joint sample return project with the French, with the results that will be described below.

With the introduction of the *Scout* programme that allowed American scientists to propose small investigator-led, institute-managed Mars missions, the network concept was revived with an eighteen-lander network mission called *Pascal*. This recognised that Mars global network science had failed to get off the ground in the past because the payloads had tried to address too many disciplines and this led to complex, heavy landers that could not be flown in sufficient numbers within cost constraints. If the payload were focussed on a single discipline, in this case meteorology, and restricted to the most important measurements, namely surface pressure, temperature and wind speed, water vapour concentration, and panoramic images, this would lead to small, light, landing systems that allow the number needed for global coverage. In the USA, *Pascal* or similar projects have a regular chance to gain flight status through the *Scout* opportunities that come along every two years or so outside the main Mars exploration framework, constrained inside a budget cap, currently $475 million.

While *NetLander* has been superseded by a rover mission, *ExoMars* – see later – in ESA's programme, there are still many network enthusiasts in Europe who keep looking out for a way to fly some version of the mission. ESA itself has studied a concept called *Mars-NExT* that would insert a new mission to Mars in its *Aurora* programme, after *ExoMars* and before attempting sample return. The mission description envisions a spacecraft carrying three (or four) lander probes launched in 2018 on board a Russian *Soyuz* rocket from Kourou to establish a network of stations on the surface of Mars, and so is essentially *NetLander* or *MarsNet* under yet another name.[3]

Meanwhile, a network mission focussed on atmospheric science called *MetNet* is being advanced by a consortium including the Finnish Meteorological Institute in partnership with the Babakin Space Centre and the Space Research Institute, the last two both in Russia. The Finns have long been somewhat unlikely champions of Mars exploration, devising the concept of *MetNet* as long ago as the 1980s and eventually starting development work in the year 2000. The goal is eventually to deploy twenty or more semi-hard landers on the Martian surface, beginning with a small number, perhaps just one, on a precursor mission presently scheduled for launch in 2011.

8.3   The sample return saga

NASA began studies for a sample return mission even before the successful conclusion of the *Viking* mission. The report prepared by the Jet Propulsion

---

[3] Once a mission has been proposed for flight, and rejected for any reason, it is virtually impossible to re-propose it successfully under the same name, no matter how good it is. This explains the long string of names for similar concepts, each intended to convey the message that the project has been updated and improved since the last, unsuccessful attempt.

Laboratory in May 1978 recommended the name *ARES*, for *Automated Return of Extraterrestrial Samples*. It also recommended that the sample-gathering machine should be able to traverse the surface of Mars, with sufficient mobility to ensure intelligent and effective selection of the returned material. The cheapest concept, a static *Viking*-style lander with 'grab-sampling', had too high a potential for a disappointing scientific return. The samples could be returned to Earth either by Mars orbital rendezvous and sample transfer, or by landing the entire return vehicle (and its fuel) on Mars and flying directly back. The latter option involved putting down a landed mass of nearly six tons to return a sample of a few (minimum two) kilograms. The study group noted, 'with some passion', that sample return was 'a significant and most defensible goal' for NASA, and one that, at an estimated cost of $1,192 million for two launches in 1988, 'far from being excessive, provides a bargain', lasting 'in its significance, for generations'.

However, the US government and people did not share this passion sufficiently, following what they saw as a disappointing return from *Viking* on the key life question. Still, the studies powered ahead, most significantly with the *Mars Rover, Sample Return* concept that explicitly recognised that finding the right sample required a mobile robot on the surface, and one with some degree of intelligence. By 1988, when the post-*Viking* sample return could have been on its way, the *Mars Rover, Sample Return* team, now targeting a 1998 return, considered three separate launches for: (1) a 'big eye' imaging orbiter, similar to what eventually became the 2006 *Mars Reconnaissance Orbiter*; (2) a rover/gatherer, similar to the mission now known as *Mars Science Laboratory*; and (3) a sample return vehicle. This could all be achieved for a combined cost approaching seven billion dollars. The 1988 strategy noted that the Office of Exploration at NASA's headquarters in Washington was calling for a human landing as early as 2005, with more conservative voices in government putting it in 2013. Considering what it called 'a real uncertainty among planners about when the first landing is feasible or likely', the Jet Propulsion Laboratory planners opined that the most realistic date was somewhere in between, 'perhaps 2007 or 2009'.

This optimism collapsed when policymakers decided that space projects, including Mars missions, could not be afforded at such a high level. The appointment of Daniel Goldin as NASA Administrator in 1992 gave control to those who criticised NASA's expenditure plans and favoured a new, slimline approach. In May 1992, just a month after Mr Goldin took office, a NASA official reported to the *MESUR* science definition team, meeting at JPL, that a change was imminent: this was rumoured to be 'the aggressive emergence of small missions'. This duly happened: missions would be characterised by firm cost ceilings and limited development time, in contrast to the traditional approach that Mr Goldin's supporters described as 'spend what is needed, take as much time as necessary'. *MESUR* had already replaced *Mars Rover, Sample Return* as the flagship Mars mission under study, as JPL management, smelling the breeze, looked for a cheaper way forward. However, even at a fifth of the cost, it too would soon be ditched.

Mr Goldin's strategy called for three relatively low-cost missions to bring selected Martian samples back to Earth. These were to obtain:

1. Some bedrock, nearby loose soil, and atmospheric gases from an ancient sea or lake bed, identified from orbital photography and other data. Sedimentary material was considered much more likely to show evidence for climate change, and for life if any, than the igneous rock of the SNC meteorites, which had been forged in volcanoes.
2. A sample obtained by drilling into layers where groundwater is expected to be present, possibly near the 'seepage' features seen in some crater walls.
3. A sample from a modern reservoir of water, such as one of the polar caps.

Under the Faster, Better, Cheaper strategy prevailing at the time, the sample return missions were to follow one another rapidly, with the first launch in 2003, or even earlier. It was agreed by almost everyone that this was the way to take Mars exploration rapidly forward; inferences like those made from ALH 84001 and other advanced, life-related experiments were only possible in a well-equipped laboratory on the Earth. Not only that, but the samples could be shared around many laboratories and the results cross-checked, leading not only to more thorough conclusions, but also hopefully the scientific consensus that *Viking* notoriously failed to achieve.

The renewed emphasis on sample return served mainly to highlight the difficulty of the task. Following the many problems with relatively simple orbiters and landers at Mars, taking off again and returning to Earth more than doubles the complexity, cost and risk of failure. Such is the case even for the simplest sample return concept, the so-called 'ground-breaking' sample return mission, in which the sample is grabbed by a static lander from the initial landing site and returned to Earth, with only perfunctory attention to what sort of material it is. The trouble with this is that the sample could turn out to be boring – perhaps just soil with the characteristics of ground-up material from the same source as the SNC meteorites, which is what much of the soil on Mars appears to be. Then the mission, even if technically perfect, would be deemed to have failed, and NASA's programme would be left in danger of a hiatus like that which followed *Viking*. If sample return is to be done, it must be done properly, with mobility on the surface of Mars to seek out the best and most promising samples, and enough equipment on Mars to perform a preliminary analysis to check that this is indeed the 'right stuff'.

In Mr Goldin's plan, the first part of a two-part *Mars Sample Return* mission would be launched in 2003. This would deliver the rover that would find the samples and an ascent vehicle to carry them up from Mars, while in 2005 a larger payload would include copies of the 2003 vehicles but also carry an orbiter and a return vehicle. The latter would capture the sample canisters from both '03 and '05 missions and return them to Earth, arriving here in 2008. The renewed enthusiasm and optimism in NASA that made sample return seem feasible after years of fruitless planning was not just the Faster, Better, Cheaper processes in management and engineering that so reduced the cost, but also the partnership that NASA formed with the French space agency, CNES.[4] This became possible because France's Prime Minister had appointed a geologist and sample analysis scientist, Claude Allègre, to be the Minister of Education and Research. M. Allègre was as enthusiastic about obtaining Mars samples as Mr Goldin (he used his influence to press the European Space Agency, ESA, into technical studies of the feasibility of sample return missions to Mercury and Venus, too) and they soon got together and made a deal for cooperation on sample return. France would provide the orbiter for rendezvousing, capturing and returning to Earth with the sample that NASA promised to collect with a rover and launch into Mars orbit.

At the time, American and European scientists were already working together on a less ambitious programme of collaboration that revived the *MarsNet* mission to place a number of small automatic observing stations around the globe of Mars. The European agency was carrying out a joint Phase A study with NASA in the USA for a 2003 launch of the mission, now called *InterMarsNet*. If approved at the next meeting of ESA's council in April 1996, a single European *Ariane* rocket would carry a European orbiter and three US landers to form a science network on the Martian surface. This coordinated measurement approach offered important advances in many key disciplines, including seismology, meteorology and geochemistry, and was considered relatively low-risk since it had been studied before on both sides of the Atlantic. Getting together, it was agreed, was a good idea, to share the costs and the benefits, and to make a larger network than either could afford separately. The plan was going swimmingly, with enthusiastic teams of scientists, engineers and managers meeting regularly on both sides of the Atlantic to develop the details. Although under ESA auspices, the largest representation within the European contingent was from France, and it seemed clear that the same would apply to the mission itself with the seismological network, in particular, a big priority for the French.

The joint team was at work at ESA Headquarters in Paris one day when word leaked through to the senior NASA people involved that a deal had been struck between France and the USA to ditch the network mission and proceed directly to sample return. Most of the people in the working group were blissfully unaware of this and, although no announcement was made and those in the know remained tight-lipped, the mood of the meeting changed dramatically and it was soon clear to everyone that something had happened. Instead of earnestly solving problems, as the group had been doing for months, the NASA managers began to make flippant remarks and commitments were replaced by jokes. With the majority still in the dark, the meeting dragged on to the end of its agenda and the bemused participants returned home. When, some time later, they learned what was happening it was hard to know whether to laugh or cry.

---

[4] CNES stands for *Centre National d'Etudes Spatiales*.

On the bright side, of course it was wonderful to have sample return under-way, and on such an immediate schedule. However, there were plenty of people in all aspects of Mars exploration, scientific, technical and managerial, and at all levels, who thought that *InterMarsNet* was at least as important as sample return, and not only that, it made sense that it should be done first. They also doubted whether something as difficult as sample return was feasible on the budgets and time scales embraced by Mr Goldin's optimism, and feared losing everything. They were right: as the failures of *Mars Climate Orbiter* and *Mars Polar Lander* unfolded, the Jet Propulsion Laboratory was deep into the design of two sample return missions to be launched in 2003 and 2005, using the same faster, cheaper approach. In 1999 the artificially constrained cost and schedule of a sample return programme comprising two missions launched to Mars, each with a sample-collecting rover and one return capsule containing Martian rocks, had been judged by NASA and independent reviewers to be feasible. The total budget allocated by NASA for both components of the sample return mission was $650 million, in real terms just twice the cost of the much simpler *Pathfinder/Sojourner* mission (approximately $250 million in 1996).

The aftermath of the double failure in 1999 saw reality take hold of the budget and schedule. The cost for sample return rose rapidly to two billion dollars and more and the launch dates slipped from 2003 to 2005 to 2007 and then, briefly, to 2009, before the NASA/CNES partnership on sample return collapsed com-pletely. NASA was no longer confident that it was ready for sample return at any price, and France could not afford it anyway. Backing off to a CNES, rather than ESA, led version of *InterMarsNet*, one that might still carry some American instruments to Mars, seemed a possibility for a while. It was one that many would have welcomed in the Mars community, but the momentum behind the network mission had been destroyed, mutual trust was in short supply, and in the general mayhem the idea was short-lived. In March 2000, Allègre left office, followed by Goldin in November 2001. The failures of the '98 Faster, Better, Cheaper missions pushed Mars networks and sample return into an uncertain, but probably distant, future. NASA's programme architecture for Mars did not feature it at all for a time (Figure 8.2).

8.4 Developing a new strategy: Mars Exploration Pathways 2010–20

ESA could still look forward to *Mars Express*, and NASA to the *Mars Exploration Rovers*, with both missions scheduled for the same launch window in June 2003. What would happen beyond that was less certain, with the previous front-runner as flagship mission now discredited, but what remained clear was that neither was giving up on Mars. Ambitious planning cycles were initiated by both agencies but, while they moved forward on parallel tracks in terms of scientific thinking and goals, joint Mars missions were no longer on the agenda.

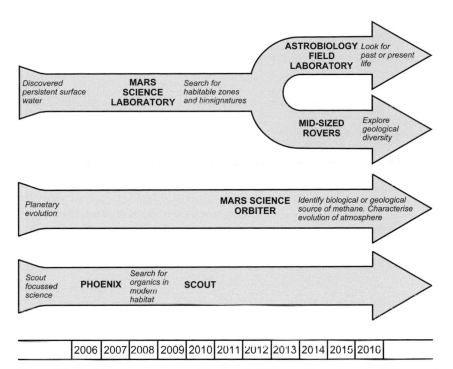

Figure 8.2 The Mars programme architecture in NASA as it was in early 2006: sample return does not feature, and neither does a Mars network, except as a *Scout* candidate.

NASA's approach was to develop a multi-branched Mars exploration road-map, the idea being that they would look ahead nearly two decades to 2020, and there would be a series of decision points where the branch or 'pathway' taken would depend on the outcome of the previous mission or investigation. The goals to be reached were defined at the highest level in NASA as *Life, Climate, Geology* and *Preparation for Human Exploration*, and they laid down some ground rules. None of these themes (in particular, not *Life*) was to dominate; the pro-gramme had to achieve a balance. It was not to assume the warm wet hypothesis for early Mars; the pathways were to include as one option the postulate that the past Martian climate was similar to the present. Finally, at least one of the options should *not* include full-up, expensive Mars sample return before 2020, while consideration should be given to whether a lower-cost, scientifically productive sample return mission is feasible in at least one of the other branches.

After about a year of working group meetings, the four alternative pathways summarised in Figure 8.3 emerged from the efforts of the scientists, engineers and managers involved. They intended that NASA would initially choose to pursue one of these, and everything would evolve from there, depending on many factors including not only advances in science and technology, but also budgets and enthusiasms among politicians and the public. Here is a précis of how their report described the pathways.

Pathway	Lines of Scientific Enquiry
1. Search for Evidence of Past Life	If Mars was warm and wet: Locate and analyse water-lain sedimentary rock Search for evidence of past life
2. Explore Hydrothermal Habitats	If active or dormant hydrothermal deposits exist: Explore for evidence of past and present life
3. Search for Present Life	If there is evidence for modern habitats and political/programmatic support: Sample Return with mobility
4. Explore Evolution of Mars	If Mars was never globally wet: Study the history of volatiles on Mars Determine initial conditions and evolution Understand Earth, Venus and Mars together

Figure 8.3 The lines of scientific enquiry that defined the four 'pathways' for Mars exploration by the United States in the early 2000s. (NASA)

Pathway 1: Search for Evidence of Past Life

Sites with sedimentary deposits that show evidence of either deposition or alteration by liquid water in the past resemble those where microbial life is found on Earth and so can be identified as having the highest potential for revealing evidence of past primitive life on Mars. The *Mars Reconnaissance Orbiter*'s planet-wide search for evidence of persistent liquid water on the surface in the distant past, two to four billion years ago, can identify the most scientifically productive landing sites for future rovers, landers and sample return missions in this and other pathways, including the landing site of the *Mars Science Laboratory*. This large rover was designed to investigate the carbon chemistry in near-surface rocks and soil and provide a rigorous and definitive examination of their mineralogy, as well as the extent to which they were formed or altered by water. Later missions in this pathway build upon this foundation and carry out further definitive tests for evidence of past life, for instance, advanced *in situ* instruments are expected to conduct biomolecular chemical analysis and higher spatial resolution examination of samples. The technology to acquire difficult-to-access samples, such as those buried several to tens of metres below the surface or at large distances from safe landing sites, can also be developed.

Pathway 2: Explore Hydrothermal Habitats

Hydrothermal activity on Mars may have provided water, heat and nutrients for the evolution of life. On Earth, subsurface heat from residual volcanism, in contact with water, has left behind readily detectable chemical and mineralogical deposits that often preserve evidence of habitats and signatures of microbes. There is reason to suspect that active or fossilised hydrothermal deposits exist on Mars, for instance from recent studies of Martian meteorites in laboratories, although none have been positively identified to date. Were an active vent to be discovered, perhaps by detection of hot spots in thermal images from an orbiter, such a vent could be an abode of present life. Landing the *Mars Science Laboratory* at such a site to examine the mineralogy and chemical composition of vent deposits would be an obvious way to initiate this pathway. A search for bio-signatures (organic compounds, isotopic signatures and microscopic textural indicators) and past and/or present life would then follow.

Pathway 3: Search for Present Life

The search for present life on Mars is thought by many to be a high-risk venture, since the *Viking* mission when expectations were too high. Not wanting a repeat of this experience, most scientists would not wish to begin the Search for Present Life pathway unless it is first shown that environments exist on Mars today that can support life. Such habitats include liquid water in the near-subsurface, active hydrothermal systems, or source regions of melt-water of the type that apparently formed the crater gullies recently detected by *Mars Global Surveyor* imagery. If a relatively warm, water-rich subsurface environment exists it will also be protected from harsh surface conditions, e.g. ultraviolet and cosmic rays.

Another possible event that would motivate a concerted effort to find living organisms is an impending landing of humans on the Martian surface, since this would be almost certain to contaminate the planet with Earth life so that detection of indigenous life would become even more difficult, if not impossible. Following an announcement of plans to send humans to Mars, there might be a rush to find Martian life first using clean robotic explorers. This would also help ensure that any Martian life forms found were not hazardous to the explorers.

This pathway would emphasize *in situ* life detection, address complex sample access problems, and require a very substantial near-term technology investment in instruments. Spacecraft cleanliness, for the purposes of planetary protection and the avoidance of false-positive detections of life, would be mandatory for surface missions in this pathway, as would confirmation in Earth's laboratories through measurements of returned samples.

Pathway 4: Explore the Evolution of Mars

This pathway would be a plausible choice if the currently favoured hypothesis for a habitable Mars – that the planet has at some time enjoyed liquid water at its surface for long periods – is proven to be incorrect, so that NASA's current focus on the search for surface habitats would become untenable. This new picture of Mars seems rather unlikely, but if it did come about it would pose a number of new questions. In particular, there would be fewer parallels between Mars and Earth, requiring a new paradigm in which the terrestrial planets evolved very differently.

8.5   Missions to implement the pathways

The first practical outcome of these deliberations was the decision to develop the large and versatile 'mega-rover', *Mars Science Laboratory*, to be much larger than the 2005 *Exploration Rovers*, *Spirit* and *Opportunity*, and originally intended to launch in 2009, as shown in Figure 8.4. The question of what to do about sample return without courting bankruptcy was addressed by resurrecting the well-worn plan for a relatively simple 'ground-breaking' sample return mission that could be launched in 2013. The concept was that which had been rejected nearly three decades earlier in 1978: a static lander, similar to an updated *Viking*, which would grab a sample from its landing site and then take off again. It was understood that this might not be wonderful scientifically, but the rationale was that it could be seen as the first step in a series of progressively more ambitious sample return missions in which experience and technical capabilities were developed progressively. This is an eminently sensible approach in fact, although not one that is compatible with the modern appetite for instant gratification, which applies to agencies as much as it does to individuals.[5]

Mars *Scout* missions appear where they do in the plan for two reasons. Firstly, since *Scouts* would be selected from competitive proposals by research labs and

Pathway	2009	2011	2013	2016	2018	2020
Search for evidence of past life	Mars Science Laboratory (MSL) to low latitude site	Scout	Ground-breaking sample return	Scout	Astrobiology field laboratory or deep drill	Scout
Explore hydrothermal habitats	MSL to hydrothermal deposit	Scout	Astrobiology field laboratory	Scout	Deep drill	Scout
Search for present life	MSL to North Pole or active vent	Scout	Scout	Sample return with Rover	Scout	Deep drill
Explore evolution of Mars	MSL to low latitude site	Scout	Ground-breaking sample return	Aeronomy	Network	Scout

Figure 8.4 The missions needed to conduct NASA's Mars Exploration Pathways programme, starting with *Mars Science Laboratory* (*MSL*) in each case. This plan, dating from around 2003, needs revision to allow for the selection of *MAVEN* for the 2011 (now slipped to 2013) *Scout* opportunity, pre-empting the aeronomy mission shown here in 2016, and for the fall (back) into disfavour for *Ground-breaking Sample Return*. These updates make the second pathway, Explore Hydrothermal Habitats, the only one that looks viable in 2009, and even that might benefit from the addition of *Mars Science Orbiter* before the astrobiology mission to locate the recently discovered sources of methane more precisely.

---

[5] A common complaint, containing a grain of truth along with an element of hyperbole, made by those writing research proposals these days is that the funding agency will not consider any application that does not specify in advance what its outcome will be. Of course, if one could do that it would not be necessary to do the research.

universities, the community would get regular chances to fill any gaps in the programme of larger missions defined and selected by NASA's centralised planning process and provide balance across science disciplines. Secondly, since the Principal Investigator's institution is given responsibility for implementing the flight hardware, while NASA retains control over the flow of money, the Agency finds this an admirable way of smoothing expenditures of the programme over time. The recently selected *Scout* mission, the upper-atmosphere explorer called *Mars Atmosphere and Volatile EvolutioN* (*MAVEN*), although relatively cheap at a cost of less than five hundred million dollars, is already under threat of delay or cancellation due to the cost overruns on *Mars Science Laboratory*.

MAVEN had already been delayed, before it was even selected. As Figure 8.3 shows, the second *Scout* (after *Phoenix*) was to have launched in 2011, but it was delayed to 2013 in December 2007 to keep the overall budget on track. This would have meant no Mars launch was planned for the 2011 window, making it the first launch opportunity without a mission since the gap between *Mars Observer* and *Mars Global Surveyor* in 1994 (until *Mars Science Laboratory* slipped into it, of course).

8.5.1   Mars Science Laboratory

Mars Science Laboratory is a rover with a mass of seven hundred and seventy-five kilograms, more than seventy times that of *Sojourner* and nearly as much as a small car.[6] Having decided that such a heavy vehicle could not be landed safely using airbags, and recognising that a more precise landing technology had to be developed in any case if future plans were to pan out, the mission designers came up with the concept of a 'sky crane' to lower the vehicle onto the surface of Mars. This employs a platform controlled by rockets, in the sideways as well as the vertical dimension, which can hover long enough to winch the rover down on a tether and place it at the desired point on the surface safely, as pictured in Figure 8.5. After the rover is down and free of the tether, the crane, relieved of its weight, is allowed to shoot off into the sky and crash some safe distance away.

The *Science Laboratory* earns its name by the large and sophisticated payload of instruments it carries. These include three cameras, several spectrometers, radiation detectors and a meteorological station. Some of these, and indeed the mission itself, have been under threat from budget overrun problems. The *Descent Imager*, designed to take images at a rate of five frames per second during about two minutes of descent and landing, looked for a time to be a definite casualty of mission cost growth, NASA arguing that the high-resolution camera on *Reconnaissance Orbiter* should be able to provide all the landing site context images needed for the rover mission. Its twenty-centimetre resolution could

---

[6] A current-model BMW Mini Cooper weighs one thousand, one hundred and thirty-two kilograms; the original Morris Mini produced from 1959 to 2000 just six hundred and seventeen kilograms.

Figure 8.5 NASA's *Mars Science Laboratory* making a controlled descent under a sky crane, which hovers above the surface while the rover is winched down.

provide the location of loose debris, boulders, cliffs and other features of the terrain for understanding the larger geological context surrounding the landing site and planning the path of exploration after landing. However, the cost saving turned out to be small and the camera was reinstated. The publicity value alone of taking colour video during the rover's descent toward the surface, and of recording any difficulties the novel and ambitious landing system might encounter, is worth the price (which, in any case, according to some estimates was zero, most of the work having already been done, and the cost of removing things already designed in to the project being far from negligible).

The other instruments on board are:

- *Mast Camera*, which takes colour stereo pictures of the surrounding terrain as the rover travels around, and helps it to navigate. It can operate at up to ten frames per second, to make movies, and has a ten-times zoom lens and multiple filters to make the images useful for compositional analyses.
- *Hand Lens Imager* to provide colour images, but on a much smaller scale, down to the diameter of a human hair. It carries white and ultraviolet light sources so it can work both day and night, and induce fluorescent emissions that help to detect carbonates and other water-deposited minerals.
- *Alpha-Particle X-Ray Spectrometer*, a new version of those on *Spirit* and *Opportunity*, which measured the abundance of chemical elements in rocks and

Figure 8.6 *Mars Science Laboratory*, pictured sampling a rocky outcrop with its laser ablation tool. The laser beam vaporises some of the rock, creating a flask of gas and plasma, the light from which can be analysed by a spectrometer on the rover.

soils, showing how the material formed and if it has been altered by wind, water or ice.

- *Laser-Induced Remote Sensing for Chemistry and Micro-Imaging* fires a laser at rocks and soils from a distance of a few metres and analyses the composition of the vaporised material. Figure 8.6 gives an impression of how it will look.
- *Chemistry and Mineralogy* analyser directs a fine beam of X-rays through powdered material to measure the abundances of igneous and aqueous minerals. Targets include olivine and pyroxene, two primary components of the basalt that forms when lava solidifies, and jarosite (potassium iron sulphate hydroxide, $KFe_3(SO_4)_2(OH)_6$ to the chemists), a mineral salt that is uncommon on Earth but seems to be plentiful on iron-rich Mars, having precipitated out of water made acidic by volcanic sulphur.
- *Sample Analysis at Mars*[7] is a suite of three instruments, including a mass spectrometer, gas chromatograph and tuneable laser spectrometer, which accounts for more than half the science payload on *Mars Science Laboratory*.

---

[7] SAM and DAN are further examples of the constructive use of words to give a catchy acronym.

Together they search for organic compounds, including methane, and measure the abundances and isotopic ratios of all the common light elements, including hydrogen, oxygen and nitrogen, in atmospheric gases such as methane, water vapour, carbon dioxide, nitrous oxides and hydrogen peroxide.

- *Radiation Assessment Detector*, designed specifically to prepare for future human exploration by measuring the effects of high-energy cosmic rays that penetrate to the Martian surface. These consist of protons, energetic ions of various elements, neutrons and gamma rays, and include not only direct radiation from outer space, but also secondary radiation produced by the interaction of cosmic radiation with the Martian atmosphere and surface rocks and soils.
- *Dynamic Albedo of Neutrons*[7] experiment, uses a pulsing neutron generator and detector to detect water content as low as one part per thousand and resolve layers of water and ice up to two metres beneath the surface.

Finally, *Science Laboratory* has a weather monitoring station attached to a vertical mast to measure atmospheric pressure, humidity, ultraviolet radiation from the Sun, wind speed, wind direction, ground temperature and air temperature. At the time of writing, the mission development team at the Jet Propulsion Laboratory reported a two year delay in the launch and another large cost overrun, taking the total over two billion dollars and provoking a spate of recriminations from NASA's budget managers in Washington.

The technical problems that caused the extra spending have to do mainly with the complicated mechanical subsystems that drive the heavy vehicle across the Martian surface and those that it uses to manipulate samples for analysis. Hopefully, neither the engineering nor financial hardships common in this sort of ambitious project will derail it completely, although a slip to a later launch date becomes unavoidable. The next opportunity after October 2009 is in December 2011, the window originally occupied by ESA's *ExoMars* rover, before it slipped to 2013, and then 2016, as a result of broadly similar exigencies to those afflicting *Mars Science Laboratory*. Of course, delays do not reduce the overall cost of a mission, but always increase it, although they do reduce the spending in the short term and allow time to solve technical problems. The losers are the scientists, who have to wait for their data, and the participants in the missions following on, which themselves have to slip to make way for the extra effort and expenditure in the workshop on hardware that should have already been on its way to Mars.

The choice of landing site is crucial if the powerful capabilities of *Mars Science Laboratory* are to be fully exploited. The possibilities were reduced after much discussion to seven: by name these are Mawrth Vallis, Nili Fossae, Jezero Crater, Miyamoto Crater, South-west Meridiani, Holden Crater and Eberswalde Crater. At one stage the clear favourite was Meridiani, which has been shown to be interesting, safe, and full of sedimentary material by years of exploration with the small *Opportunity* rover. On the other hand, the massive, large wheeled and long-range *Science Laboratory* rover has excess capability for Meridiani, and could deal with somewhere with rougher and possibly even more interesting terrain. Gale Crater or the canyon landscape of Valles Marineris would fit this category, but possibly take the trade-off between interesting complexity and safety and engineering concerns too far in the other direction. The workshops to debate the issue will go on until the last minute, possibly for years if the launch of the mission is delayed, particularly as new evidence continues to come in from *Mars Express*, *Mars Reconnaissance Orbiter*, and ground-based planetary astronomers.[8]

8.5.2 *Mars Science Orbiter*

Europe's debut with *Mars Express* made two contributions towards pointing the direction for future Mars research that even NASA's seasoned Mars veterans quickly took on board. One is the evidence from the *OMEGA* near-infrared spectrometer of widespread deposits of sulphates, similar in composition to those *Opportunity* found on the surface in Meridiani Planum. Sulphates are always deposited by water, so their occurrence over very large areas supports the hypothesis that water was once persistent over the planet. Even more significant, potentially at least, was the announcement in April 2004 by the long-wavelength *Planetary Fourier Spectrometer* team on *Mars Express* that they had detected methane in the Martian atmosphere, and that it might be spatially variable. Although non-biological sources of methane are likely to exist as well, from residual volcanic activity, for example, or as part of the influx of small, icy meteorites and comets, the estimated amounts of these are too small to account for the new observation, implying that it may be a biosignature, as it often is on Earth. Even if it is not, the presence of methane suggests that the atmosphere is communicating with the subsurface and, potentially, with sources of geothermal heat and water, indicating in turn the possibility of a modern habitat.

Earth-based observers also pitched in with detections of methane on Mars, but like the spectrometer on *Mars Express*, their sensitivity, and especially the evidence that they may be localised, is close to their detection limit. The potential importance of methane on Mars, not only in its own right but as a signpost to the most interesting places to land, quickly led the planners at the Jet Propulsion Laboratory to plan a new mission to fly a more sensitive spectroscopic instrument. This is the *Mars Science Orbiter*, available to be grafted into the pathways approach right after *Mars Science Laboratory* should NASA decide to take this route. *Mars Science Orbiter* is essentially another *Mars Reconnaissance Orbiter*, but with a different payload, focussed on high-resolution atmospheric spectroscopy capable of detecting small traces of methane and formaldehyde, rather than

---

[8] In December 2008, the team analysing data from the imaging spectrometer on *Mars Reconnaissance Orbiter* announced the detection of substantial fields of carbonate minerals, almost certainly deposited in a warm, wet environment around three to four billion years ago, in rocky outcrops in the Nili Fossae region. This was one of the top choices as a landing site for *Mars Science Laboratory* even before this key discovery.

Figure 8.7 Nili Fossae, one of the seven locations considered as possible sites for the landing of *Mars Science Laboratory* in September 2010. A recent interpretation of spectra from telescopes in Hawaii and Chile suggests that plumes of methane gas are being released in this area; spacecraft observations have already shown that clays and other interesting minerals are present on the surface.

landing site photography. The insertion of a new spacecraft in orbit around Mars in 2014 or so would also provide a valuable additional communications relay to any assets operating on the surface. ESA, in particular, would be delighted to have back-up for its plans for *ExoMars*, particularly as its launch date slips into the future, but NASA has yet to make up its mind about the mission.

In October 2008, one of the ground-based observers who first reported the detection of methane, a gas with strong biological implications, in the Martian atmosphere, claimed to have refined the observations to show that there are large concentrations of the gas over Nili Fossae, shown in Figure 8.7, a region that has already been shown to be rich in clays and other minerals that point to a watery past. If confirmed, this factor is likely to be decisive,[9] but it is unclear at the time of writing whether confirmation and pinpointing of local sources of methane requires *Mars Science Orbiter* to fly before *Mars Science Laboratory* or whether existing observatories can do the job.

8.5.3 *Astrobiology Field Laboratory*

If a vent emitting methane (and probably other interesting gases, including water vapour) is located by high-resolution cameras and spectrometers in orbit and then explored *in situ* by rovers on the surface, and confirmed as a promising location for life-related studies, NASA has waiting in the wings a concept for a yet more advanced rover called the *Astrobiology Field Laboratory* that could be

---

[9] The discovery was announced orally by Michael Mumma, a planetary scientist at NASA's Goddard Space Flight Center in Greenbelt, Maryland, at a meeting of the planetary division of the American Astronomical Society at Cornell University. The Project Scientist for *Mars Science Laboratory* commented that he would wait for the finding to be written up in a scientific journal before deciding its impact on landing site selection for the mission – 'talks don't count'. The paper duly appeared in the journal *Science* in January 2009, to the predictable media storm about signs of life.

launched in 2018 or later, a larger follow-on to ESA's *ExoMars* shown in Plate 35. *AFL* would be the first NASA mission since *Viking* to be explicitly focussed on the search for evidence of past or present life, using the kind of advanced payload focussed on biological assays and analyses discussed in Chapter 7.

8.5.4 *Deep Drill*

There is a good chance that the mobile *Science Laboratory* and *ExoMars* explorers, and the subsurface sounding radars on *Mars Express* and *Reconnaissance Orbiter*, will confirm that the most interesting potential habitats need samples from well below the surface. In that case, NASA would not begin work on *Astrobiology Field Laboratory*. Instead, the large rover would be replaced by a mission with less surface mobility but the capability to explore below the surface, *Deep Drill* (shown in Plate 36). Even if *Astrobiology Field Laboratory* does go ahead, a mission to drill is sure to follow at some stage, since it is likely that the key ingredient for life, liquid water, is to be found in significant quantities only at a considerable depth, in most locations at least. Another job for an advanced drilling capability is extracting cores from deep within the layered deposits at one of the poles. Not only is water ice exposed at the surface here in the summer months, a probe of the layering beneath the surface is sure to be a priority as a way of looking back in time.

Achieving the desired depth – set at fifty metres by some study groups – is a considerable challenge in either case and requires an ingenious engineering approach. The relatively simple drill developed for *ExoMars* has a goal of just two metres depth. One of the challenges is how to lubricate and cool the drill bit, and to flush out the detritus, as it goes ever deeper, when it is too cold to use water or oil as would be done on Earth. The material displaced not only has to be removed, it needs to be collected in some sort of a coherent way so it can be analysed. Yet another problem is that the type of bit to be used depends on the material to be drilled through,[10] which is not known at the outset and anyway may change on the way down. Even if the same bit material can be used for the entire drilling, the bit itself will almost certainly wear out and need to be changed.

Considerable infrastructure is also required on the surface. A sufficiently large and massive derrick to handle the drill and the samples, power supplies, analytical equipment (some of it fairly sophisticated, to include detecting microscopic life) and so forth. The debate about how far robots can go in this direction before human engineers need to be on-site will not be resolved for some time.

---

[10] A drill bit made of diamonds embedded in tungsten gives a good performance for sandstone and most other sedimentary materials, but in igneous material like basalt, which is harder, a drill made of a softer material such as bronze as the matrix holding the harder diamonds has been found to work much better, because it exposes fresh diamonds as others wear or fracture.

8.5.5 *Ground-Breaking Sample Return*

Ground-Breaking Sample Return was not long on the active mission list, but the basic idea has always been important, and remains a recurring part of the exploration agenda. As the most affordable and reliable approach to sample return, many planners, particularly the engineers, think it should be attempted first as a technology 'pathfinder'. The criterion for success would be simply to bring back samples from Mars, taking 'pot luck' on what sort of material is returned and not expecting major scientific dividends on the first try. The really interesting samples would follow later once techniques for finding and recovering them had been honed and a reliability record built up. This sort of 'sensible' approach characterised the early space programme, but has gone out of fashion in recent years, with agencies, the public and the media tending to demand everything at once and expecting instant success in difficult quests.

NASA's earlier concern about the practicality of sample return led to the mandatory inclusion of some programmatic pathways that could proceed without it, and also produced the directive that in any case sample return of any description should not occur before 2020. The gloom was deepened by the announcement of President George W. Bush's vision for Mars in 2004, which is discussed in detail in the next chapter. While this superficially gave top-level support to Mars exploration, its practical effect was to divert NASA's available resources – which were not significantly enhanced as part of the announcement – to developing a space shuttle replacement and a manned Moon base. This meant money and effort actually moving away from Mars, in the short term at least. The 'ground breaker' had the huge advantage over other sample return missions of being affordable in the relatively short term, leaving it with a chance that it could be kept in the programme.

As conceived *circa* 2003, the mission would consist of two identical landers, each carrying an extendable arm with very simple sampling devices (e.g. a combination of a scoop and a sieve, and a miniature rock corer) and a context camera, supported by an orbiter that would receive the samples by a rendezvous in Mars orbit and return them to Earth. Only about five hundred grams of fines, rock fragments and atmospheric gas were to be obtained by each lander, directly from the material within reach at the landing point. The absence of mobility was not only to keep cost and risk down, but could be rationalised by arguing that the landers would target sites that had been previously characterised by earlier missions as regions known to contain aqueous deposits. This would help guarantee the material returned to Earth would contain the first sedimentary rock and soil from Mars to be analysed in our laboratories, although few went so far as to predict that material within reach of the sampling arm would contain any evidence of life.

It was always intended that ground breaking would be followed by a mission with a rover that could gain access to a bigger range of samples and target less accessible locations. However, it became clear a few years later, when the properties of the terrain traversed by the *Mars Exploration Rovers* had been analysed, that maximising mobility and selectivity is essential from the outset. This, plus the increased confidence that came from recent successes, giving rise to a feeling that NASA could proceed directly to more ambitious missions, led to the groundbreaking concept losing scientific support and slipping out of contention again.

This took sample return of any kind off the agenda, consummating the fear that had led to support for ground breaking in the first place. After a meeting in April 2006 of the Mars Exploration Program Analysis Group, a large forum made up of the entire American Mars science community, including NASA managers, and some Europeans, Russians and Japanese, its chairman was moved to record that the 'community expressed concern that network science and sample return were no longer even shown in the long-term plan' of the Agency. The Mars Exploration Program Analysis Group recommended that NASA team up with foreign partners to make these missions happen, but, still remembering the *InterMarsNet* fiasco, European scientists were not responsive this time, and had grand plans of their own instead.

8.6 Pathways at work: discoveries change the plan

The *Exploration Rovers* showed that early Mars was wet and had an active hydrological cycle that involved groundwater that ebbed and flowed in response to periodic climate change. This sent NASA along the pathway that pursues Search for Evidence of Past Life, to search for a probable habitat and to begin there to seek evidence of past life through studies of carbon chemistry and complex organic chemistry in sedimentary deposits. This led to *Mars Science Laboratory* being assembled for 2009 launch, while the current *Reconnaissance Orbiter* searches for its best landing site, possibly followed by *Mars Science Orbiter*, which would do the same job for *Astrobiology Field Laboratory*.

The essence of the Pathways approach is that the points can be switched to another direction in the network if new discoveries dictate the need. The *Astrobiology Field Laboratory* would only get the go-ahead if it can be landed near an extinct, or preferably active, hydrothermal vent or hot spring, or a localised source of methane (which may well be the same place, Nili Fossae for instance). Such a location would have to be identified in the next few years, and the technology to land there with precision would have to be developed. Both of these are major challenges. The site is likely to be small in extent, while the error ellipse for landing packages on Mars is currently measured in tens of kilometres. The capability to hover, and winch down a rover or other package gently, is being developed for *Mars Science Laboratory*, and will soon be put to the acid test; for its successor, the sky crane may have to navigate a fairly large distance until it finds and recognises just the right spot. The analytic measurements then performed, in 2018 or later, will be capable of determining whether life ever existed at the identified habitat. A positive answer would drive exploration beyond 2020 to widen the search and to look for survivors wherever they might be.

The possibility must be faced that the worst-case scenario will be found – that there are no vents on the surface, and yet deep drilling is not viable because the interesting layers either do not exist or are too deep, even for a fifty-metre drill. What then? The strategic plan calls for a shift to the Explore the Evolution of Mars pathway. With an aeronomy mission already underway as a result of the second *Scout* selection, and ground-breaking sample return out, the next mission after *MAVEN* could be a network, similar to the old *NetLander*, but now a NASA-only version. Multiple (ten or more) long-lived stations distributed globally about the planet would focus on geochemical surveys, seismological measurements and meteorological monitoring. Atmospheric scientists and geologists have long sought such a mission to characterise the behaviour of Mars as a terrestrial planet, regardless of its potential for life.

In this way, the Pathways concept had all bases covered. Politicians, managers, and scientists alike appreciated the flexibility to respond to changing circumstances, especially to new discoveries about Mars, whether serendipitous or as part of each planned step in the programme. The philosophy was soon put to the test, with plenty of new insights to respond to, as the results from the *Mars Exploration Rovers* and *Mars Express* started to flow in. As it has turned out, not just the choice of pathways but the selection of missions within each of the four pathways needs revision, just five years after they were designed to cover every possibility, but the philosophy behind the approach remains sound.

8.7 Back on the agenda again: sample return

So far, it looks like NASA will stay on the Search for Evidence of Past Life pathway after the approved missions *Mars Science Laboratory* and *ExoMars*, assuming nothing they find forces a change. In that case, the return of samples of sedimentary rock to Earth laboratories comes next. The pathway in its original form would have NASA undertaking ground-breaking sample return in 2013, but that went out of the window when the *Spirit* and *Opportunity* rover results conjured up the nightmare of a static lander with only a short arm to try to collect the aqueous sediments that might be tantalisingly out of reach. This means going beyond 'ground breaking' and using mobile vehicles that have sufficient reach to access and collect the most interesting samples from the outcrops that were once immersed in standing water. These sites would be the same as the destination for the *Science Laboratory*, so they would already have been sampled *in situ*. A short-range rover carrying only cameras and sampling tools could make the original design somewhat less expensive and still be capable of reaching the most desirable samples *Opportunity* found distributed throughout Meridiani Planum. A range of one or two kilometres during two to four weeks spent on the surface would likely be an adequate compromise between collecting enough diverse samples while avoiding degradation in the solid rocket fuel in the ascent vehicle due to the extreme cold.

What might make even more sense is to use the *Science Laboratory* rover, with its expected range of at least six kilometres and a sample cache 'shopping basket', to hold the samples, perhaps for as long as a decade, until the 'postman' arrives to collect them and take them back to Earth. A change of management that put Dr Alan Stern in charge of the Science Mission Directorate at NASA led to the request to the team getting *Mars Science Laboratory* ready for launch in 2009 to add the capability to cache samples for collection and return to Earth by a new mission in 2018. This new sample return mission would probably displace both the *Astrobiology Field Laboratory* and the *Deep Drill* project. He said, 'I want to get serious about Mars sample return and this is the way to do it. This has been going on all my life, waiting for Mars sample return and it never gets there. We're going to do a pragmatic, but competent sample return.'

But then Alan Stern left NASA in April 2008, putting the sample return mission in doubt once again. The addition of a sample carrier to *Mars Science Laboratory* went ahead for a while, but now it seems the samples may never be collected. In any case, the difficulties of retrieving the samples from a ten-year-old existing cache are proving greater, under detailed study, than previously appreciated. It might be better just to collect new samples. The 'postman' would have to have such a capability in any case, to cover the possibility that it cannot find or cannot reach the cache, for instance if *Mars Science Laboratory* ends its life in an inaccessible place.

8.8   Europe goes to Mars with *Aurora*

While NASA was developing the Pathways approach, which involved working cautiously in a forward direction, switching and now revising the schedule of missions as they went, ESA developed its own plans, and decided on almost the opposite philosophy. They wanted to see Europeans walking on Mars in a finite time, and although it was not possible to put a definite date on it without knowing what budget they could count on in the next few decades, they knew they wanted the first landing to be within the lifetimes of most people around now, and 2040 was informally adopted as a target. They then worked backwards from this overarching goal to decide what to do next, stopping short of including an actual human landing in the programme so as not to price it out of consideration before it had got started. An imaginative publicity campaign was launched using evocative artwork like that in Figure 8.8 to catch the attention of the European public, which mostly was not aware that its own governments and space agency were ready to make such a large commitment. *NetLander* was an obvious candidate for the first step after *Mars Express*; it had been extensively studied, was affordable, and had very wide scientific support. It also was supportable as a 'precursor' for the planned manned expeditions, since it would extend human experience of Mars 'close-up' from a few landing sites to

Figure 8.8 The masthead of the European *Aurora* programme.
Close cooperation within ESA, as well as collaboration with European and Canadian industry (Canada being the only non-European member of ESA) and academia, is a key aspect of *Aurora*. While its overarching goal is to see European astronauts on Mars as soon as practicable, officially ESA is not committed to this and describes its goals as:
– Explore the Solar System and the Universe
– Stimulate new technology
– Inspire the young people of Europe to take a greater interest in science and technology
– Remote sensing of the Martian environment
– Robotic exploration and surface analysis
– Mars sample return missions
– A robotic outpost on Mars

perhaps as many as twenty (from one crash, with *Beagle 2*, to twenty landings if only European spacecraft are counted). It would also greatly aid in landing site selection and in gaining key meteorological data to improve landing safety.

However, ESA is nothing if not ambitious. The community it represented became hungry to get engaged with the life question, just as NASA was cautiously backing off. The ESA selection process is far more capricious than that of NASA; where the latter has many layers of management and reviews, making decisions carefully and gradually, European space scientists of a particular persuasion can lobby and get their way in one fell swoop. At a key planning meeting in Birmingham, England, in April 2005, those who happened to be there ignored the arguments for network science and brushed aside concerns that ESA had no experience of landing successfully on Mars, and decided to go straight to a rover mission. Not only that, the payload was to be boldly focussed on the search for life. They even named it *ExoMars*, a contraction of *Mars Exobiology*

Mission. It was to be launched in 2011, later revised to 2013, and then to 2016. In terms of its scientific ambitions, *ExoMars* is similar to NASA's *Mars Science Laboratory*, although the step of seeking evidence of life on Mars is a larger stride than NASA was prepared to advertise. However, the size of *ExoMars* at two hundred and ten kilograms is more comparable to the 2005 *Exploration Rovers*, *Spirit* and *Opportunity* (at one hundred and eighty-five kilograms each), than it is to the *Science Laboratory*, which weighs in at a formidable seven hundred and seventy-five kilograms.

The original plan for *ExoMars* called for the parallel development of a Mars orbiter and data relay spacecraft, a descent module using an inflatable aeroshield and/or a parachute system for braking, and the Mars rover itself. The orbiter was an early casualty of the development process, when the engineers designing the mission realised it was not affordable within the budget of only six hundred million euros (again closer to the eight hundred and fifty million dollars NASA spent for two *Mars Exploration Rovers*, than to *Mars Science Laboratory*'s two billion dollars, even allowing for the fact that NASA and ESA count their mission costs differently). The European rover now depends on the American *Reconnaissance Orbiter* – which will, by then, have been on station for ten years – to act as a data relay, and has no supporting observations from orbit. The launch slip followed for similar reasons and budget growth continues. From time to time, ESA debates upgrading the mission with a new budget, a larger *Ariane-5* launch vehicle replacing the Russian *Soyuz*, and commissioning an orbiter after all.[11]

The *ExoMars* rover shown in Figure 8.9 has been described by ESA as a field biologist on Mars, a natural successor to the robot geologists (*Mars Exploration Rovers* and *Mars Science Laboratory*) sent by NASA. A key feature justifying this claim is the drill it carries in order to acquire samples from depths down to two metres below the surface. Biological material is likely to be better preserved in these samples than anywhere near the surface, so the mission can aim to characterise the biological environment in preparation for sample return missions and then human exploration. The drill system is being designed so that it can penetrate rock, within or underneath which the best samples may lie undisturbed, whereas soft material such as sand or clay may have been mixed and sterilised to a considerable depth. The drill also has to be designed to generate a minimum amount of frictional heating that might destroy any interesting specimens as it uncovers them.

8.8.1 Goals for *ExoMars*

ESA's stated goals for the 2016 rover are:

1. To land on, or be able to reach, a location with high astrobiology interest for past and/or present life signatures, i.e. access to the appropriate geological environment.

---

[11] The latest news is that there will be an orbiter in 2016, and the rover will slip yet again, to 2018.

2-m depth

Figure 8.9 The European *ExoMars* rover, shown deploying its drill to obtain samples from two metres below the surface.

2. To collect scientific samples from different sites, using a rover carrying a drill capable of reaching well into the soil and surface rocks. This requires mobility and access to the subsurface.
3. At each site, to conduct an integral set of measurements at multiple scales: beginning with a panoramic assessment of the geological environment, progressing to smaller-scale investigations on interesting surface rocks using a suite of contact instruments, and culminating with the collection of well-selected samples to be studied by the rover's analytical laboratory.
4. To characterise geophysics and environment parameters relevant to planetary evolution, life and hazards to humans.

Using conventional solar arrays to generate electricity, the rover will be able to travel a few kilometres over the surface of Mars, navigating and operating autonomously, during its nominal six-month lifetime. Included in its approximately forty-kilogram astrobiology payload along with the lightweight drilling system will be a sampling and handling device, a microscope to scrutinise the sample, and spectrometers to analyse its composition, before and after vaporisation with a laser. Then there are some sophisticated life detection devices, the simplest of which is a gas chromatograph–mass spectrometer similar in principle to that on *Viking* but of more advanced design. NASA will contribute a device to detect and measure oxidants and an organics detector sensitive at the parts per trillion level to the presence of amino acids and nucleotide bases in the samples. Finally, there is a remarkably ambitious device called a life-marker chip, which is under development in European labs. This 'lab on a chip' uses fluorescence measurements to detect any of a range of about twenty different molecules or classes of molecules associated with the antibodies that are a reliable signature of past or present life.

8.9 International collaboration takes a back seat

Thus, with a certain panache, the European *ExoMars* sets out to achieve the goals of NASA's *Astrobiology Field Laboratory* within a mass and cost budget that is less than NASA's *Mars Science Laboratory*, which precedes both in the current plan. This decoupling, or dislocation, between the two programmes is friendly yet firm, and can seem puzzling in view of earlier enthusiasm for joint missions as a way of reducing costs. In fact, the change of heart reflects not only the disappointments of past attempts at a coordinated programme, especially the recent network vs. sample return debacle, but also the realisation of the added cost of doing business on an international scale. The latter involves travel and documentation that can more than offset the savings that are more obviously available from splitting mission components between two budgets.

A major factor is the imposition on NASA of the *International Traffic in Arms Regulation* technology exchange regulations, a key component of the US 'war on terror' philosophy that embraces technology of every kind. Whatever good this may achieve, its side effects include the exclusion of friendly European scientists from meetings at which their own ideas are discussed by their American counterparts, and even a recent request by NASA for a European expert to participate in the selection of a new space mission for flight without being allowed to read the proposal or the mission description ('just comment on the science'). The technical and political difficulties of preparing a mission jointly, which rule whenever Mars is mentioned, seem to be less onerous for the outer planets, where ESA and NASA are actively discussing a follow-on to the hugely successful *Cassini-Huygens* mega-mission to Saturn and Titan, albeit with strictly compartmentalised hardware contributions.

After *ExoMars*, ESA has pencilled in plans to grasp the nettle of Mars sample return for itself. A Mars sample return project remains essential if ESA's seventeen participating governments stay resolved to implement ESA's ambitious plan of sending humans to Mars. How this mission differs from the NASA sample return is not yet clear, nor is when the political tide might turn so that the two agencies once again wish to pool resources of money and talent on a shared venture. International cooperation makes a great deal of sense for a number of reasons, and its return seems inevitable somewhere between the current large rover projects *Mars Science Laboratory* and *ExoMars*, being undertaken separately, and human exploration, which can scarcely be imagined without it. A senior NASA administrator recently said on a visit to England 'Can you imagine US astronauts landing on one side of Mars and Europeans on the other, without having worked together to get there? I can't.' Those Europeans who have had to deal with the US visa system may not be so sanguine.

ESA placed contracts for the advanced development of sample return subsystems with various parts of European industry, thus nailing a proclamation to the mast that it would follow *ExoMars* with a unilateral sample return mission as soon as the budget allowed, perhaps in 2015. This was at about the same time as Alan Stern announced similar plans for an independent US mission, and before *ExoMars* had its latest cash crisis. Whether the challenges and opportunities will bring the two together remains a hot topic.

8.10 Return of the Russians: *Phobos-Grunt*

The Russians, too, are making plans to return to Mars, for the first time since the failure of *Mars-96* a decade ago. While they could have chosen a better name, to Western ears at least ('grunt' in Russian means ground or soil, and refers to the fact that this is a sample return mission) the *Phobos-Grunt* project has been given a

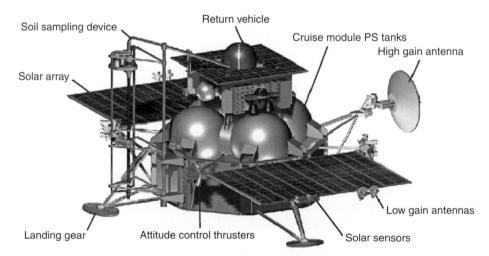

Figure 8.10 *Phobos-Grunt*. 'Grunt' is Russian for ground, and refers to the fact that the spacecraft will land on the larger of the Martian moons and obtain a soil sample for return to Earth, possibly as soon as 2015.

high priority for funding by the Russian government. The spacecraft, shown in Figure 8.10, has the goal to bring back to Earth soil samples for laboratory analysis, not from Mars itself but from the Martian moon Phobos. While in the vicinity, the mother craft will undertake orbital studies of planet Mars, including atmospheric dynamics and dust storm development, the near-space plasma and dust components, and the radiation environment.

In post-Soviet Russia, the space mission component development is led by an industrial company, NPO Lavochkin, with the soil sampling science overseen by the Vernadski Institute of Geochemistry and Analytical Chemistry and the other scientific studies of Phobos and Mars by the Russian Space Research Institute in Moscow. Project management, and the competitive selection of payload instruments, comes under the new Russian Federal Space Agency. Launch had been planned in the September 2009 launch window originally targeted by NASA's *Mars Science Laboratory*. Now it looks as if the return of the samples to Earth will take place in 2015 at the earliest, after spending about a year at Mars and on Phobos. The departing spacecraft weighs nearly two tons; the returning portion will be just fifty kilograms, including a twelve-kilogram sample container with the material from Phobos, which will be parachuted down to the surface of the Earth and recovered.

Further reading

The Case for Mars: The Plan to Settle the Red Planet and Why We Must, by Robert Zubrin and Richard Wagner, Pocket Books, 1998.

Chapter 9
Towards human expeditions

In Part II, the motivation for continued exploration was summarised in terms of understanding the origin and evolution of Mars, understanding the climate history of the planet, and finding out whether there are any traces of past or present life. In Part III, the previous chapter looked at how these aspirations are being addressed with unmanned probes, and now we turn to the preparatory work being done in anticipation of a manned mission. Many lines of scientific enquiry will probably require this eventually, notwithstanding the advances in robotics that will certainly come. While science is the subject of this book, it is also important to acknowledge that planetary exploration, indeed any kind of exploration, is not just about scientific topics. There are strong political and cultural drives to leave the Earth and investigate, perhaps even inhabit, other worlds, driven by feelings that are shared to varying degrees by almost everyone and not just scientists. Indeed, going to Mars is one of the oldest dreams of the human race, symbolising our survival and advancement, and we must assume that, if we endure as a technically advanced species, a human Mars expedition will happen someday.

Affordability will be crucial to any plan for moving beyond the planning stages of a journey to the Moon and Mars. In the USA, with Europe probably following, if not actively participating in, NASA's strategy, there will be no landing on Mars before there is a permanent base on the Moon. This will be the case whether or not the Moon is used as a staging post to Mars. If the drive to the Moon drains rather than creates funds or enthusiasm for Mars, the longer expedition may disappear into the distant future, well beyond the thirty-year horizon where it now sits with most of the pundits. Much depends on one particularly uncertain factor: the degree of international cooperation. Missions enabled by an international consortium of agencies are far more likely to progress in a timely manner and survive the inevitable crises without being cancelled than any independent plan by a single space agency, especially if they are very expensive. NASA studies as long ago as 1989 set the cost of a human mission to Mars at something in excess of three hundred billion dollars, or five to ten times the final cost of the *International Space Station*, a daunting total for any country working on its own.

Both NASA and ESA have pronounced aspirations for sending manned missions to Mars in the next thirty to fifty years, and both have indicated that they expect to cooperate with the other in due course. This will remain no more than a dream, however, until some of the yet unsolved practical problems of manned interplanetary space flight are overcome, leading to the development of the key technologies and skills needed for the establishment of manned bases and perhaps even colonies on Mars. There are probably no technological show-stoppers that will not yield to the kind of research and development that put men on the Moon in the 1960s; the problem is going to be in making the larger venture affordable and sustainable. Present-day propulsion systems are crude in the extreme, conceptually not much more than massive versions of the combustion-driven rockets the Chinese developed a full millennium ago. Where are the real versions of the propulsion systems that, in science fiction since the time of Verne and Wells, have been allowing large spacecraft with no discernable propellant tanks to cover huge distances at high speed without refuelling? Reality is supposed to overtake fiction in a shorter time than that, and in many other fields it has.

The need for new technology is, of course, linked to the budgets invested. Better propulsion and other advances might make space flight cheaper, and engender valuable spin-offs to other industries. It also may mean tangible paybacks, for instance, mining precious metals on Mars on a scale that makes it profitable despite the distance from the market on Earth. Fast, cheap space flight that pays for itself in the end is a dream, but neither of these scenarios is completely unrealistic, indeed they may be inevitable, although they still lie well in the future.

For the present, the mechanical means of getting to Mars relies on chemical rockets carrying huge amounts of dangerous fuel and large outlays of cash by governments motivated by enhancing their prestige or achieving strategic advantages. Present plans in the United States, still the front-runner, although maybe not for long, call for the use of the *Space Station* and a new base on the Moon as stepping-stones or service stations. To this end, NASA is replacing the elderly shuttle fleet with new launch vehicles and manned spacecraft that can service not just the station in low Earth orbit, but also the lunar base. Advanced versions will then go on to Mars, with the same or a different vehicle, carrying very heavy payloads. The sort of plan they will follow is illustrated in outline in Figure 9.1. Next there are the humanitarian questions of maintaining physical and mental health in an artificial, unfamiliar and possibly hostile environment, and humdrum but difficult matters like food provision and waste management. Key decisions have to be made, such as how many people should go, in order to have enough effort and the right skill mix? How can the crucial question of redundancy be addressed when the possibility of crew members dying en route or on Mars is so sensitive it is scarcely allowed on the agenda?

When all of that is sorted out, there remains the advance work that can be done on practical questions associated with Mars itself. To what extent is Mars a potential source of resources such as building materials, water, food and rocket fuel? Other than trying to survive, and looking around, what should crews do

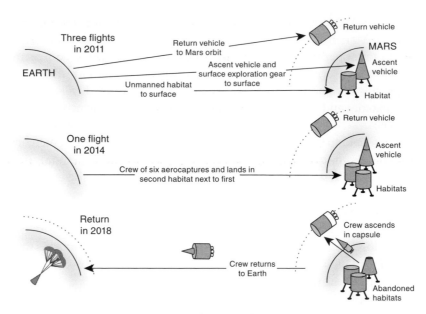

Figure 9.1 A study published in June 1998 by NASA of a manned mission to Mars required four launches of a newly developed, heavy-lift launcher, with about four times the capability of the space shuttle. The first three unmanned flights delivered the crew habitat and the ascent vehicle to the Martian surface, and the return vehicle to Mars orbit. The crew of six followed in the next launch opportunity more than two years later. The *Ares* launcher now under development could make this sort of mission feasible in the decade after 2020.

while they are there? They will need a good, versatile science plan and the tools to do proper research; the first expedition will have no trouble getting into the history books, but the subsequent ones had better do something the folks at home think is exciting and worthwhile if they do not want to achieve an *Apollo*-like relapse into isolation and introspection on a crowded Earth.

9.1   Finishing the *International Space Station*

Space stations have long been considered an obvious stepping-stone that would be utilised for any human mission to the planets. Von Braun's plans in the 1950s and 1960s took the availability of an Earth-orbiting base for granted, as the crews that would have to assemble the large Mars-bound ships in Earth orbit would need somewhere to live and store equipment. As early as 1971, while *Apollo* astronauts were still travelling to the Moon, the Soviet Union established the first long-term human presence in low Earth orbit with the launch of the *Salyut* space station, while NASA followed in 1973 with *Skylab*. The Soviet Union followed the success of *Salyut* with the first large station, *Mir*, assembled over a decade, beginning in 1986, and the USA started on space station *Freedom* in 1988. Ten years later this became the *International Space Station*, with additional participation by the Russian, European, Japanese, Canadian and Brazilian space agencies.

Originally the station was to be a place where humans would go to perform scientific, medical and industrial research and manufacturing that could not be done on Earth. Delays in completing the space station and shuttle problems have led to cost growth and delayed the project and greatly limited its utilisation. After the *Columbia* accident took the shuttle out of service, the number of astronauts resident on the space station was reduced to two, since there were not enough flights by the smaller *Soyuz* to supply any more. Without the manpower to press ahead with the non-essential jobs, like science, the space station's mission stagnated. Since the shuttle returned to flight in 2006, matters have improved, and the space station's sustaining partners have reiterated their support for completing and using the facility.

President Bush's 'Vision for Space Exploration' contained little new money but put pressure on NASA to replace the shuttle and get on to the Moon and Mars. This, and the fact that the lunar base will probably be serviced directly from Earth and not via the space station, put the Agency's commitment to the station under strain. NASA made deep cuts in its budget for research in Earth orbit, hitting especially hard that portion devoted to understanding and mitigating the harmful effects of space travel on humans. Meanwhile, modules built by Japan and Europe have to be ferried up to the station by NASA to satisfy its international agreements. A succession of shuttle flights is underway, with plans to complete the space station by 2010, when the shuttle can be retired and further crew and supplies launches provided by Russian spacecraft, until the new *Ares* launcher starts flights to the station in 2015 or so.

9.2 Developing the Moon base

President George W. Bush can be seen in Figure 9.2 announcing his 'vision' on 14 January 2004. This revamped and rescheduled President Bush senior's 1989 'Human Exploration Initiative', directing NASA to 'Extend human presence across the Solar System, starting with a human return to the Moon by the year 2020, in preparation for human exploration of Mars and other destinations.' In Europe, ESA had already announced in November 2001: 'the ESA Council at Ministerial level approved the *Aurora* Programme dedicated to the human and robotic exploration of the Moon, Mars and asteroids. The ultimate goal of *Aurora* is human exploration of Mars foreseen in the 2025–2030 time frame.' Russia says that it has plans for human exploration of Mars and Phobos. The latest entry into human spaceflight, China, also threw its hat into the ring, describing goals for Chinese astronauts to travel to the Moon and Mars.

These were all proclamations rather than plans, but NASA's was fundamentally different from ESA's in that it involved the political establishment imposing goals on the space agency rather than the latter making plans and then seeking political support and funds. The Vision sought to put humans on Mars, but only after successfully demonstrating sustained human exploration missions to the Moon. The high projected cost of the lunar stage of the programme, which

Figure 9.2 On 14 January 2004, President George W. Bush announced his Vision for Space Exploration and listed its principal features:
– Spend $12 billion on new space exploration plan over next five years. $1 billion will be new money, the rest reallocated from existing NASA programs.
– Retire shuttle program by 2010.
– Develop new manned exploration vehicle.
– Launch manned mission to Moon between 2015 and 2020.
– Build permanent lunar base as 'stepping stone' for more ambitious missions.
– Complete commitments to *International Space Station* by 2010.

necessarily includes replacing the old shuttle fleet with a more modern vehicle capable of flying between Earth and Moon with a substantial cargo, was bound to affect plans involving missions to Mars beyond *Mars Science Laboratory*, then planned for completion in 2009. While President Bush generously charged NASA with conducting robotic exploration of Mars to search for evidence of life, to understand the history of the Solar System, and to acquire adequate knowledge about the planet to prepare for future human exploration, NASA soon found that going back to the Moon was virtually the only priority it could afford, and that budget was very tight.

9.2.1 The Moon as a test bed for Mars

Using the Moon as a test bed for human missions to Mars entails numerous technical compromises, due to the fundamental differences between the two destinations. For instance, the gravity on the Moon is only half that of Mars (at about one-sixth of that on Earth), while the length of day is nearly thirty times longer. Descent onto an airless body like the Moon has to use rockets, whereas descent through the Martian atmosphere can use atmospheric drag to reduce velocity through dissipation of energy as heat, and one or more parachutes. Of course, the final stage of the latter can also use propulsion, especially useful near the surface. For small payloads that do not require accurate targeting (not recommended for humans), a semi-hard landing with airbags works well. Once on Mars, astronauts face a communications delay with Earth of up to thirty-eight minutes, compared to just two seconds on the Moon.

Whether the Bush vision will endure long enough to send NASA astronauts to Mars, or just to the Moon, only time will tell.[1] However, it is clear that for almost everyone in politics, agencies, science or just in the street, Mars is the glittering prize and the Moon is the way station. Some NASA insiders opined that getting to Mars would be quicker and cheaper without going via the Moon; the riposte was generally that risk reduction, and not short-term economies, was the reason for going to the Moon first. Neither the Earth nor the space station provided the environment needed to learn how to survive and build substantial structures on a smaller, low-gravity, virtually airless body, and what experience there was suggested these tasks are much more difficult and dangerous than anyone had imagined. It is likely that a lot of the heavy lifting and digging will be done by robots, and these need to be developed and tested. Unpiloted cargo missions will transport infrastructure to the Moon, including habitats, power systems, transport and consumables. This is undeniably a good rehearsal for Mars, although not perfect – important differences in gravity, communications delays with Earth, and the effect of the still significant Martian atmosphere on landing and ascent techniques, all limit the relevance of practicing on the Moon for a Mars landing. The long journey time – remembering that humans have to go both ways – is probably the biggest challenge for a Mars expedition.

9.3 Transportation systems

Several different space agencies are working on the problem of getting to Mars and back, all with different ideas and approaches. The mandate to go via the Moon is the obvious constraint imposed on NASA. ESA is keeping its options open on this, but is virtually certain to team up with NASA at some point. Concepts for the vehicles that will carry the crew and its support equipment take various forms, most of which exist so far only in drawings and paintings, like the splendid-looking European trans-Mars vehicle shown in Figure 9.3. What they are all likely to have in common is a large ship for the out-bound and return journey to Mars orbit, and a lander that descends to the Martian surface and ascends back into Martian orbit for rendezvous with the mother ship, prior to return to Earth.

---

[1] President Obama, before his election, expressed broad agreement with NASA's exploration goals as they stood at the time.

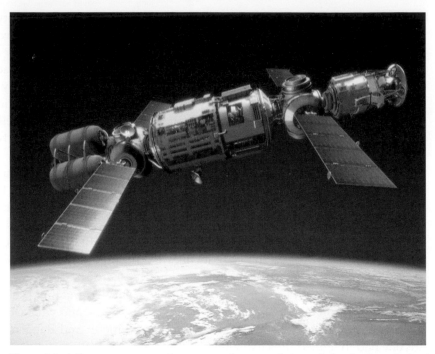

Figure 9.3 A European concept for a manned spacecraft, intended to be the culmination of the *Aurora* program started in 2003, that could be assembled in Earth orbit and ready to carry its crew to Mars by about 2043.

It seems most unlikely that a flight directly from the surface of Mars back to the Earth will be attempted, at least not until major technical facilities and fuel supplies have been established on Mars in the more distant future. More likely, crews will transfer twice for each leg, once in Earth orbit and once in Mars orbit. Both the interplanetary ship and the surface-to-orbit shuttle will come in two versions, one for crew and the other an unmanned cargo ship. The assumption in many past scenarios that the interplanetary spaceship will be assembled in Earth orbit must now be questioned: it might now have to take off from, and return to, a launch pad on the Moon, or possibly be assembled in lunar orbit by technicians based on the Moon.

9.3.1 From the Earth to the Moon

After the last *Saturn V* lifted off in 1973 carrying *Skylab* into orbit, NASA was left with a very limited array of launchers, and nothing of comparable size. This situation was to last for nearly forty years. The shuttle could carry 27,500 kilograms of payload, in addition to the mass of the vehicle itself and that of the crew, whereas the *Saturn V* lifted nearly five times as much useful mass. Even if it were not too small and too technically dated, the shuttle was designed to fly into low Earth orbit and not to fly to the Moon, much less to Mars. The best that could be done with the shuttle would be to make a large number of lifts into Earth orbit with pieces of Moon and Mars spacecraft, and assemble them there for their onward journey, using the space station as a base. At least six very expensive shuttle launches would be required per journey to the Moon, and about twenty for Mars, along with the inconvenience, mass inefficiency, and risks associated with on-orbit assembly.

The first stage in implementing the Bush vision, then, is to retire the shuttle in 2010, leaving only mid-sized commercial launchers with even less capacity than shuttle until a replacement in the *Saturn V* class can be developed.[2] To minimise the development costs and quicken the schedule, NASA elected to build both a man-rated crew launcher vehicle and a heavy launch vehicle for cargo, utilising shuttle-style solid rocket boosters, together with a liquid-fuelled upper stage using an upgraded *Saturn V* engine. The *Ares I* will take a payload about the same size as the shuttle carried, enough for a crew of four to six, and be supplemented by a lunar cargo launcher that carries 125 tons to low Earth orbit, comparable to the capacity of the old *Saturn V*.

The heavy launch vehicle will later be adapted as a manned vehicle for the journey to Mars, since the mass that has to be carried for even a small human crew, on a short-term *Apollo*-style mission to the surface of Mars, is at least ten times that of the most massive robotic mission. When the details are worked out, it may be much more, well beyond the capability of a single *Ares I* payload. For a crew of six planning to erect a habitat and use transportation on the surface, the mass that needs to be sent to Mars will exceed the total mass of the *International Space Station* at completion, about 500 tons, possibly by a considerable margin. This mass has to be taken from Earth to Moon first, requiring perhaps as many as fifty of the new *Saturn V*-class *Ares* launchers. Clearly there is scope for contemplating investments in new technologies and the development of even larger launchers before going on to Mars.

9.3.2 Transfer to Mars

Once reliable vehicles with heavy lift capabilities from Earth to Moon are fully operational, the even bigger challenge of the transfer to Mars will be tackled in earnest. Achieving acceptable levels of safety and comfort for a crew of six for a five-month each-way cruise is a task that goes far beyond those for the two-day trip from Earth to the Moon. The long duration requires a much larger volume be available for the crew, including the provision of living quarters, toilets and a galley, and of course all of their supplies must be carried, including, in the first instance at least, the fuel to return to the Earth–Moon system. When contemplating new technology for this part of the journey, the most promising

---

[2] According to reports in the media, President Obama's team asked the Bush administration to try not to do anything to preclude additional shuttle flights after 2010, should they decide to approve these to plug the gap. Military rockets might also be used.

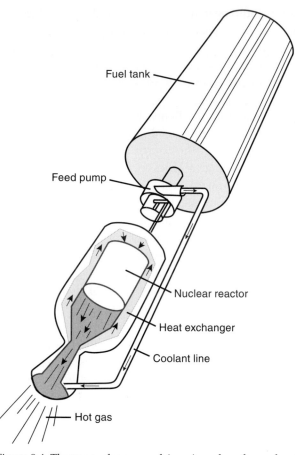

Figure 9.4 Thermonuclear propulsion: A nuclear thermal rocket operates by the same basic principles as chemical rockets, that is, the expansion of hot gas through a nozzle to provide thrust. The gas concerned, probably hydrogen because it has the lowest molecular weight, is supplied as the fuel for the journey. In flight it is passed through the core of a nuclear reactor, where it is heated to a very high temperature before being expelled; more than three thousand degrees centigrade must be achieved to obtain sufficiently high performance.

candidate looks to be thermonuclear propulsion. Figure 9.4 shows how this works.

Nuclear thermal rockets were first developed and tested in the early 1960s, but then abandoned because there was then no really exciting application that would justify the development costs. The advantages for travel to Mars include trip times shortened by as much as thirty per cent because the thrust levels achievable are nearly thirty thousand newtons [3] and, unlike chemical propulsion, this impulse is applied continuously. After the development cost has been written

---

[3] This can be compared to about ninety thousand newtons, the thrust of one of the Rolls-Royce *Olympus* engines used to power the *Concorde* supersonic airliner.

off, lower costs per trip should result, mainly from the more compact nature of the energy source, which saves the need to move large masses of liquid fuel around.

The main development problem is obtaining a compact, lightweight nuclear reactor that will not add too much to the mass of the vehicle. With current technology, reactors produce a high flux of energetic neutrons, which is powerful enough to damage the actual structure of the motor over a period of time if it is not bulky and strong.

9.4 The danger from radiation and weightlessness

Protection of the humans on board from the radiation generated by the reactor is also a key issue, of course, generally addressed by an elongated spacecraft design with as much distance and structure as possible between the rocket and the crew. In addition to separating the humans as far as possible from the reactors, shielding will be required between the two to keep the radiation dosage to an acceptable level. The shield can be fairly small if the geometry of the layout is carefully chosen, given that the crew quarters will necessarily be quite compact. The same shield might protect the crew from bursts of energetic particles from the Sun, if it could be oriented appropriately when such a burst was detected. However, shielding from galactic cosmic rays is much harder, since these come from all directions, and this could require a large and massive shield. Crew exposure to radiation occurs throughout the round trip flight, as well as on the Martian surface. Without adequate shielding, there is a cumulative increase in the risk of radiation-induced cancer and damage to the central nervous system.

Designing protection into the spacecraft means first obtaining reliable measurements of the flux of galactic rays and solar energetic particles to which the astronauts would be exposed, to help to decide the scale of the shielding that will be needed. A radiation monitor was flown to Mars on board the 2001 *Mars Odyssey* orbiter, collecting data along the same route and for the same time as the trip to be taken by humans. Once in orbit about Mars, the instrument monitored the radiation outside the planet's atmosphere, including that produced at the eleven-year maximum of solar activity. Although typical, this limited set of data is not enough. Some of the most dangerous exposure comes during relatively brief energetic solar particle events that, in extreme cases, may induce debilitating nausea and vomiting, and in very extreme cases death, among members of the crew. Statistics on those are essential.

The last time humans had prolonged exposure to energetic particles from the Sun was during the *Apollo* missions to the Moon in the early 1970s. (Low Earth orbit, including trips to *Space Station*, do not count because the astronauts are still inside the protective shield offered by the Earth's magnetic field.) Something that was not well understood at the time of *Apollo* was the extent to which dangerous particle emissions are confined to relatively infrequent events called coronal mass ejections, when tens of billions of tons of material, principally protons (hydrogen nuclei stripped of their electrons), are ejected from the Sun at speeds of around three hundred and fifty kilometres per second. There was one such event in August 1972, in between the last two *Apollo* missions (*Apollo 16* returned on 27 April, and *Apollo 17* launched on 7 December, 1972). This was extremely fortunate: if either of the crews had been in interplanetary space when this particularly severe burst struck their spacecraft, it is possible that they would all have been killed, despite being inside the protective hull. It was not until October 1989 that there was another burst as deadly as that one.

The message is clear: astronauts travelling to Mars will need two kinds of protection, one against their own nuclear power plant, cosmic rays and the frequent, smaller bursts from the Sun, and another against the rare but deadly major coronal mass ejections, if one should occur during their excursion. Fortunately, the arrival of a dangerous burst can be predicted (reliably, for a ship on its way to Mars, since the particles have to pass the Earth first, where they will be observed by satellites and ground stations) in time to allow the crew to take shelter in the most protected part of their ship for the limited period – a day or so – that the level is high. Reducing the radiation hazard by the provision of shielding comes at a serious cost in terms of additional mass and propellant. If the reactor is kept at a sufficiently large distance, the crew quarters can be in the shadow of a geometrically shaped shield, possibly a tank filled with water or a smaller volume of a dense metal, or even an electromagnetically generated magnetic field, mimicking the Earth's protective magnetosphere. These can be aligned towards the Sun when the arrival of a burst has been forecast. Probably only a small part of the crew's habitat will be fully shielded from solar and cosmic radiation during the cruise to and from Mars. During solar storms, the crew must shelter in the shielded part of the spacecraft, but since these are infrequent, taking cover need not be overly disruptive to operations or lifestyle.

The negative effects of extended periods of weightlessness have been known for decades and extensive experience has been accumulated. The crew will use exercise machines and dietary supplements, including a carefully controlled fluid intake, to minimise muscle atrophy and maintain cardiac output, knowing that they will suffer some degradation but that they will return slowly to normal upon landing on Mars. Bone loss, particularly in weight-bearing bones, is a more serious problem since it increases the chances of a catastrophic injury while on the mission. Addressing this requires long periods of quite aggressive and unpleasant exercise, including compressing and jarring the bones to simulate normal stresses; experience has shown that the hardest part is persuading the astronauts to keep up the grinding routine for sufficiently long periods.

9.5 Landing and return vehicles

Landing on Mars is a risky business at present, and the success rate has certainly not been as good as a potential crew member on the first expedition would wish. Anyone going to Mars later this century will want to do so knowing that the risks associated with soft landing have been reduced substantially from the current fifty per cent or so. In the longer run, this will be achieved routinely by building spaceports, with flat, instrumented landing fields offering laser guidance to automatic pilots and boasting safety records comparable to present-day airports on Earth. Even then it will still be desirable to make pinpoint landings on rough, previously unvisited sites, either from a base on Mars or directly from an orbiting Mars space station. This will require advanced terrain recognition systems, using lidar for example, and techniques for hazard avoidance during final descent and soft landing. Reliable facilities for predicting the weather, especially high winds, turbulence and dust storms, will also need to become routine. These are likely to include a network of weather satellites and surface weather stations, as well as balloon launches and continuously airborne dirigibles. The floating stations are the best way of detecting local disturbances, measuring winds and turbulence in particular. Finally, as for Earth, numerical forecasting models will fill in the coverage and identify potentially unstable atmospheric situations based on a continuous flow of local and global data from the measurement systems.

Whether for the pioneers who land first, or for future colonists and holiday-makers, the most important uncertainties are during entry into the atmosphere, and just before landing. The former must take into account the huge variations in atmospheric density that are found at very high levels, determined mainly by what is happening deep below. Dust storms have the largest effect, expanding the lower atmosphere and changing the density by more than a factor of two at a hundred kilometres above the surface, where the ship starts aerobraking and the human or robot pilot is calculating angles of approach and parachute deployment times. The region between the surface and twenty kilometres altitude contains the highly variable planetary boundary layer itself, threatening the now slowly moving craft with side winds that can push it off course or even topple it over.

During daytime, the surface heats the lower atmosphere creating convection as the warm air rises to replace the cooler air above. This process, when vigorous, induces instabilities that manifest themselves as clear-air turbulence and wind shear (large changes in wind speed or direction over a relatively small distance). Katabatic winds – those caused by the local topography – can also create unpredictable wind shear. As the incoming vehicle reduces speed and steers towards its landing site, it may be exposed to drifts and tilts that are difficult to control. The *Spirit* rover came close to having this problem when it opened its parachute and found its rate of descent too rapid, due to lower-than-expected middle atmosphere densities, and experienced unexpected oscillations that nearly exceeded safe ranges. The airbags would not have delivered the rover safely if it swept across a sharp rocky outcrop as it touched the surface, to give just one example of many such hazards. In the fullness of time, operators at bases on the ground on Mars will release balloon sondes to check the weather conditions before a landing or takeoff, much as they do at airports on Earth. Until then, pilots and their support equipment will have to be intelligent, adaptable and cautious.

The weather on Mars will be known and understood in a general sense, well before a manned landing is attempted, from the environmental data acquired by robot precursors at the same time as they search for evidence of climate change or of indigenous life on Mars. While in some sense every mission to Mars is a human precursor, as the time for a manned flight gets closer there will be an increasing number of robotic missions wholly or partially dedicated to obtaining the practical knowledge of Mars needed to send a crew to land with acceptable cost, risk and performance. Detailed surveying will establish whether a proposed site will support a spacecraft and, eventually, a building. Their final task will be to prepare the landing site itself by clearing or charting obstructions and any other steps that are necessary to see that it is certified for safety and suitability before the first manned craft sets down. Then they can help the pilot on the incoming ship to locate the prepared site and make a precision landing by positioning themselves around the safe area and turning on radio beacons.

Supplies can be sent separately to reduce the size and mass of the landers with people on board. These will include prefabricated structures, possibly erected in advance by robots, to provide a home for the astronauts that is available immediately they land. This will make it easier to deal with the thin atmosphere and low temperatures on Mars, and provide better protection than their landing craft from the radiation environment at the Martian surface. The supplies will need to land safely and accurately too – it is no use putting a crew down safely if they then starve to death because their supplies crash a hundred miles away. However, cargo ships do have an option not available to manned vessels, namely the possibility of sending more than are needed with the expectation that some will be lost. For reasons of economy, they are likely to be very massive, and to come in directly from Earth, rather than from orbit, which means high approach speeds. As well as using the same advanced radio and optical flight navigation and surface location techniques as the manned landers, they will need aerocapture, entry and manoeuvring systems that work at speeds estimated in some studies at up to ten kilometres per second. Deploying parachutes at that sort of speed – around Mach three – is well beyond the qualification of current systems and probably not achievable in any completely reliable design even with developments in materials and configurations. This points to the need for a lot of aerodynamic braking first. In the thin Martian air, that means a long, flat and very accurate trajectory towards the landing site, and a sophisticated heat shield design. If it can be made tough enough, the shield might be inflatable, in order to increase its area and provide more drag for a given mass.

9.6 Setting up the Mars base

Once safely down, the explorers will need some sort of habitat on the surface. They will have to stay longer than the *Apollo* astronauts did on the Moon, partly because the journey is so long and partly because time is needed to find and collect interesting samples, perhaps including drilling, and to conduct some preliminary experiments involving launching weather balloons or exploding seismic charges. There are also constraints due to planetary alignment that affect Mars journey times and fuel requirements much more than they do on trips to and from the Moon. In general, missions that involve the shortest time on the surface have longer transfer times. Expert astro-navigators, considering all factors, define a short-stay mission as about fifty days, while the long-duration class of mission has the crew on Mars for nearly six hundred days, almost a Martian year.

Even if they spend only a fifty-day Martian month on the surface, the crew will have to dedicate most of their time to establishing a safe habitat and organising their supplies, using material from the cargo ship that landed before them. For a longer stay some further supplies may arrive after the base has been set up, and have to be recovered and transported locally. Only crews on long-stay missions can move very far from their bases or carry out the most important scientific research. These will gradually become more ambitious as experience is gained, and later flights will benefit from a pre-existing habitat, set up by earlier expeditions and gradually developed to provide better facilities.

The immediate concern of the first crew, beyond mere survival, will be a detailed appraisal of their immediate surroundings to confirm the suitability of their landing site as a base in the longer term. Factors such as the bearing strength, penetration resistance and cohesion of the soil are all factors in deciding. Quicksand of one kind or another may be common on Mars, as demonstrated by the spectacle of the *Opportunity* rover stuck in a sand dune, now named 'Purgatory'. This episode led engineers, scientists and even the project manager at the Jet Propulsion Laboratory to build a sand box and try different mixtures of sand, clay and ground-up shells to simulate the likely properties of the soil underneath the rover, in order to understand how it would respond before commanding it to dig itself out. Finally, after nearly five weeks, they succeeded.

The material of which the surface is composed is an asset as well as a possible threat, of course. Sand, clay and rock will be good building materials there, just as they are on Earth. Figuring out how to acquire and use the full potential is another challenge for the intrepid explorers. Location will be important – obviously, a water supply would be wonderful, just as nomads seek oases in the Sahara. It may be better to set up a base against a cliff, in a valley or a crater, but it is undoubtedly easier to land on a relatively bland plain. Finding the right compromise will not be easy, but could be crucial.

9.6.1 Human health on Mars

Once landing safety is achieved and living quarters are set up, more subtle considerations come into play, including the potentially serious health hazards the team will have to face from factors like weather, marsquake, radiation and infection during a long stay. The top priority for research on the surface is likely to be learning more about how cosmic rays and high-energy particles from the

Sun affect the environment at the new base. Experience of this on Earth or even in space is of limited value to the new Martians, who need to explore the limited shielding properties of the thin atmosphere and the consequences of the absence of a global magnetic field. The scope of the problem further requires an understanding, not only of how the particles from space propagate through the Martian atmosphere, but also the extent to which the regolith releases a flux of secondary particles when struck. These heavier and slower particles can actually do more damage to astronauts and equipment than the fast, light particles coming in from space that tend to pass straight through human tissue unnoticed.

The particle threat will have been evaluated in advance by the deployment of radiation monitors on the surface by robotic landers. The *Mars Science Laboratory* rover will start this project, testing existing calculations of the shielding effect of the atmosphere with reliable measurements of the numbers and energies of all types of charged particles made simultaneously at the surface and in orbit. These data need to extend over the long term, in order to find the accumulated absorbed dose at the surface over time as the output from the Sun varies with the solar cycle and with solar storm events. The thickness of the atmosphere itself depends on the seasonal cycle of surface pressure and on the atmospheric dust loading, which is very variable throughout the year. It will also be important to identify the separate contributions of protons, neutrons and heavier particles, since they inflict different types of damage. Finally the natural radioactivity of the planet's surface materials (soil and rocks) may make an important contribution also, and this will vary with location.

Provided, of course, they are not themselves significantly radioactive, rock, soil and dust from the Martian surface offer a readily available source, not just of building material, but of additional mass to screen the crew from the deadly rain of energetic subatomic particles. Permanent habitats will probably be at least partially buried, with soil piled around the exposed part as a radiation shield; this will also improve the thermal insulation. Calculating the thickness of the required cover will depend upon better knowledge of the characteristics of the Martian regolith. Its key physical properties include the particle size distribution, and particle shape, hardness and density; its cohesion, adhesion and dielectric characteristics; magnetic properties, elemental composition and chemical stability or reactivity. The thermal conductivity and heat capacity of the surroundings will affect the control of temperature in the habitat during the extreme Martian day–night cycle, which can exceed a hundred degrees centigrade. Soil properties will vary with depth and need to be known down to at least ten metres, and as a function of location, season and nearby geological environment.

If natural material cannot do a good enough job, then insulating and shielding material will have to be flown in from Earth. Since this is likely to be massive, it is particularly important that it is well designed and based on a good understanding of the size and scale, and variability, of the hazard. Another kind of brute force mitigation for all sources of radiation damage can be accomplished by flying an older crew, aged forty or over. In middle age, radiation exposure is very unlikely to become a significant health issue within the expected remaining span of a normal human life.

The fine wind-blown dust will be a problem everywhere, affecting visibility and clogging, abrading or covering vital equipment, especially during local or global dust storms. The seals on the airlocks leading in and out of the explorers' habitat could be compromised by dust contamination, and mechanical bearings inside the vessel will be exposed to dust damage if enough of it gets into the moving parts. The iron-rich dust is magnetic and conducting, and could cause short circuits in vital electrical equipment. Just how dangerous it turns out to be depends on the detailed properties of the dust, and how local meteorological conditions cause it to move around. The pioneers on Mars will need to keep a careful watch for problems while making long-term measurements and analysis using sophisticated equipment. These will be needed at every site and season; as the weather varies, so may the nature of the dust. While global dust storms mix and homogenise the finer dust, the larger, heavier particles may be sourced locally and could have properties that vary from location to location.

In addition to its nuisance value, some of the dust could conceivably be toxic. The finer particles will stick to astronauts and equipment and, once transported inside pressurised modules, will become airborne, and may be inhaled. This happened to the *Apollo* astronauts on the Moon, although the only serious problem they reported then was that the lunar dust irritated their eyes. The Martian dust is different and could contain potentially serious biological hazards; for instance, one study suggested that hexavalent chromium, an unpleasant substance that is rare on Earth, could be present in hazardous amounts (more than one part in ten thousand in this case) in the highly oxidised surface material on Mars. It is really too late to find this kind of problem after arrival, and each landing site will have to be visited in advance by robots equipped to collect and analyse dust, rock and soil samples, returning some of them for full risk characterisation on Earth. They will look for the toxic trace elements such as arsenic, lead, beryllium, cadmium, chlorine and fluorine, which are harmful in very small doses, but can be hard to detect in ordinary chemical assays without full lab facilities. The effects of the dust and soil on biological material such as enzymes, lipids and nucleic acids must also be tested to identify any unexpected potential for harming human explorers.

The microscopic life that may exist on Mars, part of the motivation for going there in the first place, could itself be a hazard. The factors that make life unlikely on the surface – the thin air, the cosmic-ray flux, and the oxidising nature of the surface soil – become helpful when considered in terms of the safety of future human explorers. Conversely, dangerous bacteria or other hazardous things may lurk at safe depths in the soil (safe for them – but perhaps not so deep as to keep the future colonists from encountering them, when drilling for water, for example).

9.6.2 Finding useable water supplies

Humans on Mars are going to need copious supplies of liquid water, and they are not going to want to bring such a heavy item from the Earth if they can help it. Instead they will rely as much as possible on an assessment of the distribution of accessible water in soils, regolith and Martian groundwater systems at each location, again carried out as far as possible in advance of the manned landing. Finding useful water is a rather different question from the search for water on scientific grounds, because the 'accessible' part means it must be relatively easy to extract as well as to find. The answers, however, may be found in the same way, with geophysical investigations followed by subsurface drilling and *in situ* sample analysis to search the Martian subsurface for ice and liquid water reservoirs. If these are scarce, adsorbed and bound water may have to be recovered from rock, soil or dust. Then the question becomes, what is the water content of accessible subsurface soil, rocks, the near-surface atmosphere including frost and dew, and so on, and what equipment is needed to release and purify water from them?

Assuming the pioneering explorers do find Martian groundwater systems, they not only have to determine the hydrological characteristics of the aquifers, they also need a subsurface drilling capability that may be very demanding. Missions to demonstrate autonomous drilling operations on the Martian surface will begin with the two-metre drill on *ExoMars*, then NASA's *Deep Drill* reaching to tens of metres, followed by whatever it takes to reach depths corresponding to subsurface reservoirs of liquid water if both of those fail. While there is a high probability that they do exist, the aquifers could well lie several kilometres deep in many places. It might be easier to set up the base at colder latitudes where permafrost layers that can be mined for water are found near the surface, or even to accept at the outset that it will be necessary to develop the technology to extract useful quantities of water from the vapour in the atmosphere, if that turns out to be easier than getting it from hydrated minerals in the regolith.

9.6.3 Biohazards and forward or backward contamination

If the colonists are blessed with a handy liquid water supply, as for example at the poles where reasonably pure water ice may be exposed at the surface, they will have to test it very carefully for biohazards such as indigenous Martian bacteria. Interesting as these may be, they may be so different from the familiar types that great care will need to be taken to characterise or eliminate them before the water is drunk or otherwise brought in contact with earthlings. It is, in a way, fortunate that the current evidence suggests that any life there may be on Mars is probably confined to niches where conditions are benign, so it seems unlikely that bacteria will be present in the airborne dust or in the surface materials with which daily contact is made. On the other hand, it is unfortunate that such niches may include the water supply. A foolproof test for the presence of life will be needed, a significant challenge especially when the nature of the life has yet to be determined, although again advanced knowledge of the infections that may have to be dealt with will be sought by robotic measurements and sample return before humans are risked. Even then, positive tests are much easier than negative ones; it will probably never be possible to say for certain that there are no life forms on Mars. Ultimately, returning astronauts will have to be quarantined until their own experience on Mars has shown it is reasonably safe to drop this precaution.

Assuming any Martian bacteria that may exist are found not to be immediately harmful to the explorers, a significant risk is still associated with the possibility of transporting any kind of replicating life form to Earth when they return. It will be very hard ever to completely eliminate the risk that it will have a negative effect on some aspect of the Earth's ecosystem. This is known in the planetary protection discipline as 'back contamination' and, although the probability of an alien epidemic is considered very small, the consequences could be enormous and few would say that painstaking measures should not be taken to minimise the risk.

It is more debatable how much time and trouble should be devoted to the problem of 'forward contamination' – the risk that terrestrial microbial life will be transported to Mars, resulting in local or widespread contamination of the Martian surface and possibly leading to false positive indications of life on Mars. This could also happen if samples returned to Earth are contaminated by the crew or their equipment, even if they do not infect Mars itself permanently. It complicates the question of how much to spend protecting Mars from terrestrial bugs to know that America, Russia and the UK have already landed there, with various levels of sterilisation first. Thorough decontamination procedures that do not risk damage to the equipment are very expensive, and most project managers do not want to spend too much of their budget this way. If they do, they can soon reach the point where the planetary protection rating becomes one hundred per cent, because the mission is cancelled due to cost overruns.

Humans travelling to Mars cannot realistically expect to avoid contact with soil and dust or the expulsion of waste products of various kinds, so both backward and forward contamination are virtually guaranteed and will have to be dealt with by appropriate procedures. This is yet another reason why precursor missions will have to include sample return. Then, the full power of analytic instrumentation that exists on Earth can be brought to bear, complex sample preparation procedures can be carried out, tests involving mechanical difficulty and/or live organisms can be attempted, detailed systems of positive and negative control standards can be established, and early unanticipated results can be followed up with a revised analysis plan. If Martian life is found in any sample, it must be assumed hazardous until proven otherwise. Hazard determination may require extensive experiments, which again could realistically only be carried out in laboratories on Earth. A key corollary is whether terrestrial microbes can survive and reproduce on Mars, again a task to be addressed in terrestrial simulation laboratories before deliberately allowing terrestrial organisms onto the Martian surface. It will also be important to determine the possible mechanisms of transport whereby surface organic contaminants might penetrate into the Martian subsurface, and in particular, into a Martian aquifer.

9.6.4 Getting around on the surface

It is hard to see how explorers will achieve all of their scientific goals unless they can cover distances on Mars that are much too large to walk. Wheeled or tracked vehicles will need to be used extensively, and these must be reliable since it will not usually, at first at least, be possible to mount rescue operations if a vehicle gets stuck far from base. Therefore it will be necessary to get to know and map the traction and cohesion properties of Martian soil and regolith in various widely dispersed regions, with emphasis on hazards such as quicksands, crevices, cliffs and dunes. When the *Mars Exploration Rover Opportunity* was mired for weeks in soft dune material, the wheel slippage was 99.5%, an unacceptable predicament for a human mission, especially as the latter will be heavier and more slippage-prone. Problems of clearances over rocks and cracks and other hazards are more predictable but not always avoidable. Landing sites and locations for bases would need to be particularly well surveyed for slopes, rock properties and abundances, and soil textures, something robots can do very well in advance. Remote measurements also help, for example by mapping the thermal inertia (that is, measuring how quickly a surface cools – dust bowls do it much faster than rocks) at high resolution from orbit or from survey aircraft.

Sooner or later, humans on Mars will probably want to travel around the planet by air. NASA recently considered a design for deployment on Mars in the relatively short term, following successful test flights in Earth's atmosphere. This was an unmanned aircraft, carrying cameras to survey the terrain, rather than humans, but it made a start on understanding how powered flight might work on Mars. Takeoff and landing are particular problems because of the thin air even at the surface, and of course the uncertain terrain and capricious meteorological conditions. The rarefied air dictates a very large wingspan and a propulsion system capable of high thrust in order to get airborne. NASA's unfulfilled plans for unmanned survey aircraft avoided this problem by deployment directly into the atmosphere on arrival, without landing, and when it did finally come down the mission would be over. This meant a short mission lifetime, possibly of only an hour or two, and a design that did not address the difficulties of takeoff and landing for routine manned flights on Mars. There is a lot of work still to be done, therefore. The choice is probably between using vertical ascent and descent under rocket power, and constructing runways and using launch catapults like those on aircraft carriers.

9.6.5 Storms and lightning

The probability of a person, a building, or a landing or departing spacecraft being struck by lightning on Mars is very small, although it may not be negligible in severe dust storm conditions. Research is needed to understand the role of electrical discharges and electrostatic effects in atmospheric processes involving dust, and all of the potential electrostatic hazards when working on the surface. Experiments will need to be landed in advance to measure the voltage required to produce electrical breakdown in the air as a function of temperature, pressure, wind and dust load in the atmosphere, and all electrical equipment for the expedition designed accordingly. The dust in the atmosphere is very dry and may hold an electrostatic charge, as may the soil particles, electrified by turbulent motion in clouds and dust storms. On the other hand, some components of the dust may be highly conducting, which would suppress lightning but might bring other hazards such as short-circuiting batteries and power supplies. The air itself may be ionised near the surface, even under normal conditions, to some extent that remains to be determined and, if so, this will influence what types of electrical discharges do, in fact, occur on Mars.

9.6.6 Growing plants on Mars

To support long visits, permanent bases and colonies, something will need to be done to the Martian soil to enable it to support plant life. It seems too much to hope that anything useful as food will ever be grown in the natural Martian environment; the addition of fertiliser and the provision of an artificial atmosphere will be essential, and probably artificial lighting, temperature control and a water supply as well. The main unknown is the quality and variability of the soil, which will determine which plants are likely to thrive given this support and encouragement. Following studies of returned samples on Earth, the acid test will be by *in situ* plant growth experiments, carried out in various artificial environments on Mars through the full growth, seed and re-germination cycle.

Perhaps plants can be selected and genetically engineered so that a completely Earth-like environment will not be needed before they will grow. Ideally, some of the natural or artificial fertiliser that will have to be added to the soil will itself be retrieved or manufactured from local mineral deposits near the base, although organic material such as peat and other products familiar to the terrestrial gardener will need to be transported from Earth. One of the more attractive aspects of human waste disposal on Mars is the possibility for recycling material as fertiliser; the Martian colonists will be expert recyclers, carrying out research to determine the rate of reaction of typical materials exposed to the Martian environment, and the dispersion rate of unwanted waste for various methods of disposal. Some plants make good fertiliser for others; the scope for the plant biologist and food scientist is almost endless.

9.6.7 Refuelling on Mars

Some time after the first landing, probably well after, when the inventories of raw materials are done and the production and storage techniques have been well tested, future explorers are likely to rely on *in situ* propellant production to return to the Earth. The most likely fuels to be exploited first are methane and oxygen, with fuel cells using oxygen, water and buffer gases such as nitrogen and argon to provide power in spacecraft cabins and in surface vehicles. Skills must first be developed for locating and acquiring whatever local resources there are, processing them and storing the end products, and then using the latter effectively in efficient motors and machines.

Assuming a supply of water from aquifers, hydrated minerals or the air, as already discussed in terms of human needs, zirconium oxide cells can be used to generate hydrogen and oxygen by the electrolysis of water. The oxygen is reasonably easy to store by liquefying it at low temperatures and under pressure. The hydrogen produced in this process is a useful fuel, although probably not for spacecraft until there are very sophisticated facilities at the Mars base, because it is much more difficult to store and use safely than methane or oxygen. This may not be the case for long, as a lot of research is being done on Earth to handle hydrogen safely as a pollution-free fuel for cars. In any case, it can be used for methane production by reacting it with carbon dioxide from the Martian atmosphere in the presence of a suitable catalyst, so the propulsion engineers will not need to find sources of methane for their rocket motors on Mars, nor will they have to extract it from the traces that exist in the air.

The electrical power requirements for these schemes for propellant production are formidable, considering how much fuel will be needed to support return missions to Earth and normal power consumption for heating, lighting and communications. At Mars' distance from the Sun, and at the bottom of its dusty atmosphere, the best that can be expected from solar power generation systems is a performance of three hundred watts per square metre, implying a need for very large solar power arrays spread across the surface. If the base is to be there in the long term, and home to more than a few people, it is hard at present to see any alternative to placing nuclear power stations capable of supplying megawatts of electricity on the Martian surface.

9.6.8 Calling home

Finally, the communication infrastructure required to support robotic missions and eventual human exploration needs high data rates and essentially continuous contact. The sort of performance that will be required is at least a million bits per second bandwidth at the maximum Earth–Mars distance, with correspondingly higher rates when the two planets are closer. An exception to the virtually continuous availability of the communications link will have to be made at superior conjunction, when the Sun is between Mars and the Earth. Until a relay station can be built to overcome this, in Venus orbit for instance, a communications shutdown lasting ten days or so will have to be endured. Past and present-day exploratory missions have had this experience regularly. Once again, research will be needed and experience gathered on the effects that atmospheric and ionospheric disturbances might have on communications activities, for example during dust storms, especially if they generate lightning.

To collect the data from the roving explorers on the surface and relay them on to Earth, and to communicate between manned and unmanned bases on different parts of Mars, networks of relay satellites, similar to those that now orbit the Earth, will be deployed around Mars. These will also provide GPS-like satellite navigation capabilities for human and robot drivers on the planet. Schemes like this are not far off being implemented, since it has long been clear that even the relatively unsophisticated current missions, like the new generation of large, long-distance rovers, could make good use of them. Their successors that will carry out pre-human exploration and sample return are certain to require multiple relay satellites, for quasi-continuous communication, high data rates, and redundancy in the event of failure. They could also double as a weather satellite network, using radio links with each other to measure high-resolution atmospheric temperature profiles as well as the more conventional imaging and infrared sounding techniques.

9.7 Ready to go?

All of these preparations sound formidable, but little of it is very far beyond our present capabilities, given foreseeable developments and sufficient political, financial and popular support. The greatest hurdle to be surmounted before humans will venture, even once, to Mars is marshalling the will of the international community of space-competent nations to begin the adventure. A total of around half a trillion dollars, or perhaps more, will be required to realise the goal of putting humans on Mars and returning them safely to the Earth.

Strategies that depend upon 'pay as you go' schemes, linked to claims that no additional money over current space budgets is needed, tend to be disingenuous. If all of NASA's current nearly fifteen billion dollar annual budget were spent on the Mars programme it would still be forty years before the first human mission could take off. The current plan to return to the Moon first will further delay going on to Mars, even if everyone agreed that returning to the Moon is a step that necessarily improves the prospects of getting to Mars. The technical challenges for the Moon and Mars are very different, and the rigours of human space travel are different for the two destinations. On the other hand, lunar missions do spur the retooling and financial motivation of industry that is required to recreate the essential infrastructure of launch vehicles, radiation-hardened electronics and many other fundamentals that are essential for either Moon or Mars journeys and that have ebbed away since the days of *Apollo*.

Now and again, schemes that require a fundamental change in cultural attitudes toward the risk to human life are offered as non-technical solutions to the challenges of human space exploration. The thought of adopting an approach that saves money and time by accepting substantially increased risk is attractive to some, even among those who must personally take the risk, but unless the world changes in unforeseen ways this is not realistic. Among the general public, space exploration may be seen as just as compelling as those other goals for which the loss of life is routinely accepted, military adventures being the obvious example. Certainly, the loss of astronaut lives during *Apollo*, and the *Challenger* and *Columbia* shuttle accidents, did not result in the public turning away from NASA's programmes for long. There was a temporary downturn in the number who said they were personally interested in tourist trips into Earth orbit following these incidents, but it is clear that space travel, in sub-orbital flight at least, will be a real industry in the next decade, despite the risks. Market soundings have been made, and *Virgin Galactic* was taking bookings in 2008 at a cost of two hundred thousand dollars for a flight 'as early as 2009'.

Much depends on what governments, in thrall to a well-entrenched safety industry, will find agreeable. Currently, placing lives at risk by knowingly cutting corners on safety is unacceptable. 'Generic' risks, those that are seen as unavoidable if the venture is to take place at all, are more palatable. The record for Space Shuttle flights is a mortality rate of about two per cent, while for expeditions to the summit of Everest it is around ten per cent, so a small percentage might therefore represent the upper limit on what could be countenanced for the first expeditions to Mars. A guess for the lower limit, the one that planners would have to be seen to be aiming for, and actively working to achieve in the longer term, might be the current risk of death through driving on the roads each year, which is about a thousand times lower. There may be some gain for proponents of human flight programmes in rigorously balancing cost versus risk so long as the balance achieved is acknowledged and made visible to the participants and to the public. One or more catastrophic mission failures are bound to occur in a programme of several decades duration – it will be essential to admit that fact in advance in order not to face big delays and higher costs, if not loss of the entire venture, when the inevitable happens.

Sending humans to Mars is possible; no insurmountable problems have been identified in the dozen or more studies that have been conducted by space agencies, industry, and a few individuals and enthusiast groups. However, the challenges are enormous and the costs are unprecedented for a non-military venture by a single nation. Hopes remain alive that some fraction of the costs of destructive wars might someday be applied by the nations of the world to cooperating on space missions to Mars and beyond, in search of a collective destiny.

Further reading

For the official account of NASA's plans for human flight to Mars, including a detailed exposition of the Bush 'Vision', see:

NASA's Exploration Systems Architecture Study, NASA TM-2005–214062, November 2005.

For an expert and technically detailed critique of this plan, including a prediction that NASA's plan will not result in a manned mission to Mars before 2080, see:

Human Missions to Mars: Enabling Technologies for Exploring the Red Planet, by D. Rapp. Springer Praxis Books, 2008.

Chapter 10
The first footfall on Mars

Few would argue that human presence on Mars is, and has always seemed, inevitable. But when will it occur? The previous chapter concluded that there do not seem to be any 'showstoppers' so far as technology is concerned. This world just needs to bite the bullet on the costs, and get on with it. When will the conditions be right for that to happen?

On many occasions over the past sixty years, the goal of sending humans to Mars seemed to be within reach, only to slip steadily into the future and stay as far away as ever. It was probably as close in the post-*Apollo* days of von Braun, in the 1970s, as it has ever been. Now, the largest space agencies, in the USA, Europe and Russia, have begun to speak openly of their individual plans for human exploration of Mars, each saying they will launch manned missions in twenty-five to thirty years from now. It is a sign of the remoteness of the goal and of the lack of serious progress that these plans are largely independent. Everyone likes the idea of blazing a trail to Mars; nobody, at least nobody with their hands on the purse strings, wants to pay to get on with it. Even the USA, while declaring a national vision to put men on Mars, has effectively settled for the lesser goal of re-establishing a presence on the Moon. Europe has enthusiastically declared its own vision for manned Mars mission, but shows no sign of putting up the funds it will require. Vision, in the context of Mars bases with men and women in them, apparently does not have its usual meaning of an ability to see the future, but its secondary meaning of an imaginary concept with inspirational significance.

The biggest obstacle to progress is, of course, the three hundred billion dollar plus price tag, combined with modern sensibilities that lead to an aversion to cheaper, high-risk strategies. Informally, representatives of the agencies say it would be great to carry out the project jointly with other nations, pointing out the benefits of cost saving and the sharing of facilities and expertise, leaving implicit the idea that this will also be good for world trade and even world peace. On the negative side looms the dark spectre of an Earth that will soon be unable to undertake any grandiose ventures in space, either because of political instability and war, or because ecological disasters like global warming force nations to turn inwards to deal with crippling problems at home.

Taking an optimistic view, we can conduct a thought experiment in which a joint US–European mission lands on Mars thirty years from now. Perhaps by then Russia and Japan, and maybe China and India, will be involved as well, but so little is known of their plans, even inside the countries themselves, that what they will do cannot be described with any degree of confidence. For the two western agencies, however, nearly all the essential elements of the first piloted mission to Mars, as well as the necessary precursors, can be described with considerable present realism, if little future certainty, from a knowledge of the technical studies and programmatic plans already completed. Unlike the rest of this book, this part is set out as a fable, to emphasise that it is based, hopefully soundly, on the best of present knowledge, but still is almost certainly quite wrong.

10.1 The *Mayflower 3* and its crew

The year is 2038. An international expedition with a crew of six, three from PASA[1] and the same number from ECSA,[2] departed Earth for Mars five months ago and their spaceship is now parked in an orbit three hundred and thirty-four kilometres above the red planet. They will spend a total of three years away from Earth, including two months in orbit about Mars and forty days on the surface.

The mission commander, aged forty-three and a military test pilot, trained on the Moon with the rest of the crew for a year prior to the launch towards Mars. In addition to himself, his crew consists of a pilot, a navigator/mission specialist, a medical doctor, a geologist specialising in sedimentology, and an electrical engineer with a PhD. On the Moon they learned how to do the tasks needed on the voyage, but just as importantly they learned to work together, while a team of psychologists took the opportunity to plumb the depths of crew compatibility. Although dozens of astronauts had been part of the training programme that led to the choice of these six, during that last year they had hardly seen anyone except each other as they maintained a gruelling training and fitness schedule.

The main structure of their spacecraft was assembled in Earth orbit by technicians housed in the *International Space Station*. The station itself, after a difficult history of technical problems, cost overruns and fluctuating political support, had finally been completed in 2019. Its main use is as a hotel in space: *Virgin Galactic* leases two-thirds of the area on the station, running a profitable business taking paying passengers there in their own space planes for one- and two-week breaks. The spectacular views, unusual weightless sports and extra-vehicular activities have proved popular attractions. *Virgin* also provides accommodation

---

[1] The Pan-American Space Association, with its Headquarters in Mexico City, assimilated NASA and the Canadian and South American space agencies in 2035.
[2] The European Community Space Agency took over from ESA in 2015.

and recreation for the crews of the space agencies, whose labs take up the rest of the space, currently dedicated mainly to supporting the assembly of the interplanetary space ship moored nearby. Watching the technicians work and seeing the gigantic vessel take shape before their eyes is part of the thrill for the tourists.

Fitting out, including the installation of the nuclear power units and the crew quarters, took place in lunar orbit. The completed hull was transferred from its orbit near the space station to a low orbit around the Moon using strap-on boosters and a piloted space tug. During orbit insertion at the Moon, it passed over the heads of its future crew, training in the base that had been established on the lunar surface a decade earlier. The flight crew made many visits to the *Mayflower 3* as it grew to completion, working with the technical team on installations on which their lives would later depend, until they knew every detail. They also learned to do each other's jobs to some extent; it could not be assumed that everyone would survive the journey, although the risk analysis showed very plainly that most of the anticipated failure modes, all designed out as far as possible but never completely eliminated, were more likely to destroy the whole crew than to take out individuals.

10.2 The journey through deep space

Three days before the departure towards Mars, the crew joined the completed interplanetary spacecraft by a short shuttle ride from the Moon base, having completed their training in the lunar environment. The *Mayflower 3* is an impressive sight: one hundred metres long and weighing over a hundred and fifty tons, only the five-hundred-ton *Space Station* beats it to the title of largest free-flying spacecraft ever built. Except, of course, for the two nearly identical sister craft which preceded it. The *Mayflower 1* was a robotically operated prototype, which flew to Mars in 2029 and tested out the landing systems successfully before being lost following rocket motor failure, that cut short the burn as the spacecraft was attempting to leave Mars orbit and return to Earth. A salvage mission would successfully retrieve the inert spacecraft from its orbit around the Sun in 2047. Of *Mayflower 2*, more later.

Once on board, a long final checkout of all of the systems begins, and each of the crew members, except the medic and the geologist, takes a turn to don a spacesuit and check the exterior of the ship, from the crew quarters at the front to the massive power module at the rear. The latter is already dissipating enormous amounts of heat.

The commitment to employ nuclear propulsion for the first human mission to Mars was made before there had been any testing in space beyond that performed in 1965, when a reactor was launched into Earth orbit to prove that it could operate in the space environment. The United States made the political decision to develop nuclear thermal drive technology early on in the Mars programme, and without much difficulty, given the rapid growth in dependence on nuclear power for domestic use on the Earth that followed the global warming trends and oil crises that peaked in the 2020s. PASA undertook the programme to design and build the clustered engines that could achieve the required two million newtons of thrust for the voyage to Mars.

Space demonstrations of nuclear drive had been tedious, in part because in NASA days each launch of nuclear material into space used to require approval from the President of the United States. This could not be freely given, or seen to be so, because many elements of the public saw the use of fissionable material in space as tantamount to deploying weapons there. However, as people got used to the spread of safe nuclear power stations, some not too far from their neighbourhood, the opposition relaxed and the demonstrations gradually ceased. An engine suitable for the return trip to Mars was demonstrated in space for the first time in 2023. Its performance justified the cost: it would cut as much as thirty per cent off the time needed for the journey. Within five years, unmanned nuclear-powered precursor missions were dispatched to Mars with a twin purpose: firstly to prove that the engines were sufficiently reliable for humans, and secondly to deliver equipment that the trailblazers would use on arrival. Each flight tested one or more of the more risky technical aspects of the journey – aerocapture, the use of atmospheric drag to slow the spacecraft enough to enter into Martian orbit, powered descent to the surface, and navigation to the desired landing site. Well before the crew of the *Mayflower 3* left lunar orbit, their landing site had been mapped by rovers, and capsules containing provisions, and navigation beacons and automatic weather stations had been left near the point where they would actually touch down. The autopilot in *Mayflower 1* had landed at the site and taken off again with pinpoint precision and in total safety.

The manned vehicle uses the same engines and techniques as the freighters, but requires a different interior design to ensure the safety of the crew during deep space travel. The layout draws heavily from lessons learned from experience on the unmanned flights to Mars, which had monitored the radiation environment every step of the way, and from the *International Space Station*, which had provided experience with long-exposure weightless conditions and cramped quarters. For the trip to Mars, private quarters are provided for four crew members to sleep simultaneously, and a larger communal area for six provides additional workspace for use during the cruise to Mars. Radiation protection is best in the galley, at the centre of the ship; everyone will huddle in there for short periods in the event that a burst of energetic particles from the Sun is detected by the array of spacecraft, some orbiting Mercury and Venus as well as the Earth, the Moon and the Sun itself. The life support system on the *Mayflower 3* recycles air, water and waste and is equipped with critical backup systems, including independent power handling, communications and air quality monitoring.

10.3 On the voyage

The ship flew itself to Mars on autopilot, leaving the crew with very little they absolutely had to do, other than keep themselves fit and rehearse the arrival procedures. Fitness was a question of exercising in a cramped space with machines that used elastic straps, rather than weights, to provide the stresses that gravity no longer provided. Plans to spin the craft to provide artificial gravity had been abandoned long ago, in the days of the early space stations. The practical difficulties of spinning a relatively small object fast enough, and the nausea induced by the moving background through the window ports and the Coriolis forces on everything that moved, so far outweighed the advantages that this cherished feature of science fiction voyages was a delusion.

For mental health they played a lot of games with each other, mostly chess and cribbage, and talked long and often to family and friends back on Earth. Formal reports, including a tacit mental health examination, were relayed from each crew member daily. And they read a lot, with their choice of books, movies and music stored on tabloid-sized thin, flat personal electronic readers still called iPods, constantly refreshed by the high-speed uplink from Earth. Perhaps surprisingly, perhaps not, the most popular choice for books and movies was science fiction of varying antiquity. Alone in his sleeping quarters, the doctor was reading:

What is this spirit in man that urges him for ever to depart from happiness and security, to toil, to place himself in danger, to risk even a reasonable certainty of death? It dawned upon me up there in the moon as a thing I ought always to have known, that man is not made simply to go about being safe and comfortable and well fed and amused. Almost any man, if you put the thing to him, not in words, but in the shape of opportunities, will show that he knows as much. Against his interest, against his happiness, he is constantly being driven to do unreasonable things. Some force not himself impels him, and go he must. But why? Why? Sitting there in the midst of that useless moon gold, amidst the things of another world, I took count of all my life. Assuming I was to die a castaway upon the moon, I failed altogether to see what purpose I had served. I got no light on that point, but at any rate it was clearer to me than it had ever been in my life before that I was not serving my own purpose, that all my life I had in truth never served the purposes of my private life. Whose purposes, what purposes, was I serving? … I ceased to speculate on why we had come to the moon, and took a wider sweep. Why had I come to the earth? Why had I a private life at all? … I lost myself at last in bottomless speculations.[3]

10.4 Entering the realm of Mars

There were a great many exciting experiences for the crew of the *Mayflower 3*, but few as dramatic as the strategy used to get into Mars orbit. Arriving at a speed of more than ten thousand miles per hour, propulsion and navigation systems were used to start the deceleration and to align the trajectory of the spacecraft so that it passed through the Martian atmosphere on a path that is only twenty-four kilometres above the surface. This is low – Olympus Mons is twenty-five kilometres high – but essential, since the air at this level is thick enough so that the

---

[3] H. G. Wells, *The First Men in the Moon*, 1901.

craft is decelerated by atmospheric drag. Aerocapture is a frightening method of arrival, but is used for piloted and for large cargo missions alike because of the large savings in the mass of propellant that must be carried from the surface of Earth to Mars. It requires a massive heat shield, used at Mars, and again upon return to Earth, and very precise navigation. Aerocapture has to work at the first attempt – a second chance would require a trip around the Sun.

The navigator on board *Mayflower 3* is an onlooker as the craft bumps its way at supersonic speeds through the Martian atmosphere. The trajectory is controlled by massively redundant computing systems on board the ship, on the Earth, and on the surface of Mars. These are all linked to each other, and share and assimilate a mass of data on the speed and temperature of the incoming spacecraft, and its position and height, tracked by redundant radar systems on the ship and on the ground. Deviations from the planned trajectory, mainly due to meteorological variations in the atmospheric density, are constantly corrected using the aerodynamic lift modulated by small changes in the attitude of the blunt-nosed craft. The crew has nothing to do but ride the buffeting, feel the temperature rise, and watch the view, framed by a spectacular glow from the spacecraft's thermal shield and its halo of atmospheric plasma. It seems forever, but after twenty minutes they emerge into the blackness of space on the other side of the planet, with the spacecraft now captured into a loose elliptical path around Mars.

The atmosphere is used again to change the orbit to one that is lower and more circular, more suitable for supporting and observing the crew on the ground, and for their eventual rendezvous and return. This is aerobraking, a less fraught experience than aerocapture, because the spacecraft passes through the upper atmosphere repetitively, gradually converting the energy of the vehicle's orbital velocity into heat. Errors are unlikely to be fatal, and can be corrected on subsequent orbits. Also, the minimum height reached on each dip into the atmosphere is still more than a hundred kilometres above the surface, so the buffeting and heating is much less than it was on arrival.

Small burns of the low-thrust conventional rocket motors on the *Mayflower 3* near apoapsis, the most distant point on the orbit from the planet, control the height of periapsis, the closest point. The length of the burn is carefully calculated to optimise the trade-off between maximum drag and maximum safety. Early experiments with aerobraking using unmanned spacecraft, when there was little current information about the state of the atmosphere below, had to be conservative, and took up to six months to decelerate into a close, circular orbit. Pushing the apoapsis down too fast could have meant taking too big of a bite of the atmosphere, de-orbiting the spacecraft to crash on the surface. *Mayflower 3* does not have that luxury – the crew needs to get on with its mission. On the other hand, the atmosphere is much better monitored by satellite and surface meteorological stations in 2038 than it was in the early years of the twenty-first century, and the outcome of each pass through the atmosphere much better predicted and controlled.

During aerobraking, a constant watch is kept for the sudden onset of dust storms, although the season when they are most common has been deliberately avoided. If unusually large amounts of dust become airborne, whipped up by strong surface winds, the heating of the atmosphere as the dust absorbs solar energy expands the lower atmosphere and increases the density encountered at the aerobraking level. In the event, there are no such problems, and the *Mayflower 3* makes just three aerobraking passes and then makes the remaining orbit adjustments by chemical rocket burns at apoapsis until the desired final orbit is achieved: circular, at a height of one thousand, three hundred and fifty kilometres, inclined at twelve degrees to the equator.

10.5   Preparing to land

The landing site was determined years before *Mayflower 3* left the Earth. Starting in 2004, the rovers *Spirit* and *Opportunity* had resolved one of the longest standing questions in planetary science when they produced convincing evidence that early Mars was warm and wet. By 2015, the *Mars Science Laboratory* had discovered several locations that looked like habitats for ancient life. Furthermore, these were on Meridiani Planum, west of Vallis Marineris at two degrees south latitude, six degrees west longitude, a region that has many level sites where safe landings could be made, alongside interesting geological structures like hills, cliffs, and craters (Figure 10.1). Based upon the results of a search by orbiters, principally the *Reconnaissance* and *Science Orbiters* in 2006 and 2016, no alternative was identified to be superior to Meridiani as being readily accessible for the search for evidence of past life and offering safe landings.

Two sample return missions in the 2020s confirmed that Meridiani is an ancient shallow sea, having last been inundated by groundwater more than a billion years ago. Initially, the layered evaporites and sediments were the primary evidence for the persistence of liquid water on the Martian surface. Then the first returned samples showed conclusively that sedimentary rock at this site was deposited 2.8 billion years ago by a large body of briny water. Estimates from ancillary data such as morphologies of tributaries and outflow channels and the chemistry of the sediments, suggest that water was present for an extended period, at least tens of millions of years. The evaporites left behind by liquid water from the subsurface came and went with climatic events that punctuated the history of Mars. The depth of that record is still unknown, although craters explored by rovers revealed the stratigraphy down to a depth

Figure 10.1 The view at Meridiani Planum obtained by the *Mars Exploration Rover Opportunity* near its landing site. The first astronauts to land will see something similar.

of nearly fifty metres. The landing parties that follow today's pioneers will set up drilling stations to probe even deeper, and to find water.

The crew of *Mayflower 3* examine their landing site from orbit through powerful telescopes. What they see is familiar from their training, using imagery from unmanned orbiters and landers. Meridiani is an ideal site for landing, since it is one of the most rock-free areas on Mars. Slopes are small and craters are very few. It is very near the equator, which brings the most temperate, or least extreme, climate, with temperatures that are chilly but Earth-like during the day. Meridiani is susceptible to dust storms, although never so bad that surface operations by robotics had to be interrupted.

Studding the surface of Meridiani Planum are the outcrops of the sedimentary rock that form the expedition's objective. Scientists using the best analytical instruments on Earth analysed dozens of samples of these rocks and confirmed that these particular sediments are capable of preserving chemical or morphological evidence of life, if it ever existed at this site. However, after the third set of samples was returned, still no biosignatures had been found. This was interpreted both ways by supporters and opponents of manned missions to Mars: the former said humans must go and do the job properly, while the latter said there is no proof there is anything to find. It was just as well that the human exploration programme had been pressing ahead in the meantime with technology development and launcher evolution during this period, and the momentum had become almost unstoppable.

10.6 Descent to the Martian surface

With a good weather forecast obtained, and the landing beacons and life support hardware on Meridiani activated and verified, all preparations are complete for the descent to the surface. Four of the crew will make the short trip down, the pilot taking charge of the descent module they named *Kingfisher*, while the mission commander and the navigator remain in orbit for the long wait until they return. For these two, the whole focus is getting to Mars, following a pre-arranged plan as closely as possible, and getting safely back. From their orbiting station, they follow every step of the team on the surface below, and survey them and their surroundings with high-resolution optics. Any deviation from plan, whether as a result of exciting discoveries or of any kind of system failure, including physical or mental illness among the crew, is first referred to them, and they remain the command, as well as the communications, link between the explorers and mission control back on Earth.

In terms of the environment and the technique required, landing on Mars is intermediate between landing on the Earth and on the Moon, and the pilot has trained on both, as well as in advanced Mars simulators. The near vacuum and relatively low gravity of the Moon mean that a solely propulsive system is required for landing there, while on Mars, as on Earth, the atmosphere can be used to reduce the entry velocity, saving a lot of fuel but requiring the addition of a heat shield and parachutes. The *Kingfisher*, incorporating temporary quarters and the return-to-orbit vehicle, is a heavy spacecraft that needs to use all of the available options: propulsive deceleration to leave Mars orbit, then frictional deceleration using the heat shield, followed by moderately sized parachutes, and finally a controlled landing under power.

The moment that *Kingfisher* encounters the top of the atmosphere, friction heats the leading surfaces of the vehicle, protected by its inflatable heat shield, to temperatures of nearly fifteen hundred degrees centigrade. The shield also provides a significant fraction of the deceleration on the drop toward the surface, following a shallow figure-of-eight pattern to increase the flight path and to reduce the g-forces on the crew. Half-way down, the speed of the ten-ton spacecraft has fallen to a fraction of the original Mach three, and the first of a series of parachutes snaps open. Near the surface, the parachutes are discarded and rockets are used for the final deceleration, and to provide active control for the approach manoeuvres.

The landing itself requires advanced optical navigation using terrain recognition techniques, and a propulsion system that can hover and move considerable distances sideways. Despite the greater size and mass of the spacecraft, and an even higher reliability requirement, this is not qualitatively different from the techniques used for the 2011 landing of the robotic rover, *Mars Science Laboratory*. The vehicle must not only avoid hazards, but also rendezvous with the buildings and supplies that are already on the surface. It needs to avoid dropping parachutes and discarded fuel tanks on these assets below, and needs to get close without landing on top of them. The pilot's extensive training on simulators means that she recognises the terrain instantly as the lander closes in on the spot selected for touchdown, and stands by ready to steer manually if required. In fact, it is not needed: the *Kingfisher*'s computers and sensors, linked with the beacons and radars on the ground, take the heavy craft down automatically to touch down without any detectable jolt well within the twenty-metre-radius error ellipse, two metres from dead centre, and eighty metres from the waiting Mars habitat.

The surface operations plan calls for the crew to remain in the lander overnight before transferring to the habitat, which they activated remotely and which has been fully functioning since they entered Martian orbit a week ago. Once they leave the *Kingfisher*, it will remain on standby, as a lifeboat in case of need, for a further thirty-six hours, and accessible for habitation within ninety minutes throughout the surface mission. All systems are in a nominal state. The fuel for return to orbit has a residual margin of thirty per cent post-landing. The automatic deployments outboard from the lander were completed on schedule, including the three-metre-diameter high-gain antenna pointing directly to Earth. Communications with Earth and with the command module in orbit checkout as nominal: the crew send short messages to their families. The light-time delay to Earth is currently thirty-five minutes and fifty-nine seconds.

Their first impressions of the landscape are dominated by the softly undulating grey bleakness of Meridiani Planum. Hundreds of thousands of images acquired over many years have someway failed to capture its utter emptiness.

Only two features, each barely discernable, interrupt the surface beyond the habitat's structures: the rim of a nearby small, shallow crater, and the tracks of a dozen rovers that have plied the surface. The grey-blue cast of the ground comes from the millimetre-size spheres of haematite precipitated from the ocean that existed here three billion years ago.

10.7 First Foot

After system checkout by the crew, in contact with the orbiting mother ship and mission control on Earth, followed by a four-hour rest period, two of the four pioneers get ready to step onto the surface. The geologist won the debate about who should be first by the expedient of declaring himself the least important for the survival of the others and therefore the success of the basic mission. He is also deemed the most eloquent and suitably educated choice to speak about what he sees as he looks around to the billions of people glued to their televisions back on the Earth.

'The dream is alive,' he says, 'we come in peace for all mankind.' The reporters on their news websites write that this is an unoriginal but appropriate and safe thing to say. The geologist looks around and describes the low hills on the horizon and the nearby rocky outcrops, the sky and the weather. The medic joins him, and they walk together to the nearby habitat and enter the airlock. While they are doing this, the pilot and engineer are changing the vehicle configuration for the return trip, following procedures that have much in common with the *Apollo* missions to the Moon seventy years earlier. Within twenty-four hours of the landing, all four pioneers are making themselves at home in the habitat module, prior to the commencement of exploration.

10.8 *In memoriam*

On their second excursion from the habitat to the surface, the explorers broached a subject that had remained tacitly off limits, for themselves, the controllers in Houston, and even the press legions following every step of the mission. This was the loss two years earlier of the *Mayflower 2*, the first manned mission to Mars, along with all of its crew. They had made it as far as Mars successfully, but during the crucial aerocapture manoeuvre a small navigation error had increased the stress on the hull of the spacecraft beyond nominal limits, and parts of it had broken away. The result was a loss of directional control and the spacecraft hit the surface at high speed, ironically not far away from the base in Meridiani where they were later to have landed, in the canyon known as Ganges Chasma. The Mayflower class of space ship was never intended to land, and so there was never any chance of a safe landfall for any of the crew once they failed to achieve Mars orbit. Photographs from orbit showed the wreckage strewn over a wide area, most of it up against a cliff at the edge of the canyon.

The crew of *Mayflower 3* held a memorial service in their spacesuits, and erected a plaque to their predecessors on a temporary stand. A later mission would make the trek to Ganges Chasma and install the plaque on the wreckage. The whole event was televised back to Earth. Once it was over, the crew would not mention *Mayflower 2* again, publicly or privately, until they were safely back on Earth.

10.9   At Meridiani Base

Over fifty tons of landed hardware had preceded the crew over a period of ten years. The habitat's main housing was built in Europe and delivered to Mars by a direct flight using ECSA's large *Ariane VII* launcher. First named Meridiani Base, which was later revised to Coradini Base after an influential administrator who drove the early European planetary programmes, it consists of a four-metre-high by twelve-metre-long cylinder on support struts, entered by a short wide ladder. Five hundred metres from the main habitat, and partially buried to reduce its contribution to the radiation hazard, a small nuclear reactor is currently producing fifty kilowatts of electricity. Reservoirs and reprocessing plants for oxygen, water and waste are connected by a network of pipes and cables, all put in place over the preceding ten years by robotic constructors descended from the car-building robots first used in the mid 1990s. From them, tactile sensors in robotic hands on robotic arms transitioned very rapidly from laboratory benches in universities to major development projects in industry worldwide. Mobile robotic platforms were developed to lift and move massive pieces of equipment on the Earth and on the Moon, and those awaiting the crew of *Kingfisher* can operate over up to a kilometre across the Martian surface carrying large loads in the low gravity of Mars. The accurate alignment of large bulky pieces and the laying of cable and pipelines require delicate skills as well as powerful lifting, so the larger machines have to combine the skills of an engineer while being able to act as a bulldozer. They stand idle now, not to be used again until a second module is added in six year's time.

Development of robotic manipulation technology skills began in earnest during the assembly of the *International Space Station*, and continued with activities on the Moon. During the five trips to the lunar surface between 2017 and 2022, approximately five hundred kilograms of rock and regolith samples were collected from diverse sites within a ten-kilometre radius of the first lunar habitat. While exercising various Mars tools, including a manned rover, a five-metre-deep vertical core of the lunar regolith was acquired. The core was returned to Earth for study in the same labs that will receive Martian samples in a year when *Mayflower 3* returns to Earth, then the deep cores two years later. Due to its size and complexity the deep drilling capabilities will gradually be extended on the fourth and later expeditions, until depths of at least one kilometre, where the most habitable environments are expected to be found, can be accessed.

10.10 Survival and resources

Supplies stored on the surface are sufficient for the crew to survive only a few days beyond the end of the launch and rendezvous window for return to Earth. An Earth-initiated rescue is wholly impractical for either the surface crew or for the tended command module. Failure to launch from the Martian surface or to leave Mars orbit, or any of the other critical events on the journey back to the Earth environment, results in the loss of the mission and the crew.

Among the items they and subsequent teams will study is the conversion of *in situ* resources into fuel, and into oxygen and water for human consumption. However, there are no plans for them to depend on this for the foreseeable future. Manufacturing fuel on the Moon was found to be too difficult to accomplish with sufficient reliability and efficiency to be practical, and even optimistic projections of conversion efficiencies for *in situ* resource utilisation showed that a single factory established on the Martian surface would generate enough liquid oxygen for fuel for a return trip to Earth only after a full year of operation. Without new breakthroughs in technology, the cost of producing fuel on Mars or the Moon actually works out to be more expensive than the cost of carrying fuel as part of the cargo missions. The fuel factory and storage farm that would be needed would occupy several acres of Martian real estate, and the robotic construction and unattended operation of such a facility is made enormously complex by the need not only to make but to store the fuel needed for the return flight. All aspects of the preparation for the crew's return flight to Earth must be complete prior to the launch of the crew from Earth.

The story is similar for consumables such as water and oxygen – stay-times of one Mars year or less are least expensive when the consumables are flown in from Earth. However, as was the case with the Moon, construction projects can make good use of *in situ* materials, mostly soil and rock, with carbonates, phosphates and other minerals used to make cement. During successive landings, all delivered by shuttling between Earth and Mars orbit, a protective mound of rock and soil is built up around the habitat by robotic earth-moving equipment. Future long-duration missions, and likely colonies, will place new crew sleeping quarters underground. Nuclear reactor power supplies will continue to be buried beneath surface material, and kept at a distance from the habitats, just as the first has been at the Meridiani site.

A week into the first surface expedition, the outpost's power system continues to operate nominally and radiation levels on the boundary closest to the reactor are now measured only once per day. Machines controlled by the crew have laid power lines and secondary feeds to equipment and the rover battery recharging station. The length of their stay, and the intention to re-use, extend and improve the habitat, means that housekeeping is a permanent drain on astronauts' time. As they learn how to do repetitive chores efficiently in their new environment, the challenge is to keep the time spent in these tasks to about half of the total, so exploration and the scientific work can go ahead. The reliability of the habitat's systems is crucial; lessons from the early days of long-duration human activity in space, on the space station and on the Moon, showed that maintenance could easily become very nearly a full-time occupation for the crew.

The amount of work required to keep everything working smoothly is not a bad thing. Keeping very busy is a good psychological antidote to the awful loneliness and fear associated with a journey beyond the limits of human experience. So very far from home, and in constant danger from a fragile environment, even specially selected and trained humans are at risk of unpredictable reactions. Being in a team helps, as does a regular high-speed link with base and family, but not always enough.

10.11 Science operations

Testing and improving the habitat is the second of the explorers' main objectives, the first being to get there safely and call home. With that done, they can focus on their science goals, which are also twofold. The most important is to look around and choose the best samples to bring back, then to acquire and stow them. The second is to verify and support the plan for the next mission in two years time, known simply as *Mars 3*, by preparing the ground for drilling, and for an enlarged and improved habitat that would allow a five-hundred-day surface stay. While the crew of the *Kingfisher* work on other tasks, any robots not needed by them are already at work excavating sleeping quarters for the crew of *Mars 3*. A rigid roofed structure will be erected in a broad trench and covered by one and a half metres of soil, reducing crew exposure to cosmic rays by thirty per cent.

The first extended stay at the future site of deep drilling occurs ten sols after landing, with several hours spent assembling the first element of the base of the drilling tower. Pads, uprights and crossbeams of the tripod base are the first to be assembled. In the first few hours, work proceeded with about the same difficulty as had been experienced during training exercises on the Moon. This early confirmation of the validity of lunar training would bode well for scheduling future tasks. The light-load crane on the crew rover worked well in its first test; crane control was also excellent. The first upright and two cross beams of the drilling rig were interconnected, and a single footpad was attached to the upright. Once this demonstration of working on the surface is complete, further work on the rig is left to be performed by the crew of the *Mars 3* project.

The excursions suits are cumbersome, and visibility directly downwards is very limited. Gloves, especially the opposing thumb, are the greatest impediment to working quickly. Handling the wide range of tools and rover fixtures, including the trailer hitch, is not a major problem, but grasping the smaller couplings on the hard Martian surface proves difficult. Arm and leg strength and stamina are somewhat better than anticipated, a testimony to the success of the exercises on the Moon that familiarised the crew with working in low Martian gravity. Communications between the excursion crew and the outpost remain good throughout, and Earth is in continuous contact with the remote site through the telecom satellite poised overhead, stationed beyond the moon Deimos in areosynchronous orbit.

The small crater near the landing site that had been identified by the 2028 rover as being of exobiological interest, because of its exposed sedimentary rock strata with evidence of carbonates, organic material and geothermal activity, is confirmed as the target for the returned sample. Now named Squyres, after the team leader for the rover missions in the 2000s, it is visited every day for ten days by the astronauts, working in pairs. Working from the top down, they hack, chisel and drill the exposed face of the fifteen-metre high cliff until they have removed, bagged and labelled samples from each of the strata, each older than the last. Nuggets of interesting material from other locations, identified by unusual appearance or from a crude *in situ* analysis, are also collected until the mass limit of two hundred and fifty kilograms is reached. The samples are carried back to Earth in compartments external to the crew cabin, where they are stowed before leaving Mars, in vaults enclosed in the body of the return vehicle to protect the Earth's environment against a possible breach in the sample canisters.

10.12 Launch to near Mars orbit and on toward Earth

Preparations to return to Earth begin five Mars days prior to leaving the surface. Systems checks on the landing module dominate, but more housekeeping must be performed. The last of the shipping containers are filled with trash, sealed, bio-cleaned, and buried at least five hundred metres from the habitat and potential science sites. The habitat's life support systems are prepared for hibernation. Just prior to launch the power system will be taken offline.

Unlike the ascent vehicles used on the Moon, the *Kingfisher* has an aerodynamic shape, required for efficient passage through the Martian atmosphere. The ascent and rendezvous sequence it will follow has been practiced three times for robotic sample return, and takes almost a whole day to accomplish, most of this being the delicate manoeuvring required to meet up with the mother ship and mate successfully. The life support and propulsion systems have all been repeatedly checked out by the two crew members who remained on board; departure for Earth can commence almost immediately once all of the rest of the crew has joined them.

The first stage is an agonisingly slow spiralling out from Mars, using chemical and nuclear propulsion. The low thrust of the main engine does not lend itself to a quick getaway, although, by firing throughout most of the journey, the return time of six months is less than it would have been with a comparable mass of rocket fuel. Most of the conventional fuel that was carried has now been used up, for manoeuvring into and out of orbit, descent and ascent. The mass of the samples returned on board, as much as two hundred and fifty kilograms, is small when compared with this, and the spacecraft returns to Earth considerably lighter.

The flight back, with the home planet getting larger in the sky, is less stressful than the journey out, although by now the low gravity and cramped conditions are taking their toll on the crew. They broadcast to mission control, with relays to their families and to Earth's media, every day, as they have done all along, even when on the surface. One last moment of extreme tension is still to come: the aerocapture manoeuvre into Earth's atmosphere. With a better-characterised, less-variable and denser atmosphere to work with this is easier than at Mars, and the advanced servo systems that steer the spacecraft in mean that navigation errors are all but impossible. Still, the heating and buffeting of the now less-than-pristine hardware they ride means the possibility of mechanical failure always looms.

Once the flaming ride is over, the crew uses the last of its rocket fuel for its spiral descent from the capture orbit to rendezvous with the *International Space Station* in low Earth orbit. Once on board, debriefing and medical checks are carried out and quarantine begins. They will be there for some time. *Mayflower 3* orbits alongside the station, its exterior looking weathered and, in places, charred, an object of wonder to the tourists who paid a premium to *Virgin Galactic* to be there at this historic time. After refitting, it will fly again to Mars in two year's time, and the retrieved *Mayflower 1* will take its place docked near the station for refurbishment work that will eventually see it joining its sister ship on the Mars run.

10.13 Quarantining the returning crew

The *Apollo* astronauts were quarantined after returning to Earth from the first lunar landing. At the time, public awareness of biocontamination was limited and the quarantine of the astronauts was not particularly rigorous. The return capsule vented into the Earth's atmosphere at high altitude during descent and, while the crew donned protective clothing, including helmets, before transferring to a raft in the open sea, they suited up in the contaminated capsule. *Apollo 11* astronauts were isolated for two weeks in a caravan while tests were conducted, but the President of the United States, Richard Nixon, entered their quarters before the quarantine had elapsed.

Such casual precautions will not do for the first Mars expedition: extraordinary care must be taken to avoid contaminating the Earth with biological or environmental materials. Although no evidence of living organisms has been found by any measurements on Mars or by analysis of the dozens of return samples, there remains a finite chance that crew of *Kingfisher* might have been exposed on Mars to organisms or materials hazardous to the Earth's ecosystem. Unlike the uninformed public at the time of *Apollo*, today's populace is well aware of the potential for biological hazards. Therefore, the crew, together with medical and biohazard teams, will spend at least one week in the *Space Station* – longer if anything untoward shows up – before returning to Earth, and their movements will be restricted for a further month once they land. On the ground, they are held in a special facility in Utah, near the sample handling facility into which all returned samples from robotic missions are quarantined before release to the curatorial facility at Johnson Space Flight Center in Houston.

Medical evaluations of the returning crew are conducted in the hospital and laboratories that constitute much of the Utah facility, along with studies of the impacts of the long voyage and unavoidable exposure to the Martian environment. The rocks, soils and gases that the crew brought back with them are subjected to the standard procedures for assessing the safety of samples before they are released to Houston and subsequently distributed to qualified scientists, first in America and Europe, and then worldwide. Although the safety checks revealed nothing of biological significance, within a few weeks of their release the samples are making headlines all over the world. The findings dramatically alter the plans for the next and subsequent flights to Mars.

Epilogue: Beyond the horizon

Beyond the tentative plans of the space agencies for human expeditions and temporary bases on Mars, the details of what may happen begin to merge with science fiction. Advanced concepts such as major colonies on Mars, or 'terraforming' in which the climate is manipulated by planetary engineering on a massive scale to be more conducive to human habitation, have been studied, although not in much detail, and the investment in the research needed to make them plausible, if not realistic, has not advanced very far. This is partly because the prospect is so distant that scientific, technological and sociological breakthroughs that cannot be predicted at present will make huge differences, positive and negative, to what actually happens. It is also because much of science fiction (as opposed to fantasy) tries to be factually based and to make a serious attempt to predict future reality. The events described are so far ahead that anyone who knows the story so far, and has some basic knowledge of the limitations of the laws of physics and economics, can be almost as effective, or even with luck more effective, at predicting the future, as a large team of specialists at one of the world space agencies. The technology that will be available when the time really comes to build *Kingfisher* may well be virtually unimaginable by anyone now.

This effectively brings us to the end of this book, since the stated goal was to talk about the *realities* of Mars exploration, past, present and future. It will have become clear to anyone who has read from the beginning that the future of anything as complex as a Mars exploration programme is something that is subject to great and rapid change, even on a time scale shorter than a year. Thus, while it is possible to take a factual approach to missions that are approved but not yet flown, there are plenty of precedents for cancellation, modification or technical failure. These, and many other factors, not excluding success, modify plans for missions less than a decade ahead, and even change the direction and aspirations of the whole programme, as *Viking* did thirty years ago. It can easily vanish altogether, through the kind of audience fatigue that killed the *Apollo* programme in its prime, especially if Mars turns out to be perennially lifeless and the political situation in the late twenty-first century is such that it is seen as having no strategic or commercial value. Those who have worked on Mars exploration know that there is no known law that says the United States, or anyone else, has to take the path of learning more about Mars by visiting it with robots and astronauts.

Still, we may try to glimpse what may happen on Mars in the next few centuries if the age-old fascination is not dimmed by a new familiarity. Any such predictions will not be accurate – that is impossible – but still, the game must be played, if only to amuse future readers who dust off old books and read with the benefit of hindsight. After all, we laughed at Lord Nelson ('Long and careful study … has led me to the conclusion that some form of life definitely exists on Mars.') and even at von Braun ('…the large [one hundred and seventy-seven tons] glider will land, aeroplane fashion, on the sands of the Martian plains'), even while recognising his genius. So, here goes.

Mars in 2200

It seems very likely that, if civilisation on Earth does not collapse due to some great environmental or military disaster, humans will step on Mars in the present century. Mars will probably turn out to be of interest only to the specialists, the planetary geologists interested in rock formation under low gravity and the palaeontologists interested in planetary formation and evolution. To the writer and dreamer and most of the public it will soon come to seem dreadfully boring, like the Sahara desert but with no oases and no camels or spiders or scorpions.

Nevertheless, mankind will persevere with Mars. The reason will be basically pragmatic: staying in space. By 2200 hundreds of Earth-like planets will have been observed circling distant stars, some of them with the unmistakable spectroscopic signatures of water and oxygen in their atmospheres. Humans will be exploring the surface of frigid but Earth-like Titan, and robots will roam the surface of Venus and the subsurface oceans of Europa and Enceladus. The Moon will be sprinkled with bases of many nations: there will be a branch of the United Nations concerned with maintaining a protocol of harmony and mutual support between them.

Until the nearest stars are reached – in 2040, someone will be writing a book like this one but about the advanced plans the Terrestrial Space Agency has developed for such a trip in another forty years' time – Mars remains the only good place for humans to live and develop. Will they then go on to colonise the planet, which means building cities with humans living there in large numbers, perhaps born and dying there and never returning to Earth? From our present perspective, this seems most unlikely, just because of the staggering logistics that would be involved to get established. However, many well-informed Europeans used to say that about the Americas or Australia: the present perspective and attitudes simply will not apply in a hundred years' time. If by then Los Angeles has spread to cover the entire Mojave desert, including Death Valley, and conservation of water and other vital resources has been developed to a fine art to make it possible, things might be different. It is also necessary to assume that space flight becomes cheap, fast and easy, huge assumptions that may require new, undiscovered physics and not just engineering and fiscal refinement.

Expeditions similar to the one described in the previous chapter are, from the perspective of the space community in government and the aerospace industry in 2009, imagined to lead naturally to longer and longer stay-times, and eventually to colonisation of Mars. Today, it is simple speculation to suggest that the accumulation of outpost hardware on the Martian surface will bring inhabited colonies closer to reality. Because self-sufficiency is unnecessary for human expeditions of one Mars year duration, it is a wholly different undertaking to equip a colony. Planetary colonies imply nearly complete independence from Earth; apart from refurbishment of certain infrastructure and, by analogy with distant colonisation on Earth, rescue in dire circumstances. Colonisation of Mars will require a decision that is distinct from the resolve to undertake the initial exploration of the planet.

Whether future governments of Earth will have reason to support a Martian community is unknowable. Most of the various justifications expressed today for Mars colonies do not bear present-day scrutiny. One is a discovery of truly monumental proportions demanding decades of study on the surface. Probably, only the discovery of living organisms or of the remains of an advanced civilisation on Mars would qualify as monumental. However, it is far from obvious that even this comes close to providing the required motivation – the round trip is not so arduous as to overcome the increased costs and risks associated with colonisation – since outposts will suffice. The analogous condition of trans-Earth migrants was the desire to remain permanently away. Something of benefit must be at the site, such as resources of immense value back home or opportunity for improvement in the human condition. Catastrophe on Earth, e.g. environmental collapse, is sometimes cited as sufficient motivation not only to colonise Mars but to shift a significant fraction of humanity to live there. These fantasies fail to account for the risks inherent in living on a planet that is entirely inhospitable to humans. Can living in a spacesuit or dome on Mars be preferable to living in one on Earth?

Perhaps terraforming could be a solution to living on Mars in Earth-like conditions, although even on the face of it the time scales are wrong. Terraforming is not achievable on the time scales of even several generations of human life. The current, unwanted global project to alter the Earth's climate through the activity of six, or is it eight, billion humans is progressing slowly, but more quickly than anything that can feasibly be done on Mars. Even if it was clear how to begin, what would cause future governments to act on such a long-term venture? The cost of proactively responding to global warming might be comparable to establishing a substantial colony on Mars, but it is a cost that governments are so far not really willing to pay.

More likely than colonies is the continuing emplacement and expansion of scientific outposts on the Martian surface. Outposts will be viable even in the event that financial and other support is intermittent – not the continuous support that fledgling colonies would need. In this instance, the scientific stations present today on the Antarctic continent are reasonably good analogues to Martian outposts. Antarctic stations are inhabited by a transient population of scientists and support staff. The science performed has evolved, from the study of the geology, biology and climate of the location to include things like the present search for meteorites from Mars and neutrino detector arrays implanted in the ice for mapping the Universe. Antarctic outposts are in no sense independent from the rest of the world. Moreover, that link can be beneficial, as a long queue exists of people wishing to go south for a visit of months to one or even two years – not unlike the time away required for a tour of duty at an outpost on Mars.

Appendix A
Data about Mars

Technical data for the basic properties of Mars, its atmosphere, and its orbit are summarised in the following tables.

The planet

Bulk parameters	Mars	Mars/Earth
Mass	6.42×10^{23} kilograms	0.107
Equatorial radius	3397 kilometres	0.533
Polar radius	3375 kilometres	0.531
Mean density	3.93 grams per cubic centimetre	0.713
Equatorial gravity	370.6 centimetres per second^2	0.377
Escape velocity	5.03 kilometres per second	0.450
Moment of inertia	0.366	1.106
Visual geometric albedo	0.150	0.409
Solar irradiance	589.2 watts per square metre	0.431
Black-body temperature	210.1 kelvin	0.826
Day (sidereal)	24 hours 37 minutes 22 seconds	1.029
Obliquity (axial tilt)	23° 59′	1.074
Orbital eccentricity	0.093	5.593

The atmosphere

Mean surface pressure	6.1 millibars
Scale height	11.1 kilometres
Average surface temperature	214 kelvin
Diurnal temperature range (at *Viking 1* Lander site)	184 to 242 kelvin
Wind speeds (at *Viking* Lander sites)	2–7 (summer), 5–10 (autumn), 17–30 (dust storm) metres per second
Mean molecular weight	43.34 grams per mole
Atmospheric composition (by volume):	
Carbon dioxide (CO_2)	95.32%
Nitrogen (N_2)	2.7%

The atmosphere (cont.)

Argon (Ar)	1.6%
Oxygen (O_2)	0.13%
Carbon monoxide (CO)	0.08%
Water (H_2O)	0.02%
Nitrogen oxide (NO)	0.01%
Hydrogen (H_2)	0.0015%
Xenon (Xe)	0.0008%
Helium (He)	0.0004%
Neon (Ne)	0.00025%
Monodeuterated water (HDO)	0.000085%
Krypton (Kr)	0.00003%
Methane (CH_4)	0.000001%
Hydrogen peroxide (H_2O_2)	0.000001%

The orbit

Orbital parameters	Mars	Mars/Earth
Semimajor axis	227.92×10^6 kilometres	1.524
Orbital period (sidereal)	686.980 days	1.881
Perihelion	206.62×10^6 kilometres	1.405
Distance from Earth (minimum)	54.5×10^6 kilometres	—
Distance from Earth (maximum)	401.3×10^6 kilometres	—
Aphelion	249.23×10^6 kilometres	1.639
Mean orbital velocity	24.13 kilometres per second	0.810
Orbit inclination	1.850°	—
Orbit eccentricity	0.0935	5.599
Length of day	24.6597 hours	1.027
Obliquity	25.19°	1.074

Appendix B
Space missions to Mars

Mission	Country	Date	Type of mission	Notes
Mars 1	USSR	1962	Flyby	FAILURE: Lost contact before flyby
Mariner 3	USA	1964	Flyby	FAILURE: Shroud separation
Zond 2	USSR	1964	Flyby/Lander	FAILURE: Lost contact in flight
Mariner 4	USA	1964	Flyby	SUCCESS: Returned 22 images
Mars 1969A	USSR	1969	Flyby	FAILURE: Launch
Mars 1969B	USSR	1969	Flyby	FAILURE: Launch
Mariner 6	USA	1969	Flyby	SUCCESS
Mariner 7	USA	1969	Flyby	SUCCESS
Mariner 8	USA	1971	Orbiter	FAILURE: Launch
Mariner 9	USA	1971	Orbiter	SUCCESS
Mars 2	USSR	1971	Orbiter/Lander	FAILURE: Parachuted into dust storm
Mars 3	USSR	1971	Orbiter/Lander	FAILURE: Parachuted into dust storm
Mars 5	USSR	1973	Orbiter	FAILURE: Loss of pressure in Mars orbit
Mars 6	USSR	1973	Lander	FAILURE: Landing failure
Viking 1 (Lander)	USA	1975	Orbiter/Lander	SUCCESS: First lander on Mars
Viking 1 (Orbiter)				SUCCESS
Viking 2 (Lander)	USA	1975	Orbiter/Lander	SUCCESS: Second lander on Mars
Viking 2 (Orbiter)				SUCCESS
Phobos 1 (Lander)	USSR	1988	Orbiter/Lander	FAILURE: Mars orbit insertion failure

Mission	Country	Date	Type of mission	Notes
Phobos 1 (Orbiter)				FAILURE: Commanding error
Phobos 2 (Lander)	USSR	1988	Orbiter/Lander	FAILURE: Reached Mars orbit, lost contact
Phobos 2 (Orbiter)				PARTIAL SUCCESS: Computer failure
Mars Observer	USA	1990	Orbiter	FAILURE: Fuel line rupture on Mars insertion
Mars '96	USSR	1996	Orbiter/Landers/ Penetrators	FAILURE: Launch, crashed into the Pacific
Mars Pathfinder	USA	1996	Lander	SUCCESS
Sojourner			Rover	SUCCESS: First rover on Mars
Mars Global Surveyor	USA	1996	Orbiter	SUCCESS
Nozomi	JAPAN	1998	Orbiter	FAILURE: Solar flare induced failure
Mars Climate Orbiter	USA	1999	Orbiter	FAILURE: Mars orbit insertion failure
Mars Polar Lander	USA	1999	Lander	FAILURE: Mars landing error
Deep Space 2	USA	1999	Penetrators	FAILURE: No signal on landings
Mars Odyssey	USA	2001	Orbiter	SUCCESS: Ongoing mission
Mars Express	Europe	2003	Orbiter	SUCCESS: Ongoing mission
Beagle 2	UK	2003	Lander	FAILURE: No signal on landing
MER-A Spirit	USA	2003	Rover	SUCCESS: Ongoing mission
MER-B Opportunity	USA	2003	Rover	SUCCESS: Ongoing mission
Mars Reconnaissance Orbiter	USA	2006	Orbiter	SUCCESS: Ongoing mission
Phoenix	USA	2007	Lander	SUCCESS

Appendix C
Mars study groups

Discussions and output from all of the following groups, plus two with general community-wide membership, namely, **International Mars Exploration Working Group** and **Mars Exploration Program Analysis Group**, and several others not listed here, led to many of the ideas and inspiration used in the preparation of this book. All concerned are gratefully acknowledged, and absolved from any peculiarities of interpretation or error.

National Research Council Space Studies Board European Science Foundation European Space Science Committee Joint Working Group on Cooperation in Planetary Exploration, 1984

US delegates

Eugene H. Levy, University of Arizona
Donald M. Hunten, University of Arizona
Harold Mazursky, US Geological Survey
Fred L. Scarf, TRW Space Systems
Sean C. Solomon, Massachusetts Institute of Technology
Laurel L. Wilkening, University of Arizona

European delegates

Hugo Fechtig, Max Planck Institut, Heidelberg
Hans Balsiger, Bern University
Jacques Blamont, CNES, Paris
Marcello Fulchignoni, Istituto di Astrofisica Spaziale, Rome
Keith Runcorn, University of Newcastle upon Tyne
Fred Taylor, University of Oxford

National Research Council Space Studies Board European Science Foundation European Space Science Committee Committee on International Space Programs, 1998

US members

Berrien Moore III (chair) University of New Hampshire, Durham
Robert J. Bayuzick, Vanderbilt University, Nashville, Tenn.

Robert E. Cleland, University of Washington, Seattle
Bill Green, US House of Representatives (former member),
    New York City
Jonathan E. Grindlay, Harvard University, Cambridge, Mass.
Joan Johnson-Freese, Air War College, Maxwell Air Force Base, Ala.
Victor V. Klemas, University of Delaware, Newark
Donald G. Mitchell, Johns Hopkins University, Laurel, Md.
James Morrison, BDM Inc. (retired), Washington, DC
S. Ichtiaque Rasool, International Geosphere-Biosphere Program, Paris, and
    University of New Hampshire, Durham
John A. Simpson, University of Chicago
Darrell F. Strobel, Johns Hopkins University, Baltimore

European members

François Becker (chair), International Space University, Strasbourg, France
Michel Bignier, La Réunion Spatiale, Paris
A. Mike Cruise, University of Birmingham, Birmingham, England
Alvaro Giménez, Laboratory for Space Astrophysics and Theoretical Physics,
    Madrid, Spain
Robert J. Gurney, University of Reading, Reading, England
Gerhard Haerendel, Max-Planck-Institut für Extraterrestrische Physik,
    Garching, Germany
Manfred H. Keller, Deutsche Forschungsanstalt für Luft und Raumfahrt,
    Cologne, Germany
Jean-Claude Legros, Microgravity Research Center, Free University of Brussels,
    Brussels, Belgium
Dag Linnarsson, Karolinska Institutet, Stockholm, Sweden
Herbert W. Schnopper, Smithsonian Astrophysical Observatory,
    Cambridge, Mass.
Fred W. Taylor, Clarendon Laboratory, Oxford, England

ESA *Kepler* Mars Obiter Science Working Team, 1981–6

M. Ackerman, Institut d'Aeronomie Spatiale de Belgique
P. Edenhofer, Ruhr Universitat Bochum
V. Formisano, Istituto Plasma Spazio, Frascati
M. Lefebvre, CNRS, Toulouse
F. Mariani, Istituto di Fizica, Roma
S. K. Runcorn, University of Newcastle
F. W. Taylor, University of Oxford
U. von Zahn, University of Bonn

NASA *Mars Aeronomy Observer* Science Working Team, 1986

Donald M. Hunten, University of Arizona
James A. Slavin, Jet Propulsion Laboratory
Lawrence H. Brace, NASA Goddard Space Flight Center
Drake Deming, NASA Goddard Space Flight Center
Louis A. Frank, University of Iowa
Joseph M. Grebosky, NASA Goddard Space Flight Center
Robert M. Haberle, NASA Ames Space Flight Center
William B. Hanson, University of Texas
Devrie S. Intriligator, Carmel Research Center
Timothy L. Killeen, University of Michigan
Arvydas J. Kliore, Jet Propulsion Laboratory
William S. Kurth, University of Iowa
Andrew P. Nagy, University of Michigan
Christopher T. Russell, University of California Los Angeles
Bill R. Sandal, University of Arizona
Edward J. Smith, Jet Propulsion Laboratory
Fredric W. Taylor, University of Oxford
Yuk L. Yung, California Institute of Technology
Ulf von Zahn, University of Bonn
Richard W. Zurek, Jet Propulsion Laboratory

Mars Rover Sample Return Science Working Group, 1989

Michael Carr, US Geological Survey
Arden Albee, California Institute of Technology
Jacques Blamont, CNES, France
Doug Blanchard, NASA Johnson Space Center
Michael Drake, University of Arizona
Michael Duke, NASA Johnson Space Center
Fraser Fanale, University of Hawaii
Johannes Geiss, University of Bern
Matthew Golombek, Jet Propulsion Laboratory
Jim Gooding, NASA Johnson Space Center
Ron Greeley, University of Arizona
James Head III, Brown University
Hugh Kieffer, US Geological Survey
Conway Leovy, University of Washington
Eugene Levy, University of Arizona
Harold Masursky, US Geological Survey
Chris McKay, NASA Ames Research Center
Doug Nash, Jet Propulsion Laboratory

Gary Olhoeft, US Geological Survey
Toby Owen, University of Hawaii
Robert Pepin, University of Minnesota
Sean Solomon, Massachusetts Institute of Technology
Steve Squyres, Cornell University
Fred Taylor, University of Oxford
Heinrich Wanke, Max Planck Institute, Mainz
Gerry Wasserberg, California Institute of Technology

Us National Academy of Sciences Committee on Planetary and Lunar Exploration, 1991

Larry W. Esposito, University of Colorado, *Chairman*
Alan P. Boss, Carnegie Institution of Washington
Andrew F. Cheng, Johns Hopkins University
Anita L. Cochran, University of Texas at Austin
Peter J. Gierasch, Cornell University
Jonathan I. Lunine, University of Arizona
Lucy-Ann McFadden, University of California, San Diego
Christopher P. McKay, NASA Ames Research Center
Duane O. Muhleman, California Institute of Technology
Norman R. Pace, Indiana University
Graham Ryder, Lunar and Planetary Institute
Gerald Schubert, University of California, Los Angeles
Peter H. Schultz, Brown University
Paul D. Spudis, Lunar and Planetary Institute
Peter H. Stone, Massachusetts Institute of Technology
G. Jeffrey Taylor, University of Hawaii
Richard W. Zurek, California Institute of Technology
Invited Participant (European Liaison member, 1981–91): Fred W. Taylor,
    Clarendon Laboratory, Oxford

The *Mars Reconnaissance Orbiter, Mars Climate Sounder* Investigator Team, 2006[1]

D. J. McCleese*, Jet Propulsion Laboratory
J. T. Schofield*, Jet Propulsion Laboratory
F. W. Taylor*, University of Oxford
W. A. Abdou, Jet Propulsion Laboratory
O. Aharonson, California Institute of Technology
S. B. Calcutt, University of Oxford

---

[1] The asterisks denote members who were survivors of the investigator teams for the *Pressure Modulator Infrared Radiometer* experiments, a direct precursor of the *Climate Sounder*, which flew and were lost on *Mars Observer* (1992) and *Mars Climate Orbiter* (1999).

P. G. J. Irwin, University of Oxford
D. M. Kass, Jet Propulsion Laboratory
A. Kleinböhl, Jet Propulsion Laboratory
W. G. Lawson, California Institute of Technology
S. R. Lewis, Open University, Milton Keynes
D. A. Paige*, University of California, Los Angeles
P. L. Read, University of Oxford
M. I. Richardson, California Institute of Technology
N. Teanby, University of Oxford
C. B. Leovy*, University of Washington
R. W. Zurek*, Jet Propulsion Laboratory

Glossary

I have tried to make the text as accessible as possible to non-scientists, with clear definitions of technical terms where the use of these was unavoidable. This glossary collects some of the more widely used definitions for the further convenience of the non-specialist.

Aerobraking and aerocapture   Aerobraking and aerocapture both involve steering a spacecraft into the upper atmosphere in order to use the resulting drag to reduce speed and change its trajectory. Aerobraking lowers an existing orbit, whereas aerocapture slows an incoming spacecraft so it goes into orbit.

Aeroshield   A protective structure at the front end of a spacecraft that absorbs most of the frictional heating during atmospheric entry or re-entry. May be used repetitively for aerobraking.

Albedo   The reflectivity of an object or a surface; the ratio of the reflected to the incident radiant energy at all wavelengths.

Amazonian   The geological epoch on Mars dating from the present until 2.9 billion years ago.

Apoapsis   The point on a non-circular orbit that is farthest from the planet (cf. periapsis).

Astrobiology and exobiology   The study of all life in the context of the Universe (astrobiology), the search for indigenous extraterrestrial life (exobiology). Often used interchangeably.

Astronomical unit   The mean distance of the Earth from the Sun, approximately one hundred and fifty million kilometres.

Bar and millibar   Units of pressure, defined so that one bar is the mean pressure at the surface of the Earth. One bar is equal to one thousand millibars.

Biosignature   A fossil, chemical or anything else that can be interpreted as a sign of past or present life.

Celsius   *See* Temperature scales.

Centigrade   *See* Temperature scales.

Clathrate   A compound in which a gas (here usually carbon dioxide) is physically, rather than chemically, captured within a frozen solid (water ice). The gas is released on warming to the melting temperature of the ice, rather than at its own, much lower, freezing point.

Doppler effect or Doppler shift   The change in the wavelength of light produced by the motion between the source and the observer.

Electromagnetic spectrum   The shortest wavelengths are gamma rays, then X-rays, then ultraviolet, then visible, then infrared, then microwaves and radio waves.

ESA   The European Space Agency, which has its headquarters in Paris.

Exobiology   *See* Astrobiology.

Fahrenheit   *See* Temperature scales.

Fly-by   A type of mission where a spacecraft flies close to a planet in order to observe it, but carries on past instead of landing or going into orbit.

Geology, geochemistry and mineralogy   The study of the solid body of a planet, its composition and the reactions that produce different kinds of rock, soil etc. A list of the different types relevant to Mars appears in the caption to Figure 4.7.

General circulation model (GCM)   The most detailed kind of computer model of the atmosphere, used for weather and climate forecasting.

Hesperian   The geological epoch on Mars dating from 2.9 to 3.7 billion years ago.

HiRISE   The *High Resolution Imaging Science Experiment,* a very powerful camera on *Mars Reconnaissance Orbiter,* in operation at Mars since late 2006.

Igneous rock   Solidified lava from a volcano.

Infrared   Radiation with a wavelength longer than visible light, but shorter than microwave or radio waves.

JAXA   The Japanese space agency.

Kelvin   *See* Temperature scales.

Laser altimeter   A device used on orbiting spacecraft to measure the height of the terrain below by firing laser pulses downwards and timing their return to the spacecraft. Sometimes called a 'lidar' or optical radar.

Latitude and longitude   The convention used on Mars is based on that for the Earth, so the equator is zero degrees of latitude and the poles are at ninety degrees (90° S and 90° N). There are three hundred and sixty degrees of longitude; the zero is in Airy Crater by international agreement.[1]

A system with 'planetocentric' latitude (one with coordinates derived from the angle measured from the equator to a point on the surface at the centre of the planet and longitude increasing to the east) is used for making Mars maps and imagery.

---

[1] The story of how this came to be so is related by the European Space Agency as follows: Earth's prime meridian was defined by international agreement in 1884 as the position of the large 'transit circle', a telescope in the Greenwich Royal Observatory's Meridian Building. The transit circle was built by Sir George Airy, the seventh Astronomer Royal, in 1850. For Mars, the prime meridian was first defined by the German astronomers W. Beer and J. H. Mädler in 1830–2. They used a small circular feature on the surface, which they called 'A', as a reference point to determine the rotation period of the planet. The Italian astronomer G. Schiaparelli used this feature as the zero point of longitude in his 1877 map of Mars. It was subsequently named Sinus Meridiani ('Middle Bay') by French astronomer Camille Flammarion. When *Mariner 9* mapped the planet at about one kilometre resolution in 1972, a more precise definition was needed. Merton Davies of the RAND Corporation was analysing surface features and designated a 0.5-kilometre-wide crater, subsequently named 'Airy-0' (within the larger crater Airy, named to commemorate the builder of the Greenwich transit) as the zero point.

Meteorology   The study of a planet's atmosphere, its weather and climate.

Micrometre   Abbreviation for a millionth of a metre, sometimes called a micron, written μm, used for instance when describing the size of dust particles, or the wavelength of light (the visible spectrum ranges approximately from 0.38 to 0.75 μm).

NASA   National Aeronautics and Space Administration, the American space agency, formed in 1957.

Network   A mission type that involves placing a number (usually three to twenty) of identical small stations around the globe of Mars.

Noachian   The earliest geological epoch on Mars, dating from formation of the planet about 4.5 billion years ago to 3.7 billion years ago.

Orbiter   A spacecraft designed to become an artificial moon of a planet such as Mars (as opposed to a fly-by or a lander).

Parallax   The apparent displacement of an object when viewed from different directions. It may refer to the apparent change in the position of Mars relative to the background stars when viewed from two locations on the Earth, or in principle when the Earth is at the two extremes of its orbit, either side of the Sun. This small angle is not simple to determine (particularly since Mars moves along its own orbit during the time of the measurement) but once it is available, the distance to Mars can be determined from simple triangulation.

Penetrator   A probe dropped onto the surface of a planet to strike at high speed and force its way to some depth below the surface, where it survives to make measurements.

Periapsis   The point on a non-circular orbit that is closest to the planet (cf. apoapsis).

Phase A   The first phase of an approved space project, focussing on verifying the technical feasibility and the likely run-out cost.

Precipitable micrometres   A common way of measuring the water vapour in a column of atmosphere, equal to the thickness of the liquid layer that would result if all of the water in the atmosphere condensed on the surface.

Radioisotope thermal generator (RTG)   A device for generating electrical power on a spacecraft using radioactive isotopes. Generally used only when there is insufficient sunlight to use solar panels, such as in the outer Solar System or under the surface of Mars.

Redox reactions   Chemical reactions that can provide energy to microorganisms, involving the exchange of electrons, achieving both reduction and oxidation.

RKA   The Russian space agency (Rossiyskoe Kosmicheskoe Agentsvo).

Sedimentary rock   Precipitated out of water to form clays that subsequently harden into rock.

Seismology   The study of earthquakes, or marsquakes, natural or introduced artificially by controlled impacts or explosions, and their interpretation in terms of the interior structure of the planet.

SNC meteorites   Meteorites shown to have originated on Mars, named Shergottite, Nahklite and Chassignite for the locations of the earliest falls, in India, Egypt and France, respectively.

Solar longitude (L_S)   The conventional way of describing the season on Mars, measured relative to the northern vernal (spring) equinox, defined as $L_S = 0°$. Thus midsummer is $L_S = 90°$ and so on.

Solar wind   A stream of charged particles that is continuously emitted from the Sun at high speed (hundreds of kilometres per second) and reaches all of the planets, being responsible for producing the aurorae in Earth's upper atmosphere when focussed to the poles by the magnetic field.

Spectroscopy   The principle of spectroscopy involves dispersing the light from an object, in this case a planet viewed through a telescope, into its component wavelengths by a device such as a prism or a diffraction grating. Different materials absorb or reflect more strongly at some wavelengths than others; an analysis can show what materials are present.

Spectrum   The brightness, as a function of wavelength, of light dispersed by a prism or grating, is called a spectrum from the Latin for a ghost or phantom.

Spectral lines   The features in a spectrum that reveal the presence of a particular material are usually (especially for gases) in the form of groups sharp lines.

Telluric   Spectral lines or other features or artefacts in astronomical observations that originate in the Earth's atmosphere.

Temperature scales   Most of the temperatures in this book are given in degrees centigrade, also called Celsius,[2] since this is the most familiar in everyday terms to the general reader. Water freezes at zero degrees centigrade (0° C) and boils at one hundred degrees (100° C).

In some data and diagrams produced in the USA, the Fahrenheit scale is often used. Water freezes at thirty-two degrees Fahrenheit (32° F) and boils at two hundred and twelve degrees (212° F), so one degree Fahrenheit is equal to nine-fifths of a degree centigrade.

Scientists usually use the absolute or Kelvin scale, which is the same as centigrade except that the scale is shifted to be zero at absolute zero. Water freezes at two hundred and seventy-three degrees absolute (273 K) and boils at a hundred degrees higher (373 K).

Ultraviolet (UV)   Radiation with a wavelength shorter than visible light, but longer than X-rays.

---

[2] Centigrade is preferred to Celsius because it is descriptive, rather than the name of someone otherwise largely unknown today. Anders Celsius was a Swedish meteorologist who published 'Observations of two persistent degrees on a thermometer' in 1742. He called his scale 'centigrade', meaning a hundred steps. This originally had water boil at zero and freeze at a hundred degrees, thus avoiding negative values (in eighteenth century Sweden). The convention was reversed after his death.

Index